CHINESE CITY AND PLANNING SYSTEMS

Li Yu provides a magisterial overview and analysis of the complex and contested nature of city and regional planning in China. He traces the development of modern China and outlines the evolution of planning though history to its present role in a burgeoning market economy. Li Yu maps the institutional and organisational arrangements for planning and considers the catalogue of strategic and local issues in the country. There are wonderful insights into understanding the changing state-market-civil relations in the modern world and sobering lessons for the spirit and purpose of planning.

Greg Lloyd, University of Ulster, UK

If you want to learn more about the tremendous changes that have taken place in Chinese cities during the last 60 years, as well as the role urban planning has played in such a fascinating process, this is the one book you have to read. Unfolding in a panoramic perspective, Li Yu sets out a critical and enlightening study on the responsive mechanism and practice of urban planning through the rapid urbanization process, in respect of social change, history and the political system. It also offers the reader the main motivations behind the achievements and failures in nearly 30 years of Chinese urban development. Yu takes the reader on an enthralling journey through the history of urban planning in contemporary China.

Kai Wang, China Academy of Urban Planning and Design, China

Chinese City and Regional Planning Systems

LI YU
Cardiff University, UK

Routledge
Taylor & Francis Group
LONDON AND NEW YORK

First published 2014 by Ashgate Publishing

Published 2016 by Routledge
2 Park Square, Milton Park, Abingdon, Oxon OX14 4RN
711 Third Avenue, New York, NY 10017, USA

Routledge is an imprint of the Taylor & Francis Group, an informa business

Copyright © 2014 Li Yu

Li Yu has asserted his right under the Copyright, Designs and Patents Act, 1988, to be identified as the author of this work.

All rights reserved. No part of this book may be reprinted or reproduced or utilised in any form or by any electronic, mechanical, or other means, now known or hereafter invented, including photocopying and recording, or in any information storage or retrieval system, without permission in writing from the publishers.

Notice:
Product or corporate names may be trademarks or registered trademarks, and are used only for identification and explanation without intent to infringe.

British Library Cataloguing in Publication Data
A catalogue record for this book is available from the British Library

The Library of Congress has cataloged the printed edition as follows:
Yu, Li (Senior lecturer)
  Chinese city and regional planning systems / by Li Yu.
      p. cm.
  Includes bibliographical references and index.
  ISBN 978-0-7546-7499-3 (pbk.) 1. City planning--China--History. 2. Regional planning--China--History. 3. China--Economic conditions--1949- 4. China--Economic policy--1949- I. Title.
  HT169.C6Y777 2014
  307.1'2160951--dc23
                                                                    2013032332

ISBN 9780754674993 (pbk)

Contents

List of Figures *vii*
List of Tables *ix*
Acknowledgements *xi*

1	China: An Introduction	1
2	Evolution of the Chinese City and Regional Planning System	41
3	The Government System and Administrative Framework for Planning and Development	81
4	Regional Planning and Regional Governance Innovation	105
5	Statutory City Comprehensive Plan	137
6	Statutory Regulatory Detailed Plan and Planning Norms	171
7	Non-Statutory Plan: Urban Development Strategic Plan	195
8	Land Use System, its Reform and Planning Management	221
9	Performance of Chinese City and Regional Planning System	247

Bibliography *275*
Index *289*

List of Figures

1.1	Map of China	2
2.1	The nine major co-urban areas in China	64
2.2	Usage of foreign capital	72
2.3	Location of low carbon eco-cities in China	78
3.1	Hierarchical planning system in China	96
3.2	Organizations responsible for development and construction in municipal government	99
4.1	Map of major cities along the coast of Liaoning Province	126
4.2	Spatial development of 'Five Dots and One Belt' along Liaoning coastal area	127
4.3	Integrated urban system development plan of Liaoning coastal area	133
5.1	Location of Xiamen in China	147
5.2	Map of Xiamen	148
5.3	Relationship between GDP and export and import volume	149
5.4	Gulandyu Isle	152
5.5	City Comprehensive Plan formulation process	161
6.1	Regulatory Detailed Plan	177
6.2	Relation between City Comprehensive Plan and Regulatory Detailed Plan	184
7.1	Multiple stakeholders in the Xiamen development	198
7.2	Xiamen compared with other main cities in Fujian Province	206
8.1	Application process for using existing urban land	228
8.2	Application to transfer rural land to urban land use	229
8.3	Application process of development projects	231

List of Tables

1.1	Differences in regions	13
1.2	Total and selected items of government revenue in different regions (1999)	22
1.3	GDP by three areas	22
2.1	Number of cities in different categories and their percentage, between 1949 and 1999	61
2.2	Index of the three main co-urban areas in 2000	65
2.3	The tax categories	69
2.4	The major index for globalization	72
2.5	Structure of funded sources for total investment in fixed assets	74
5.1	Prediction of population and urbanization in Xiamen	158
5.2	Four zones and their economic activities	159
6.1	Land use planning control index in a Regulatory Detailed Plan	178
7.1	Comparison of quasi-provincial-level cities along the east coast of China	204
7.2	Percentage of FDI within total industrial output in 2000	204
8.1	Reasons of conditional approval or refusal for application of planning permission	244

Acknowledgements

The author would like to acknowledge the support from Professor Michael Bruton in the preparation of this book.

Chapter 1
China: An Introduction[1]

Planning in any country is inextricably linked with the history, politics and administrative structure of that country. Thus this introductory chapter is concerned with reviewing:

- the historical development of modern China;
- the social, economic, political and environmental issues that face the country today and with which the planning system is attempting to deal; and
- the system of governance.

This chapter will also explore the theoretical implication of the background for the planning system in China.

Introduction

The People's Republic of China (PRC), more commonly known as China, forms a major part of Asia. To the east it is bounded by the South and East China Seas, the Yellow Sea and the Korean Bay. Its land boundaries include Vietnam, Laos, Myanmar (Burma), Bhutan, Nepal, India, Pakistan, Kazakhstan, Kyrgyzstan, Tajikistan, Mongolia, Russia and North Korea, whilst it has sea boundaries with Indonesia, Malaysia, Brunei and the Philippines in the South China Sea (see Figure 1.1).

China covers an area of 9,596,960 square kilometres. Within the territory, there was a population of 1.37 billion in 2010 according to the Sixth Census. China is the fourth largest country in the world after Russia, Canada and the United States.

The country is diverse in many respects. Climatically, it can range from tropical in the south to sub-arctic in the north. From a topographical perspective, China has deserts in the west; mountains and high plateaus in the central areas and coastal plains and deltas in the east. In term of ethnic groups, whilst the Han Chinese account for 91.9 per cent of the population, there are 55 minority groups such as the Zhuang, Uygur, Hui, Yi, Tibetan, Miao, Manchu, Mongol, Buyi, Korean, all of which account for the remaining 8.1 per cent of the population. As a country

---

1   The social, economic and political development of China is occurring so rapidly that in writing this chapter the immediate news reporting media has been relied on more than might be expected in a textbook of this sort. Only through such sources is it possible to be up to date.

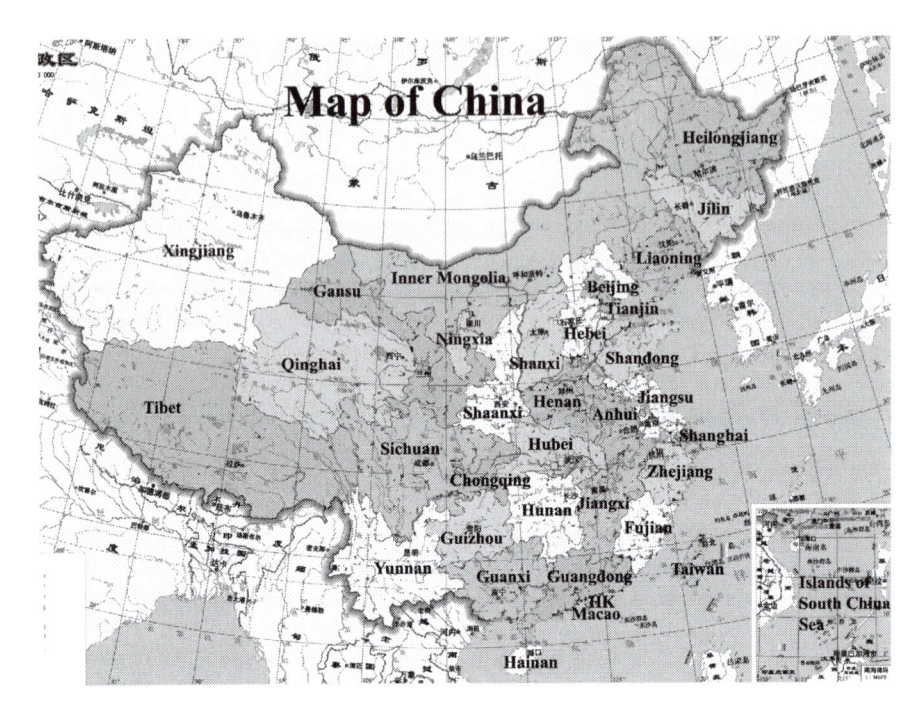

Figure 1.1 Map of China

with the largest population in the world, there are different local dialects, e.g. Cantonese, Wu (Shanghaiese), Minbei (Fuzhou), Minnan (south Fujian Province-Taiwanese), Xiang, Gan and Hakka, being spoken as well as minority languages, but Mandarin is the most popular and common language.

China is well endowed with natural resources, with plentiful supplies of coal, iron ore, petroleum, natural gas, mercury, tin, aluminium, lead, zinc and uranium, and has the world's largest potential for hydropower.

The rapid economic development of the last three decades has resulted in environmental costs – notably air pollution from greenhouse gases and sulphur dioxide particulates. These were a consequence of a dependence on coal, acid rain, water shortages, particularly in the north, and water pollution from untreated waste and deforestation. In addition it is estimated that since 1949 one-fifth of agricultural land has been lost to soil erosion, economic and urban development and the encroachment of deserts. In 1993, 10 per cent of the land was given over to arable crops; 43 per cent to permanent pastures; 14 per cent to forest and woodlands and 33 per cent to other uses with 498,720 square kilometres being irrigated.

China has a long and impressive cultural and scientific history. It was unified as a nation under the Qin Dynasty in 221 BC; the succeeding dynasties were Han, Three Kingdoms, Jin, Northern and Southern Dynasties, Five Dynasties, Song, Yuan, Ming, with the last dynasty, Qing, being replaced by the Republic of China

(ROC) in 1911. The People's Republic of China (PRC) was established on 1 October 1949 following a revolution led by Mao Zedong.

Physical Characteristics

China covers approximately 3.7 million square miles (9.6 million square kilometres), including Taiwan, Hong Kong and Macau (Special Administrative Regions – SARs). From north to south (Heilongjiang to the South China Sea Islands) it covers 49 degrees of latitude; from east to west (Heilongjiang to Xianjiang) it extends over 62 degrees of longitude.

China's topography is divided into a three-step staircase descending from west to east. The Qinghai-Tibet plateau is the top step, with an average height above sea level of 12,000–15,000 feet (4,000–5,000 metres); the second step consists of the Sichuan basins and the Inner Mongolian and Yunnan-Guizhou plateau with a height of between 3,000 and 6,000 feet (1,000–2,000 metres), and the lowest step, which is generally below 1,500 feet (500 metres), includes a belt of plain land in Manchuria, and north China; the middle and lower reaches of the Yangtze, and the Yellow, Huai and Pearl rivers all of which flow east to west. The bulk of the Chinese population and most of its important cities are located in the area covered by the third step.

Given the size of the country and its varied topography China has a diverse climate. In northern China, where Beijing is located, the climate is typically continental with warm to hot summers and freezing cold and frosty winters which bring occasional light snow, whilst in the extreme northeast winters are cold enough to freeze the rivers for up to six months, and summer temperatures are also cooler, with humid conditions and high rainfall. The annual average temperature in Beijing is 11.8 degrees Celsius, with a low of -4.6 Celsius in January and a high of 26.1 Celsius in July. The annual average rainfall is 24 inches (619 mm).

Along the Yangtze River, where Shanghai is located, it is warmer than the north in both summer and winter but it can rain at any time. Winter weather is very changeable, alternating between spells of heavy rain and bright but cold and frosty periods with occasional snow. The main rainfall occurs during the summer months whilst along the coast from July to October typhoons (tropical cyclones) occur with strong winds and very heavy rainfall. The annual average temperature in Shanghai is just over 16 degrees Celsius, whilst the annual average rainfall is 45 inches (1,135 mm).

Southern China, which includes Guangzhou, Guangxi and Fujian, is the warmest and wettest part of the country in the summer because of its proximity to the tropics whilst the winters are mild. However, the inland region of southern China is mountainous with cooler summers and rain in the highlands and sunny and warm winters with little rain. The average annual temperature of Haikou, the capital of Hainan Island, is 24 degrees Celsius, whilst the annual average rainfall in the Yangtze River basin varies between 39 and 78 inches (5,600 mm).

Southwest China, which consists largely of the high plateau of Tibet, experiences extremely cold and frosty winters and warm and rainy summers whilst northwest China (Xianjiang) is predominantly desert and has very hot summers and bitterly cold winters made even more severe by icy winds. Rainfall is light. The average annual temperature of Urumqi, the capital of Xianjiang, is 7 degrees Celsius whilst the average annual rainfall in Inner Mongolia, Xianjiang and the Qinghai-Tibet plateau is 8 inches (200 mm) (Harding 1996).

The Development of Modern China

China: 1840–1911

It is not possible fully to understand modern China without an appreciation of its historical, cultural, economic and political development. This section aims to provide a brief understanding of the development of contemporary China, concentrating on the period from 1890 to the present day. For centuries a characteristic feature of China was a long established ideology and culture with supportive institutions, including:

- centralized imperial rule;
- Confucianism;
- an agrarian economy;
- a family centred social security system;
- self sufficiency; and
- traditional feelings of the superiority of the Chinese race.

This ideology and culture developed through the control of successive dynasties starting with the Xia, Shang and Zhou dynasties of the 'ancient era'. The recorded history of China began in the fifteenth century BC when the Shang Dynasty used graphic markings that evolved into the present Chinese characters. City-states grew up in the Yellow River valley, which saw the beginnings of the development of Chinese civilization. 221 BC is commonly accepted as the year in which China was first unified as a large kingdom or empire under the Qin Dynasty. Successive dynasties developed bureaucratic systems that enabled the emperor to control an increasingly large territory that reached its maximum under the Yuan (Mongolian) and Qing (Manchurian) dynasties before the Republicans secured power in 1911.

It is generally accepted that in its development China alternated between periods of political unity and 'separatism', and at times it was dominated by foreigners, most of whom were eventually assimilated into the Han Chinese population. The influx of cultural and political influences from many parts of Asia, carried by successive waves of immigration, expansion and assimilation, eventually merged to create the Chinese culture.

Yet, despite its history of bureaucratic control and cultural development China entered the nineteenth century in decline and crisis for a couple of reasons, namely, the Manchu monarchy and the Qing Dynasty were unwilling to adapt to the rapid changes occurring in the outside world. These included adapting to industrialization; a large rural economy, although a few urban areas were beginning to industrialize, e.g. Shanghai, and the majority of Chinese being illiterate and leading their lives according customs and practices that had operated for centuries.

The First Opium War (or the First Anglo-Chinese War) between 1840 and 1842 is significant in Chinese history. Five treaty ports were opened for trade, and Hong Kong Island was ceded to the United Kingdom as a result of the 'Treaty of Nanking' in 1842 after the war. The First Opium War marked the beginning of modern Chinese history. China had declined and been invaded by the foreign powers. This greatly impacted the policies of the Chinese government after 1949, typically since China started its open door and reform policy in the 1980s. Chinese people and government have been struggling to be stronger in terms of the economy. Economic development is regarded as top priority.

The defeat of China by the Japanese in 1894–5 further established quite clearly the nature and extent of China's decline. It resulted in the loss of Taiwan. The defeat of China by Japan highlighted the success of the Japanese industrialization policies at the same time as highlighting the failure of China's attempts to modernize. As a consequence, there were attempts to introduce reforms, such as making the Manchu monarchy a constitutional democracy. The failure of China in the war with Japan in 1894–5 led to a new round of foreign powers scrambling for concessions in China.

The Boxer Rebellion in 1900 (Esherick 1987) had ultimately weakened the Qing Dynasty and paved the way for the republican movement. 'Ultimately it was a question of overhauling a civilization rather than reforming a state; a case of rejecting overpowering tradition and searching for new forms of legitimacy outside it – and therefore outside China' (Hutchings 2001: 4). Reforms were subsequently introduced which questioned the centralizing ideology that had underpinned Chinese society and government for centuries, including reforms to:

- education, when, in 1905, the Imperial Court abolished the traditional education system which had trained the elite and replaced it with a more open and questioning system based on principles that operated in the Western world;
- administration, which abolished the old centralizing boards and replaced them with ministries which took a proactive role in their areas of responsibility, such as the development of industry, trade and railways;
- the constitution, which established provincial assemblies in 1908.

Despite the fact that these reforms were substantive, they were primarily introduced in a final attempt to save the Qing Dynasty and the Manchu monarchy, and were not implemented with enthusiasm. Indeed their introduction was simultaneously accompanied by attempts on the part of the Qing Dynasty and the monarchy to reassert control over the provinces and concentrate power in the hands of the Manchu. These moves inevitably antagonized large sections of the population and led to the establishment of a republican revolutionary movement led by Sun Yat Sen, which in 1911 successfully ended imperial rule.

Revolution 1912–28

In 1912 the newly established Republic of China (ROC) appointed Sun Yat Sen as president, and established a parliament which met in Nanjing. Although the Manchus had made some progress with their reforms in the period 1900–11, including the emergence of an industrial structure in Shanghai, Tianjin, Wuhan and the northeast provinces, and the construction of a modest network of roads and railways, the newly established ROC set an ambitious agenda to create new institutions based on ideas which departed from the traditional Chinese ideology, many of which were drawn from the West, to establish an understanding of and loyalty to these new ideas, to develop industry and trade, to improve transportation and communication and to improve and extend education. At the same time the ROC was faced with both external and internal difficulties including insecure international boundaries, with Russian expansion in Mongolia, British interference in Tibet, Russian and Japanese rivalry in northeast China, and the frustration surrounding the 'International Question' with the Chinese nationalists wanting the economic privileges enjoyed by foreigners to be removed but the foreigners not being prepared to relinquish their privileges, and weaknesses arising from the existence of regional armies which were controlled by Yuan Shikai (an important warlord) and not the ROC.

In 1912 Yuan Shikai succeeded Sun Yat Sen as president of the ROC. As president, Yuan Shikai defeated the nationalists (Kuomintang) in 1913 and drove its leaders into exile, including Sun Yat Sen. Yuan Shikai attempted to revive imperial rule with himself as emperor. However Yuan died in 1916, 88 days after becoming emperor. However, his militarization of politics continued in the years after his death, with warlords effectively governing the territories they controlled, despite the existence of a central government. Indeed the lack of a credible central authority made national development impossible. In the words of Hutchings: 'China needed unity – and the strong government armed with powerful ideas and popular support required to impose it' (Hutchings 2001: 6).

Student protests on 4 May 1919 at the continued refusal of the foreign powers to hand back to China the former German concessions in Shandong led to the establishment of a broad movement concerned to promote Chinese nationalism and a 'new culture' based on Western ideas of democracy, science and Marxism. Marxism in particular appealed to China owing to the success of the Marxist revolution in Russia which suggested many parallels with China. At the same time China, with its search for unity and a new identity, was seen by the Soviet Marxists as being a potentially friendly ally protecting the eastern boundaries of the Soviet revolution. Soviet communists thus made contact with Chinese communists in Shanghai, other Chinese cities, and with Sun Yat Sen, who at this time was head of the Kuomintang, a separatist military government in Guangdong. These approaches resulted in the establishment of the Chinese Communist Party (CCP) and re-organization of Sun Yat Sen's separatist military government into a revolutionary movement.

The Kuomintang formed an alliance with the Chinese Communist Party and in the space of a few years a modern political operation had developed with military backing from the National Liberation Army and ambitions to overthrow the warlords, unify China and end foreign privileges. Following the death of Sun Yat Sen, Chiang Kai-shek became leader of the National Revolutionary Army and in 1926 moved north from Guangdong to the Yangtze. With help from the industrial workers and the peasants, who had been mobilized by the Chinese Communist Party, he defeated the local warlords in that area. Kuomintang troops entered Beijing in 1928 and Chiang Kai-shek established a new central government in Nanjing (Martin 1983).

Kuomintang and Chiang Kai-shek 1928–49

Between 1928 and 1937 Chiang Kai-shek attempted to establish control over the territory. Apart from the provinces in the lower Yangtze Valley which were dominated by the Kuomintang (Nationalist Party), Chiang's control was nominal and unity was more of an illusion than reality. In the northeast, the northwest, the southwest and Guangdong the local warlords were more powerful than the newly established central government. The countryside in Jiangxi Province was controlled by the communists.

Following the Long March, when Mao Zedong and the communists retreated from Jiangxi to Shaanxi in an attempt to avoid Chiang's efforts to overcome their resistance, the Communist Party and its army continued in northern Shaanxi.

The war between China and Japan started in 1937. Japan soon controlled north China, the Yangtze Delta and the Yangtze River as far inland as Wuhan and by 1939 it controlled every major port in China. The alliance between the Kuomintang and the Communist Party was established in the fight with Japan. It was assumed that with allied involvement Japan would eventually be beaten and consequently they conserved their resources in anticipation of renewing the domestic struggle for China once the war with Japan was won. The allied participation in the struggle with Japan had a much greater impact on China's status in the wider world. The Cairo Conference (1943) involving President Roosevelt, Prime Minister Churchill and Chiang agreed that:

- Manchuria and Taiwan would be returned to China after the defeat of Japan; and
- China would become a founding member of the United Nations with a permanent seat on the Security Council.

With the end of the Second World War a bitter civil war broke out between the Kuomintang and the Communist Party which was resolved in favour of the Chinese Communist Party. In 1949 Mao Zedong announced the establishment of the People's Republic of China (PRC) and by the end of the year the People's Liberation Army occupied every mainland province in China. Chiang evacuated what remained of his army and government to Taiwan.

The New China 1949–76

In parallel with these military victories the new People's Republic of China government initiated its social and political reforms in land, new marriage laws and nationalization of private assets. At the same time, China established an alliance with the Soviet Union and the provision of assistance from Russia.

Through the First Five-Year Plan 1953–7, China started a programme of rapid industrialization along Soviet lines, which primarily promoted heavy industry with the objective of becoming an economically strong country. A dramatic improvement in communications, which helped integrate the more remote regions, was operated. The programmes to remove illiteracy and improve health care were introduced in order to increase the quality of life of Chinese people.

In 1956 agriculture was collectivized and industry nationalized as part of the proposals for the 'Great Leap Forward', which was an attempt to instantaneously establish China's superiority to most of the capitalist world and to move China ahead of the Soviet Union as a socialist country.

Between 1966 and 1970, China withdrew from the rest of the world 'to pursue a vision of revolutionary purity at home' (Hutchings 2001: 11) through the 'Cultural Revolution'. China was isolated in its own terms but was self-sufficient. 'The Cultural Revolution' resulted in the purge of many people: doctors, engineers, planners and professionals in every walk of life; religious believers; those with overseas connections and the leaders of ethnic minorities were amongst the casualties of this revolution. The Chinese economy was in chaos.

Deng Xiaoping, Reform and Regeneration 1976–Present

The revolutionary open door and reform policy introduced by Deng was concerned to use the finance and technological resources of the capitalist world in the regeneration of the People's Republic of China, and to encourage enterprise and the profit motive, first in agriculture, then in the cities, industry, and finally in every walk of life.

Recognition that constant radical reform and revolution had exhausted the country led to radical reform within the People's Republic of China, and China started to relax its engagement with the outside world. The attraction of foreign investors to participate in the newly established Special Economic Zones such as Shenzhen, Zhuhai, Shantou and Xiamen was encouraged. Overseas Chinese were encouraged to renew their links with their motherland in order to attract investment and information and technologies from the outside world.

The People's Republic of China intended to follow the lead of Hong Kong, Taiwan, South Korea and other Southeast Asian countries in using its resources and manpower to exploit the free market economy and achieve export-led growth.

Whilst it was accepted that the People's Republic of China would engage with the rest of the world it was also established that it would not become part of the Western world. Rather it would retain its identity, while adopting characteristics of the Western world. This hybrid new identity was ideologically redefined by Deng as 'socialism with Chinese characteristics' and the intention was that it would ensure the continued political dominance of the Chinese Communist Party.

The application of Deng's reforms created opportunities for Chinese society at the same time as creating tensions. Competition from overseas spurred economic growth but at the same time undermined the basis of state socialism. In the words of Hutchings: 'A job for life, and free housing and free health care came under attack as reform of state owned enterprises gather pace' (Hutchings 2001: 13). The societal changes were immense. Urban Chinese studied abroad; millions of rural Chinese left the land and migrated to the cities; Western fashions and rock music were introduced; and activists and artists explored the new-found freedom that was available to them.

With the words of Deng Xiaoping of 'development is the absolute principle', the Chinese population dedicated itself to making money, with outstanding results. Economic growth was phenomenal; construction work dominated all cities; foreign investment, especially from Hong Kong, financed the economic boom of the 1990s. Development started in Guangdong and Fujian, then moved north to Shanghai and west along the Yangtze River; then further north to Shandong, the provinces and cities around the Gulf of Bohai and into Hebei Province, Beijing and Tianjin. However, at the same time as the benefits of this economic growth began to be appreciated so the social and environmental consequences of that growth began to be felt. However, inequalities developed between the urban and rural areas, and between hinterland and coastal provinces.

However, environmentally the impact of economic growth began to be felt in some areas, typically in loss of farm land, atmospheric pollution, deforestation and desertification thanks to the rapid industrialization and urbanization of the last 30 years.

Currently the People's Republic of China is concerned with securing economic development at the same time as preserving national sovereignty and political stability in what is a vast and large populated country. The continued transformation of the country from an agrarian to an industrial economy and from a centrally planned to a primarily market oriented economy is of paramount importance. In effect this is China's industrial revolution. These continued changes have to be achieved within a territory where there is little room to manoeuvre – territorially with the exception of the northwest, there is nowhere for the surplus population to move to; rapid industrialization has to be accommodated in an overcrowded environment; food and energy demands are met mainly through imports from overseas, which place enormous energy and raw resources risks on China.

On the economic front China is also faced with a dilemma. On the one hand, because of its size and large population, it could continue to pursue a policy of promoting and further developing its low end manufacturing economic base. On the other hand, it could attempt to promote the development of a knowledge economy based on information industries and innovation, as many developed and Southeast Asian countries have done, e.g. Malaysia and Singapore. However, such development needs a competitive service industry; a convertible currency; an established legal system; individual freedom and an educated population.

Current Issues in China

It is not possible to evaluate the performance of the planning system in the People's Republic of China without an appreciation of the issues that the system is attempting to address. For clarity and ease of understanding the current issues with which China is dealing are here dealt with individually. However, given the complex nature of Chinese society and politics these issues are invariably inter-related in a complex way and should not be thought of as freestanding. In this context this section examines and comments on the following issues:

- population;
- poverty;
- unemployment;
- economic development;
- rural–urban migration;
- social harmony;
- food security;
- sustainability and environment.

Population

China has always had a large population for the simple reason that culturally amongst the Chinese, especially the rural Chinese, there has been a concern to produce sons to work on the land as a large labour force was needed due to little mechanization in agriculture. It is Chinese culture and tradition to have sons to carry on the family name. At the same time, it is the responsibility of sons to take care of their parents in old age. In more recent years less importance has been attached to the need for sons as the economy has developed and urbanization has taken place.

Over the years China has used a range of methods to determine the size, distribution and characteristics of its population, with the result that it is not easy to be certain of: (1) the precise size of the population at any point in time; and/or (2) the rate at which that population is growing. Indeed, the first census of China undertaken on what could be considered a broadly scientific basis was that conducted by the People's Republic of China in 1953 which showed that the population of mainland China at that time was 582.6 million, unevenly distributed across the country, with 70 per cent of the population in the coastal, eastern and southern provinces.

In addition, the different conceptual views taken on population at different times by the central government further complicate an understanding of Chinese demographics. Thus, during the 1950s the overriding assumption in China was that a large population was an asset to development and essential for the 'Great Leap Forward' and national defence. The fact that overpopulation would almost certainly lead to increased poverty was ignored even though the population increased from 582.6 million to 695 million between 1953 and 1964, when a second census was taken in 1964.[2] With the general chaos arising from the Cultural Revolution population planning and control were politically unacceptable and nothing was done to address the problem until the late 1970s.

Only with the introduction of the reform era did it become accepted that China's population growth would have to be controlled if the people were to benefit fully from economic reform, even though an increase in population at this time was inevitable because of the uncontrolled baby boom of the 1950s. The effect of this baby boom was confirmed by the third census held in 1982 which showed that the population had increased to 1.008 billion with well over 100 million being of marriageable age. The government responded by raising the minimum age for marriage under the new Marriage Law of 1980, but this had little impact on population growth given the reduction in infant mortality and the increase in life expectancy at that time as a result of improved health care.

As a consequence, the Chinese government decided that population growth should be both controlled and reduced through the introduction in 1979 of the 'one child per family' policy. Incentives were offered to families who agreed to have only one child, e.g. bonuses, improved medical care, while sanctions were imposed on those who went ahead with a second child, through the system of distributing benefits offered by their work units, including the standard of accommodation provided and holidays.

One-child families became the norm in urban areas, whereas it was more difficult to operate such controls in the rural areas and amongst the ethnic minorities who were excluded from the birth control policies. Here two or more child families became the norm. The application of this birth control policy was intended to restrict China's population to 1.2 billion in 2000, and this target was reached in 1995. Subsequently, the government raised the target population for 2000 to 1.3–1.4 billion in 2010 and between 1.5–1.6 billion in 2030.

The total populations recorded at the six censuses that have been taken to date are as follows:

- 1953   0.582 billion

---

[2] For political/ideological reasons the results of this census were not made public until 1979.

- 1964   0.695 billion
- 1982   1.008 billion
- 1990   1.134 billion
- 2000   1.266 billion
- 2010   1.370 billion

This reinforces the point that China's population is extremely large and has grown significantly over the past 50 years.

More significantly, according to information from the Population Division of the United Nations (2007), China's population is growing and aging at the same time and, by 2050, China will have a total population of approximately 1.409 billion, a threefold increase over the 1950 population. Among which, 438 million people will be aged 60 and above, a tenfold increase over 1950; children below the age 5 will increase to 572 million; children and teenagers between the ages of 5 and 19 will be 218 million. They will be supported by a 'working age' population of 681 million.

Despite the achievement of population control through the application of this birth control policy, it is inevitable that China's large and aging population will place a heavy strain on food supplies, the environment, employment and the national budget, especially social services.

The difficulties arising from the size and structure of the population are further compounded by the distribution of that population. According to the research of Heilig (1999), there is a skewed distribution of population with:

- 10 per cent (approx.) of the population living in an area of only 47,000 km^2, i.e. 0.5 per cent of the total landmass of China with an average population density of 2,428 people per km^2 in the most densely populated counties and cities;
- 50 percent of the Chinese population living in an area of 778,000 km^2, i.e. about 8.2 per cent of the total land area with an average population density of 740 persons per km^2;
- roughly one billion Chinese (or more than 90 per cent of the population) living on little more than 30 per cent of the country's land area with a population density of 354 persons per km^2.

More specifically, these general characteristics are reflected in major regional/ spatial distributional variations with the bulk of the population in the mid 1990s living in the fertile plains of the east and south, including:

- 11 provinces, municipalities and autonomous regions along the southeast coast;
- Yangtze River valley (of which the delta region is the most populous);
- Sichuan basin;
- North China Plain;'
- Pearl River Delta; and

- the industrial area around the city of Shenyang in the northeast.

By contrast the population is sparsely distributed in the mountainous, desert and grassland regions of the northwest and southwest, where in 1985 the Inner Mongolia, Xinjiang and Xizang Autonomous Regions (Tibet) and Gansu and Qinghai provinces, which account for 55 per cent of the country's land area, accommodated only 5.7 per cent of China's population.

This unbalanced distribution of population is reflected in the distribution of regional economic, physical and human development as measured by appropriate regional indicators such as GDP, social consumption and capital investment in assets (see Table 1.1). This will be further explored in the section on 'Social Harmony' later.

Table 1.1   Differences in regions

	GDP by regions (thousand yuan)	Consumption expenditure (thousand yuan)	Household consumption (thousand yuan)	Government consumption (thousand yuan)	Percentage of asset projects under construction in whole country	Percentage of newly increasing fixed assets in whole country
Eastern area	515,642.2	251,526.8	188,177.2	63,349.6	40%	56%
Central area	239,743.6	139,296.4	107,354.2	31,942.2	32%	22%
Western area	121,325.5	77,149.2	59,806.4	17,342.8	28%	18%

Source: China Statistics Year Book 2000.

It also has consequences for both the densely populated and underpopulated regions. The densely populated regions are adversely affected by development pressures associated with urbanization and inward migration, and the loss of valuable quality agricultural land, whilst the underpopulated regions fail to benefit from the development process that is ongoing in China today and generally suffer from social and economic deprivation.

In summary, population is a fundamental issue for China for the reasons that:

- simple growth will inevitably produce a further 250 million (approximately) mouths to feed between 1995 and 2025, placing demands on agriculture and the fertile lands located in the south and east, which are also attractive to developers;

- uncertainty is endemic in the population forecasting process where small changes in fertility or mortality rates could result in huge differences in the number of people to feed;
- only if the adults between 20 and 49 years of age have a very low average fertility rate would it be possible to stabilize China's population below 1.5 billion;
- the unbalanced distribution of population is driving the process of rural–urban migration, which in turn is putting the fertile land of the west and south under even greater development pressure;
- social and economic deprivation in the underpopulated lands of the southwest and west as workers are laid off with the ever increasing efficiency of agricultural production.

Poverty

Poverty was endemic in China in the last century, affecting millions of families and it still exists today, although in a different form and magnitude (Hutchings 2001: 344–5). Its root causes were:

- overpopulation;
- a weak agricultural base;
- a fragile environment; and
- economic development policies set in the context of globalization.

These factors are likely to keep China poor on a per capita basis for many years to come, especially as the political, social and economic systems operating in today's society lead to an unequal distribution of any wealth that exists.

The removal or reduction of poverty, both individual and national, has been the driving force behind every movement for political change in China, including the efforts of Sun Yat Sen, Mao Zedong and Deng Xiaoping. Regardless of the fact that each of these leaders followed different policies, basically they all wanted to make China powerful and to improve the quality of life of its people.

The Chinese Communist Party through the People's Republic of China after 1949 was the first government to make any significant progress in alleviating poverty through its policies of land reform, which raised the incomes of the poorest peasants, and collectivization and communization, which introduced the principle of egalitarianism.

Nevertheless the gains achieved through economic growth and distributive policies in the early years were eroded by the Great Leap Forward in 1958–60 and the Cultural Revolution in 1966–76, which condemned the population to long-term poverty. It was only with the introduction of reform and the 'open door' policy of Deng Xiaoping in 1978 that an estimated 200 million people were lifted out of absolute poverty between 1978 and 1995 through a return to household farming, which led to a rapid increase in rural incomes and similar reforms in the urban areas, in turn leading to the rapid growth of foreign trade and investment income and a major improvement in living standards.

Given the different measures that have been used to measure poverty in China it is not clear how many people in China are poor. Using the World Bank's $1.00/day income measure, the number of poor is estimated to have dropped from about 490 million in 1981 to 88 million by 2002. Alternatively, using the World Bank's $1.00/day consumption measure, the number of poor declined from 360 million in 1990 to 161 million in 2002. According to the Chinese government, China's poor decreased from 250 million in 1978 to 29 million in 2005 (Chen 2005: 504). Whatever the actual situation is it is clear that China's achievement in reducing poverty has been significant.

Whilst much has changed in China during the 30 years of reform and economic growth at the same time little has changed in other respects. Access to social welfare, typically in education and health care, are still unequally distributed between urban and rural areas, and between coast and interior areas.

Poverty is endemic in the rural west of China, especially in Gansu, Qinghai, Ningxia and further south in Guizhou. At the same time there has been an increase in the urban poor, resulting largely from the de-nationalization of state enterprises which have had to shed jobs to survive. These new urban poor have not only lost their jobs but have also lost housing, medical care and education facilities which under the old state system were provided by their employers.

Unemployment

China has about 21 per cent of the world's population, but its labour force only occupies about 26 per cent of that globally. Employment opportunities have always been the main challenge in China. At the beginning of the reform and open door policy, the economic development policy was following the principle of high economic growth rates and high employment. However, in the transition to the market economy, especially after entering the World Trade Organization, there has been more serious competition from other countries. Increasing efficiency has been regarded as a priority by both private and state-owned enterprises. The former policy and practice of high growth and high employment will have to be changed. China is facing the risk of a serious increase in unemployment.

Urbanization may help to deal with the challenge in the rural area, i.e. the surplus rural labour force, lower income of peasants, polarization between urban residents and rural residents, and improving the living quality of the rural residents, but it also creates some problems due to the need to provide enough jobs in the urban areas where there is a serious problem of increasing unemployment, especially in the traditional industrial cities with many state-owned enterprises.

The unemployment risk has already attracted the attention of the government. According to a report by the main Chinese news agency, Xinhua News Agency, the creation of more jobs to relieve the urban poor was the major issue in the First Section of the 10th National Committee of the People's Congress in March 2003.

It was reported that according to statistics from the Ministry of Labour and Social Security, the young labour force will reach a peak during the period 2001 to 2005, averaging an annual growth of 2.9 million. This will bring the total unemployed up to 22–23 million every year, including those who are laid-off from inefficient state-owned enterprises. Moreover, the number of jobs created each year will only be 7–8 million. This leaves a job deficiency of 14–15 million. In addition, there are about 10 million rural labourers moving to urban areas every year. This leads to an increase of 300 million in the urban population in the future (Li 2012). According to former Chinese Premier Wen Jiabao China has an unemployment figure of 200 million (Wen 2010).

Some academics doubt if the cities are able to provide enough job opportunities for such a large migration, and if there is adequate available land and other necessary resources to accommodate all these potential migrants from rural areas (Cao 2002). These doubts raise the important issue of maintaining a higher rate of development in order to provide necessary job opportunities for increasing migrants from rural areas, and the increasing number of unemployed in urban areas is a problem that, unresolved, will contribute to a serious social issue challenging the Chinese government.

Deng Xiaoping stressed the concept that 'development is the absolute principle'. Development has been the government's major objective since the reform policy and during the last two decades has generally increased the living quality of the people and the GDP. The objective of achieving the economic development rate at defined stages has generally been achieved. However, both Chinese central and local government face great pressure to maintain the rapid economic development rate to provide sufficient job opportunities to resolve the problems of unemployment and mass migration. Its maintenance is an ongoing challenge.

Economic Development

Until 1978 China was undoubtedly a developing country operating a centrally planned and managed system of government based on the former Soviet system. Since 1978, it has introduced and promoted economic reforms which have moved the country away from the centrally planned economy towards becoming a market economy. An economic survey carried out by the OECD in 2005 suggested that the development of the Chinese economy can be categorized into three phases:

- communism 1949–1980s;
- communist capitalism 1985–2005; and
- capitalism 2005 onwards.[3]

3　See www.oecd.org/dataoecd/10/25/35294862.pdf.

It is found in the same survey that China's economy had grown at average 9.5 per cent per year between 1985 and 2005. It therefore resulted in China's national income doubling every eight years and this level of growth was likely to be sustained. The Chinese economic growth had contributed not only to higher personal incomes, but also to a significant reduction in poverty and improvements in the social welfare system. The Chinese economy resulted from the profound shift in government policies and reforms introduced by the 'open door' policy that had allowed market prices and private investors to play a significant role in production and trade (OECD 2005).

The transformation started in the agricultural sector in the 1980s and was extended progressively into industry and large parts of the service sector through the mechanism of the so-called 'responsibility system' (Hutton 2007). The research by Chen (2005: 505) suggests that price regulation was effectively dismantled by 2000. The transformation to a market economy changed the prior system of job security. As a consequence, unemployment increased systematically from 2 per cent in the 1990s to 4 per cent in 2002.

Throughout this period of sustained growth there has been economic and political stability. The OECD report (2005) noted that challenges remained to be addressed including the need to provide a firm framework for private sector activity, through a further modernization of the business structure and better enforcement of rules in the economic sphere, especially those relating to intellectual property rights, bankruptcy and property. The Chinese economy requires maintaining a stable macroeconomic environment, through the introduction of greater flexibility in the exchange rate which would allow the authorities to guard against any further increases in inflation. Reform in the financial system, through the expansion and further deregulation of the capital markets, is necessary. Finally, it is necessary to reduce regional inequalities, through additional expenditure on health and education in both rural areas and for migrants in urban areas, and the reduction of restraints on internal migration.

It is mentioned in the OECD report[4] that attention should be drawn to the implications for the Chinese economy of the rapid aging of the population between 2005 and 2025. The reorientation of expenditure on health and education may increase the contribution to the poorer parts of the country. In order to encourage faster growth in western China, the Chinese government may increase expenditure on infrastructure and education in its poorest western areas. To promote the urbanization process, the Chinese government will amend the restrictions imposed upon individuals changing their place of residence, thus making rural–urban migration easier. As a consequence, there will be a large-scale rural to urban migration, which must be carefully managed. Tax reforms should include the equalization of the rate of company taxation for domestic and foreign-owned companies; and the taxation should be paid fairly to domestic shareholders;

The OECD report concludes by stating that:

---

4   www.oecd.org/dataoecd/10/25/35294862.pdf.

> National income has been doubling every eight years and this has been reflected in the reduction of the poverty rate to much lower levels ... average incomes in major coastal areas are on a similar development path to that seen in other east Asian countries one generation ago. Considerable challenges face the economy, not the least of which is a rapid increase in the age of the population, but continued evolution of economic policies, especially in the areas of the allocation of capital, labour mobility, urbanization and the creation of an improved framework for the development of the private sector of the economy, should ensure this development momentum is sustained. (OECD 2005: 7)

Whilst there is a consensus that the bank credit system and the capital market in China need to be reformed there are broadly two different perspectives on the current position.

The first view held by the OECD (2005) is that the Chinese government has taken steps in recent years to stabilize the banking system, which if successful could avoid an economic crisis. For example, by taking bad loans out of the banks and placing them in specially created institutions to deal with those loans; transforming the banks into shareholder corporations, and approving foreign banks as major shareholders in the three largest banks in China.

This is not a view that is uniformly accepted, and the second perspective represented in a BBC report prepared by a multinational company (Mason 2008) argues that the Chinese economy 'is suffering a form of economic malaria with a growth rate of 9+% per annum and core inflation at 0%, which is a huge anomaly'. It further notices in the report that there is a chronic oversupply in the economy financed by state-owned banks with companies making 'too many widgets that nobody wants'. The Chinese government has controlled the retention of capital, which has been preventing the crisis. However, whilst the Chinese people save a large amount of money they have nowhere else to put their money other than the Chinese banks. If China liberalized its exchange rate there would be a flood of savings out of the country with a consequent banking crisis and an economic slowdown.

Thus, even amongst the 'experts' there is a difference of opinion concerning the ability of China to sustain its average economic growth rate of 8 per cent per annum. If it cannot continue to grow at the rate of the past two decades then the social, economic and political consequences could be serious. The future is clearly uncertain and this issue of sustained growth will dominate and underpin all other issues identified in this section.

In addition Chen (2005) identifies a number of other 'lower order' issues arising out of the economic development over the past 20 years, including:

1. ecological and environmental change, where:

    - China has 16 out of the world's 20 most polluted cities;
    - suffers extensive air and water pollution; and
    - is a world leader in the production of sulphur emissions;

2. social and cultural changes, where:

- the divorce rate has more than doubled between 1985 and 2001, from 0.9 per cent to 2.0 per cent;
- crime has increased significantly with serious crimes such as murder, robbery and theft increasing substantially since the reforms;

3. social and economic disparities, where there are significant disparities between:

- rural and urban income and welfare benefits, with China's eastern and coastal provinces and municipalities enjoying the highest per capita GDP;
- the urban middle class and the 'new urban poor';
- rural migrants in the urban labour market, where the migrant wages are 80 per cent of the wages of local citizens; where local citizens enjoy better provision of housing, health care and retirement benefits; and where rural migrants achieve less and are not advised of their employment rights. (Chen 2005: 504–9)

Collectively these issues are giving rise to unrest among significant sections of the population and have led to the government introducing its policy to foster social harmony.

Mackerras (2001: 187–208) also draws attention to other issues arising from the last 30 years of economic development and population growth.

The land best suited for agriculture is concentrated along the coast and the main river basins, which is also the area where economic development and urbanization is most rapid. The occupation of valuable arable land for urban development has then been a critical problem which may impact the food security of the 1.34 billion population. Since the open door reform there has been a return to family farming and a move away from grain to cash crops with the family farms being too small for efficient farming.

There is a mismatch between the resources and development. China has large mineral reserves in the northern and inland areas, e.g. coal, which gives rise to problems of transporting them to the developed coastal areas. China's energy is provided predominantly from coal (75 per cent), oil (20 per cent) and hydroelectric power (5 per cent). Coal, which is the largest source of energy, is generally of poor quality; often not washed and badly processed and when burned is a major source of air pollution. The difficulty of transporting coal from its point of origin to where it is needed accounts for much of the congestion on the railways and roads. Yet despite being well endowed with minerals China is heavily dependent on the import of iron ore, finished steel work, and oil to support its development programme. This creates a risk in sustainable development that Chinese development is not a self-reliant approach but is dependent upon the outside world. Given China's limited oil reserves, securing oil from abroad is now becoming a major worldwide issue.

Rural–Urban Migration: The Floating Population

China's urban population is remorselessly increasing as rural migrants are attracted by the 'pull factor' of new, well paid industrial-related jobs in the urban areas and the 'push factor' of lack of employment in the countryside. It is clear that rural–urban migration is driven by the income differential between urban and rural areas. This issue is becoming more significant in Beijing. In 2006, it was found that 38.8 per cent of rural–urban migrants came to Beijing directly for a better income and 22.8 per cent had to leave their home due to the poverty of their villages (Zhai et al. 2007). It is evident that some rural–urban migrants coming to cities for better life and some are coming out of necessity in order to survive. Correspondingly, others are seeking to improve personal skills and/or to increase their knowledge, which can be regarded as a type of 'development'. According to the 0.1 Percent of Population Sampled Survey of Beijing in 2006, 18.2 per cent of the rural–urban migrants who moved to Beijing could be categorized as moving for 'development' in skill and knowledge (ibid.). The income of rural–urban migrants varies depending upon their actual occupations. As a consequence the development of new urban areas is encroaching on previously undeveloped agricultural land at the same time as the existing urban infrastructure is being put under considerable pressure. Rural–urban migration and urbanization are inextricably inter-related.

Through research on Chinese urbanization, Bachmann and Leung (2008) argue that the Chinese government relaxed restrictions on migration to large and medium sized cities in an attempt to improve industrial growth there in the 1990s. In doing so it created a 'floating population' of rural migrants working in urban areas without '*hukou*' (household registration). This loosening of control over migration to the large and medium sized cities has had a dramatic impact on the rate of urbanization since 1995 and has reinforced the unequal distribution of population, wealth and services referred to earlier.

The parallel policy of encouraging the development of towns and small cities has had mixed results – most of the growth of towns has taken place in the eastern and southern regions where export oriented manufacturing has benefited both towns and cities. It would seem that:

> National macro economic and trade policy had contributed more to town development in coastal regions than urbanization policy ... while the pro-town policy allowed rural residents to participate in strong regional economic development, it has not catalysed large scale economic growth in towns in the absence of linkages to cities. (Bachmann and Leung 2008: 15)

Characteristically, in this rural–urban migration surplus farm workers (predominantly male) have taken low paid jobs in the cities and towns, mainly in the construction industry, whilst young female workers have taken assembly line jobs in factories that require dextrous hand–eye coordination. The construction workers are predominantly engaged in building new infrastructure and industrial and residential projects associated with the inward investment for industrial development. In particular there is significant investment into urban and new suburban development projects to accommodate the new Chinese middle class in cities such as Shanghai, Xiamen, Guangzhou, Shenzhen and Beijing.

These 'floating' workers have transferred millions of yuan back from the urban areas to their families in the countryside – an important redistributive function. Given that an estimated 800–900 million people still live in the poor agrarian provinces it is generally accepted that people will continue to pour into the major cities for the foreseeable future.

For the sake of social harmony it is vitally important for the government that the economy continues to grow and that the middle class have jobs and somewhere decent to live.

Social Harmony

The Third Plenary Session of the Eleventh Central Committee of the Chinese Communist Party in December 1978, the milestone of the open door and reform policy, formally decided that the predominant task of the Communist Party and the Chinese government should be to focus on the transition to economic development and modernization from the 'Class Struggle' (Hutton 2007). The strategy since then has been to promote economic development and to attract inward investment. Development in China started from the eastern coastal areas, using their advantages of a comparatively highly skilled labour force, convenient links with other parts of world, existing industrial bases, the tradition of the business atmosphere, and the close relationship with overseas Chinese.

This policy was a shift from the former one which attempted to reduce the regional inequalities. As a consequence of the new policy, the east coast areas have developed so rapidly that the gap and polarization have expanded between eastern China and central China, and especially western China. Data from China Statistic Book of 2000 illustrates the polarization of the regions (tables 1.2 and 1.3). In the tables, Mainland China is divided into three areas: the eastern, central and western areas. The eastern area with a total population of 511 million (NBS 2001), refers to the three mega-cities of Beijing, Tianjin and Shanghai, under direct jurisdiction of the State Council, and the provinces of Hebei, Liaoning, Shandong, Zhejiang, Jiangshu, Fujian, Guangdong, Hainan and Guangxi; the central area with a total population of 443 million (ibid.), includes the provinces of Heilongjiang, Jilin, Inner Mongolia, Shanxi, Henan, Anhui, Hunan and Hubei; and the western area with a total population of 288 million (ibid.), consists of the mega-city under direct jurisdiction of the State Council, Chongqing, and the provinces of Sichuan, Guizhou, Yunnan, Tibet, Shaanxi, Gansu, Qinghai, Ningxia and Xinjiang.

Table 1.2   Total and selected items of government revenue in different regions (1999) (10,000 yuan)

Region	Total	VAT	Enterprises' income tax	Individual income tax	Tax on real estate	Land use tax	Land value added tax	Tax on occupancy of cultivated land
Eastern area average (yuan/person)	676.07	121.43	112.22	59.52	22.71	5.84	1.06	3.73
Central area average (yuan/person)	300.15	49.66	28.79	15.75	9.01	4.70	0.18	1.99
Western area average (yuan/person)	281.12	46.40	27.86	13.48	9.51	2.92	0.21	1.78

Source: China Statistics Year Book 2000.

Table 1.3   GDP by three areas

| Region | Gross domestic product (100 million yuan) ||||
	1996	1997	1998	1999
Eastern area	39,532.08	44,453.40	48,070.92	51,564.22
Central area	19,167.58	21,642.90	22,871.37	23,974.36
Western area	9,613.22	10,728.88	11,552.05	12,132.55

Source: China Statistics Year Book 2000.

The data illustrates that although the economic development of the three areas has been growing, the rate of growth has been very different. The rate of the central area was double that of the western area and that of the eastern area double that of the central area. In the case of assets and capital construction projects, the eastern area has more projects than that of the central and the western areas, but the difference was much less than that of the GDP. The data indicates that, during the last few years, the Chinese central government has tried to develop the western area to reduce poverty as well as to promote its development. Investment in assets and capital construction projects has been the priority of the central government's policy. The central government has issued several favourable policies to encourage the investment in and development of infrastructures in the western area.

According to the China Statistics Year Book of 2001 (NBS 2001), government revenues in the eastern area were 345.5 billion yuan;[5] those of the central area were 133.1 billion yuan; and only 80.9 billion yuan were collected in the western area. The expenditure divided by the regions was 476 billion yuan, 248.9 billion yuan, and 174.2 billion yuan in the eastern, central and western areas, respectively.

There are many reasons for the inequality of development in China. In addition to its accessibility problems in some areas, and less investment and development in the past, the development concepts among the local governments and local people have had an important impact. According to Cao (2002), modernization and urbanization within the Chinese central and western areas have led to high-rise buildings and wide roads to demonstrate the so-called 'political achievements' of local politicians. The cost of all these has been paid by the local residents and income from leasing the land. Moreover, the general results of industrialization, especially the development of township enterprises in the central and western areas, have been the drawbacks of resentment and enmity. The main reason for the failure of the industrialization in the central and the western areas is the fact of the eastern area's reliance on the central and western areas as a market for their products and the suppliers of a cheaper labour force. The huge gap between the eastern area and the central and western areas in terms of income distribution and development is serious. The people of the poor areas missed the centrally planned economy in the first 30 years after the founding of the PRC.

Social and economic disparities have been a serious problem in China in the decades after the delivery of the reform policy. According to some information from the media, there are about 1,000 billionaires and three million millionaires in China. However, there are still 88 million to 161 million people of absolute poor status depending on either standard of the World Bank's $1.00/day income measure, or the World Bank's $1.00/day consumption measure.

It is evident that the urbanization process, typically the migration from the rural areas, and the disparity of development as well as economic policies, based on the principle of 'let someone become rich first',[6] has inevitably created disparity in Chinese development.

In many cities, migrants (or 'floating' workers) compose about 20–30 per cent of the local population. However, little consideration is given to supporting the living requirements of these migrants in terms of housing, infrastructure and the education of their children. Many young migrants have to live in dormitories provided by the enterprises, or have to find accommodation that they are able to afford, which is usually on the periphery of the city. These migrants normally

---

5 The exchanging rate between the pound sterling and Chinese RMB yuan at the time was that one pound could be exchanged to 13 Chinese RMB yuan.

6 The different policy of the Chinese government of 'absolute equality of all' at the beginning of the open door and reform policy for the purpose of increasing efficiency.

live together with other migrants from the same area. The settlements for these migrants are named after their home provinces or areas, e.g. 'Xinjiang Village' for those who are from Xingjiang, 'Zhejiang Village' housing those from Zhejiang Province, etc.

However, at the same time, luxury houses are popular in some metropolitan areas. The cost of a luxury house in Beijing can reach 100,000 yuan per square metre (the equivalent of 10,000 pounds sterling). The price of a house can be multiple 10 million yuan. In 2001, the price of a luxury house (villa) is around 136 million yuan in Shanghai, but the national average income was only 6,860 yuan in the urban area and 2,366 yuan in the rural area. Thus, 20,000 years' income is necessary for a normal citizen to buy such a house. Ordinary citizens with higher incomes in Beijing and Shanghai are difficult to afford such a luxury house.

The government of the People's Republic of China has recognized that increasing inequalities in society, if not addressed, could have significant negative social and economic impacts and undermine stability in society. The inequalities will lead to inefficient utilization of human capital, constrain economic growth and social development, and undermine the country's long-term prosperity, finally making reforms more difficult.

Whilst per capita GDP in China has grown at an average annual rate of more than 8 per cent over the last three decades, leading to a dramatic decline in poverty, this rapid growth has been accompanied by rising income inequalities which result from increasing income disparities between urban and rural areas and across regions. The de-nationalization of former state-owned enterprises and operations has led to the laying off of millions of people; the privatization of new initiatives has led to cost-effective operations where the wage rate is kept as low as possible through the employment of migrant workers and social benefits formerly associated with employment are no longer forthcoming.

In the view of Professor Justin Lin, Director of the Chinese Economy Research Centre at Beijing University:

> The widening income gaps are rooted in the incomplete transition from a planned economy to a market economy, with remaining ... price distortions and interventions in resource allocation inherited from the era of planning ... Increasing income inequalities further lead to increasing inequalities in the access to basic education and health care. (ADB 2007: 1)

The Chinese government has taken steps to address these increasing inequalities. It is considered by the Chinese government as a priority problem to tackle in the next 20 years. In November 2002, in the Sixteenth Central Committee of the Chinese Communist Party, the policy of 'let someone become rich first' was changed to 'a well-off society in an all-round way'. This policy emphasized prosperity of all the 1.3 billion people in China instead of only a few who have become wealthy. Mr Jiang Zemin, the former president of China, indicated that a prosperous life on average in terms of the country's whole population is still at a low level. It was decided that the government should concentrate on building a well-off society of a higher standard in an all-round way to the benefit of the well over one billion people in China (Jiang 2002).

The Eleventh Five-Year Plan 2006 made the establishment of a harmonious society in China the basic development goal. It is inevitable that a harmonious society has to be based on high and sustainable economic growth, and more jobs and development opportunities being created. It is obvious that the urgent requirements at present to establish a harmonious society need more investment in basic social services, such as education, health and other social services; physical infrastructure; and ensuring equal access to these basic services so that the whole of society benefits from growth. The development of a policy of Corporate Social Responsibility (CSR) is priority.

The Chinese government is committed to promoting CSR by improving its legal framework and strengthening enforcement, providing incentives through administrative and economic means, and building additional capacity through enhanced public awareness and training. Business should be the major contributor to CSR although other stakeholders should also play an important part in the process.

Despite these intentions there is a grumbling dissatisfaction within the working population with working conditions and the way in which local government operates. The BBC's *Newsnight* has reported on issues of dissatisfaction – about labour relations in Shenzhen and Wuhan (Mason 2005 and 2008). They are generally indicative of the problems faced by the working population and the ways in which the government is moving to create a more harmonious society.

Shenzhen   The labour force in Shenzhen is largely young, female and powerless, regardless of the fact that it is better educated and generally knows more about its rights than the previous generation of workers. They can also access to information from the Internet regarding those rights, and increasingly they are developing an image of what they might achieve by way of improvements to their living standards. The subservient attitude of the first generation of migrant workers, who were concerned solely with making a living, has gone. Yet in Shenzhen today a female migrant worker living in a dormitory, working compulsory overtime and a standard six-day week can expect to earn US$110, producing the clothes worn in the West, the toys Western children play with and the TVs and white goods consumed in the West.

Although wages and workers' aspirations are rising, workers have few employment rights. When an accident happens, even when compensation is formally awarded by the courts for an industrial injury, the factory owners often refuse to pay out. For example a woman migrant worker who damaged her fingers in a hydraulic press went to court and was awarded compensation but when she went to the factory and asked for the money for an operation to repair her hand the factory owner refused to pay and threatened the woman with violence.

However the situation is changing in Shenzhen – in 2007 one in 10 migrant workers did not return after the spring holiday. They are voting with their feet against low pay and poor conditions.

In Fuzhou similar labour relations problems are reported, where a world renowned shoe manufacturer violated China's labour laws by requiring staff to work for more than 70 hours a week when the law limits the average working week including overtime to 49 hours; underpaying staff and obstructing attempts to establish a trade union.

Wuhan   In Wuhan a different situation exists, which illustrates other aspects of the labour relations problem. Here, the heavy industries which dominate the area are still in the control of the state. These factories provide a job for life and benefits, including health care, schools and housing. However since the introduction of the market economy the workforce has been reduced by 50 per cent, creating a pool of unemployed, and the benefits privatized. The industries have a trade union but that union is led by a senior company manager who is also the leader of the Chinese Communist Party for Wuhan. The union is part of the state, and there has been no strike in 50 years. The official line is that the interests of the state and workers are the same and any problems are resolved through mediation.

The conditions in Shenzhen and Wuhan are repeated throughout the country and factory unrest has led to the introduction of the Labour Contract Law 2007 and associated regulations in an attempt to improve working conditions and improve the chances of securing a harmonious society. The previous Labour and Contract Law 1997 was vague, with employment contracts taking the form of simple templates with names and dates filled in, leaving many critical terms and unique circumstances unmentioned. Vast resources were wasted on the resolution of repeated dispute issues. Under the new (2007) legislation which came into effect in January 2008 there is a concern to secure social harmony, with:

- workers having a written contract including probationary periods and a minimum wage;
- fixed-term contracts being controlled, with a maximum of two fixed-term contracts which can only be extended by an 'open term' (permanent) contract;
- consultation on employers' conditions of service;
- exemption of certain conditions of termination, e.g. pregnancy;
- mass layoffs being permitted only in prescribed circumstances and with severance payments.

After promulgation of the new legislation, provincial regulations are following up to assist with interpretation (Eversheds 2007). With the implementation of the legislation, the business model on which Shenzhen's development was based is no longer appropriate.

Meanwhile, the official state-run union is undergoing a reformation and is changing its methods and mindset – it is recruiting migrant workers and giving them contracts; it has been engaged in a battle with Wal-Mart over the need for Wal-Mart to negotiate with the union. Under the old centrally planned economy orders were given from the top and implementation followed through to the bottom. There was no need to organize the workers. In a market economy the situation is uncertain and complex, the workers need leadership and guidance. Market forces are pushing wages up and migrant workers' lives are becoming more similar to the lives of registered workers in the cities.

The search for social harmony is likely to lead to enhanced life chances for the workers, provided that the economy continues to grow. However protest at official policies is not restricted to the workers. In Shanghai middle-class 'NIMBYs' protested in January 2008 against a £700m extension of the world's only magnetic levitation rail route, which runs through the city centre to the domestic airport. This rare show of middle-class anger against the policies is complaining about a sharp fall in property prices along the 20-mile rail route.

Food Security

China, with 7 per cent of the world's land and 21 per cent of the world's population has always had difficulty in feeding its large population and, despite tripling grain production from 113 million metric tonnes in 1949 to 454 million metric tonnes in 1993, without appropriate policies to support agriculture and protection of arable land, China may face a food crisis. A policy document, produced by Science and Education for a Prosperous China, which was summarized by the US Embassy in Beijing (1996) suggests that the issue of food security is taken seriously by the Chinese government which is concerned to provide 400 kg of grain per head of population. Since the population increases and the absolute arable land area and per capita water resources decrease, a productivity increase of 80 per cent per unit area will be required if this is to be achieved. Chinese productivity in grain production, which in the 1980s was only 50 per cent of that of the USSR; 28 per cent of the US and 20 per cent of Canadian levels, was low and needed to be improved. It is because of more people living in cities and improvement of people's living quality that general dietary habits of Chinese people have been changed. Urban people eat twice as much meat as rural people and as rural–urban migrants adopt urban patterns of consumption so the demand for meat will increase, as will the demand for additional grain to supply feed for the animals which will supply the additional meat. This has implications for food security.

However, several obstacles to achieving food security in China were identified in the report (USA Embassy Beijing 1996). They consist of natural obstacles, which include the annual loss of arable land; extensive flooding of some arable land; desertification, salination and alkalization; and drought and shortage of water. From economic perspectives, the obstacles include the lack of investment available for agriculture as a consequence of the large investments in rapid industrialization and urbanization; the family basis of agriculture which prevents China from taking advantage of the efficiencies of large-scale agriculture. In terms of technical obstacles, the report (ibid.) commends that the duplication and poor coordination of R&D initiatives at national, provincial and county levels, and the lack of 'technical' knowledge that prevents many farmers understanding and applying modern techniques, are the main problems.

Moreover, China lacks appropriate resources, and suffers from poor management and the inability to make available improved crop varieties to all those who could benefit from them.

The report (US Embassy Beijing 1996: 1–8) concludes that:

- if China is to achieve the living standard of an advanced country by 2050 Chinese agriculture must increase production by 20–40 per cent, which can only be achieved through unit productivity increases;
- without more resources and new technologies the needed food increase will be difficult to achieve;
- the protection and improvement of ecosystems will become more important;
- Chinese agriculture has great untapped potential given that it still lags some 10 to 20 years behind the advanced countries;
- agricultural R&D needs structural reform, including more investment; an increase in expenditure on the provision of agricultural extension services to help farmers; a restructuring to eliminate overlap and waste among China's 1,400 agricultural R&D institutes and a concerted effort to retain agricultural graduates to work in agricultural research;
- priority should be given to the promotion of agricultural research since the time needed to show results is often long (10–20 years) and the process of bringing results to the field is difficult;
- boosting Chinese agricultural productivity is the main task.

A slightly more optimistic view is put forward by Heilig (1999) following a detailed analysis of China's ability to feed itself. He (ibid.) suggests there was certainly a problem relating to food supply between 1995 and 2025 when the country would have to feed an additional 260 million people (roughly equivalent to the then population of the United States), but a growing preference for meat will push up China's grain demand only moderately. In addition to the trend towards a more diversified diet in China, e.g. eating more vegetables, fruit and fish, and less roots and pulses, would change the structure of agricultural land use.

Heilig (1999) further mentions that the drive for urbanization will continue, given the huge excess population in agriculture, the income gap between rural and urban employment and the labour demands of the urban industry and service sectors. The consequence of urbanization will promote the development of a commercial agricultural industry and drive its expansion. Heilig (ibid.) argues that the extent of China's cropland had been seriously underestimated in the 1980s and 1990s with a range of 95–150 million hectares. Surveys by the Chinese State Land Administration supported by remote-sensing analysis by US scientists established that in 1988 China had a cultivated land area of 132.5 million hectares which declined slightly to 131.1 million hectares in 1995. China had arable land reserves that could be brought into cultivation and would increase its area of land with cultivation potential to 197 million hectares, despite the fact that 35 million hectares of this land are only marginally suitable for grain production. Although soil degradation problems do exist, they are largely outside cultivated areas.

It is the view of Heilig (1999) that land with water is the most critical resource in improving agricultural production, with a mismatch between the distribution of water resources and demand from agriculture, such that 58 per cent of cultivated land is in the north and northeast provinces which have only 14.4 per cent of total water resources. Nevertheless, China is wasting huge amounts of water through inadequate open canals irrigation systems, whilst water pollution is also a severe threat to agriculture.

Heilig (1999) concludes that China has enough arable land and water which makes it possible for it to feed its projected population of 1.48 billion in 2025, even without technical and scientific improvements to productivity. Indeed he claims that the country has enough potential to produce 650 million tonnes of grain, even when allowance is made for: (1) 25 per cent of arable land being used for other types of agricultural production such as vegetables and fruits; and (2) the continued encroachment of urban development onto farmland. However Heilig (ibid.) emphasizes that this is only a possibility which could be undermined by basic policy decisions, such as, China might find it economically more attractive to import a certain amount of grain instead of pushing domestic grain production to its limits thus freeing some parts of the limited cropland to be used for better purposes than producing rice, wheat or maize, and for political and economic reasons which cannot be anticipated there is no guarantee that China will develop its agriculture and its economy in a way that maximizes its agro-climatic potential.

However, Heilig (1999) is adamant that China would be able to feed itself if it undertook serious efforts to solve the following problems:

- improving water use efficiency in agriculture by increasing the water-supply deficiency in deficit regions; improving water quality and irrigation efficiency;
- implementing trans-river basin water diversion from the Yangtze River to the Yellow River to provide a better water supply to the North China Plain;
- removing the transport bottlenecks in the food supply chain, e.g. increasing harbour, rail and road capacity;
- promoting the development of larger farms by gradual privatization which would eventually introduce a market for arable land and more productive farms;
- increasing grain imports to free up some arable land for high-value crops such as vegetables, fruits or nuts;[7]

---

7 In 1999, when the report was published, Heilig held the view that by slowly growing grain imports into China of up to 50 million tonnes the world market would not be

- intensifying the implementation of flood prevention measures;
- strengthening biotechnology research into genetically modified plants and animals for China's future food supply;
- continuing the effective family planning programme to ensure that demand for food does not grow more rapidly than expected.

The final two paragraphs of the report succinctly summarize the basis on which China could feed itself and be self-sufficient.

> China's farmers are currently cultivating some 132 million hectares – an area which is almost 40% larger than previously estimated. Therefore, grain yields in the 1980s and early 1990s were inflated (due to the overestimated denominator), which – on the other hand – means that the farmers actually have more room to increase productivity. In addition there are some 30 million hectares of land reserves – primarily in the north. Most of the reserves in arid regions could be cultivated, if irrigation water would [sic] be available. We believe that the most critical resource for China's agriculture is water, not land. The regional distribution of water resources does not match the agricultural (irrigation) demand, and many rivers and lakes are seriously polluted. Massive investments into the water infrastructure will be needed. Soil degradation exists but often outside cultivated areas. Its effect on agricultural productivity can also be reduced by adequate management and conservation measures as indicated by the six-fold increase in grain production since the early 1960s.
>
> Finally it should be emphasized that economic and policy measures are the key to China's Food Security. The "Great Famine" in China, which probably killed more than 30 million people during the "Great Leap Forward" (between 1959 and 1961) was not caused by shortage of land or water, insufficient agricultural technology, or higher than expected food demand. It was caused by faulty policy measures. More than anything else, it will be the decisions of policy makers that will determine whether China can feed its population by the middle of the next century. (Heilig 1999)

According to the Xinhua News Agency, a survey published in March 2006 by China's Ministry of Land and Resources showed that between 1997 and 2006 a further 6.6 per cent (eight million hectares) of China's arable land had been lost (Xinhua News Agency 2006). It was further noticed in the report that a critical challenge was the poor quality of the arable land in China. Most of the arable land lost to urbanization in China is high quality agricultural land which has led to a position where only 28 per cent of the country's remaining arable land is high yielding farmland and 32 per cent is low yielding (ibid.).

---

disturbed because it was well within the 'normal' historical fluctuations and could be met by production increases in the United States.

In 2011, the Agricultural and Rural Committee of National People's Congress indicated that 8,200,000 hectares of Chinese arable land had been further reduced since 1997. The Committee noted that China had approximately 121.7 million hectares of arable land, covering 13 per cent of its territory which amounted to 0.11 hectares per head of population, less than 40 per cent of the world per capita average (National People's Congress 2011). The prime factors contributing to this loss of arable land include the booming economy and growing urban sprawl in China and the restoration of degraded or fragile ecosystems in western China. These factors account for around 85 to 90 per cent of the losses of arable land.

The challenge facing China is that China's population comprises 22 per cent of the world total. It has been growing by some 10 million people annually; however, the country has only 7 per cent of the world's arable land.

Although China recorded grain output of more than 589 million metric tonnes in 2012, the volume of China's food imports, e.g. the rice imports, were 2.32 million tonnes in the same year, reaching a new peak (Xinhua News Agency 2013b).

It is the opinion of Chen Xiwen, deputy director of the Leading Group on Rural Work under the Central Committee of Communist Party of China (Xinhua News Agency 2013a), that there is a potential threat to food security if Chinese grain output will not speed up in the process of rapid urbanization and industrialization. China should increase grain supply owning to urbanization which may unleash newcomers' demand for more farm produce. Chen further indicates that although China produced 159 million more tonnes of grain last year than in 2003, China's grain imports hit a record high in 2012 to stand at 72.3 million tonnes (ibid.).

China's first priority is to feed its own people, thus being self-sufficient, and although it has large grain reserves to weather the current world food crisis, globally politicians are concerned that any change in China's policy towards self-sufficiency could have a major impact on the global grain markets. In 2007 China produced more than 501.5 million tonnes of grain, which is very close to the nation's annual consumption; imported 31 million tonnes and exported nine million.

However, the UN warning which was delivered at the emergency food summit held by the British prime minister on 22 April 2008 put the global position into perspective: 'Global food shortages and rocketing prices are a "silent tsunami" that must be tackled urgently' (*Metro* 2008). Whatever the reality is, it is clear that food security is likely to be a major issue, both globally and for China, in the foreseeable future.

Sustainability and Environment

Sustainable development has been a crucial challenge in China and a major issue of concern during the rapid development. China has approximately 22 per cent of the world's population but only 6.4 per cent of its land resources and 6 per cent of its fresh water. Rapid economic development has led to damage of natural resources, especially land and water, and environmental pollution will be a continuing problem for the Chinese people. China has a serious shortage of cultivated land. The total area of China is 9.6 million km^2, but the available land for agriculture or urban development only amounts to 7.7 million km^2 (Yeh and Wu 1999). Sixty-two per cent of the cultivated land is located to the north of Huaihe River and its catchment areas. Water resources in these the areas, however, contribute less than 20 per cent of the total water resources in China. Areas with good water resources exist in the south along the coast but the total cultivated land of which only occupies 38 per cent (Chen et al. 1999). The sustainability issue in China will inevitably influence the whole world given the size of its population.

Although China is a country with a shortage of cultivated land, the occupation of cultivated land for urban development and industrialization has been a common phenomenon in most Chinese cities. After decentralization and marketization, local government has emphasized that total GDP increase and rate of development are the main objectives of planning and development policies instead of efficiency and quality. This has created a result that urban plans are proposing to expand the size of the city and to increase local urban populations. The expression 'to be stronger, and to be larger', which refers to being stronger in competition and larger in city size (population and physical area) and economic scale, has been appearing frequently in many urban plans. However, to achieve the targets, the possible measures are to expand the existing built-up areas, or to let the city sprawl. The consequence is then to occupy cultivated lands. The more rapidly developing areas, which demand land for city extension, are usually those areas with available and better quality cultivated land.

In addition to the lack of land resources, the shortage of water is also serious. There are about 300 cities lacking sufficient water supply, of which 114 cities have a serious water shortage. These include the rapidly developed cities along the coast, and some mega-cities, e.g. Beijing, Tianjin, Nanjing. The average water usage in Beijing is about four billion m^3. The deficit between demand and supply is around 16 million m^3 per day (Chen et al. 1999).

Development has also brought water pollution and seriously impacted on water supply. According to the investigation by the Ministry of Water Resources in the Eighth Five-Year Plan period (1991 to 1995), 46 per cent of the total of 700 rivers and 75 per cent of the 50 major lakes in China have already been polluted. Of the 19 urban rivers running through eight cities within the Yangtze River catchment area, 18 urban rivers suffered from pollution problems, the most polluted ones being the Suzhou River in Shanghai and Qing River in Nanjing.

The decline in the quality of the water system further damages the ecology system, and more seriously creates problems of water shortage. With the increase in population, economic development and the hastening of urbanization, the water problem that is illustrated in shortage and pollution, and flooding, will greatly affect sustainability in China.

The other major challenge of Chinese development is the quality of its environment. Sixteen of the world's 20 most polluted cities are located in China.

A negative impact of rapid urbanization and the economic development during the last 35 years have greatly increased environmental pollution and reduced cultivated land, the quantity and quality of agricultural crops are decreasing due to polluted land and water. It is also crucial that solid-waste production is expected to more than double by 2016, pushing China far ahead of the United States as the largest producer.

Although economic development has been maintained at a rate above 8 per cent annually during the last two decades, the treatment of pollution has taken about 8 per cent of GDP (NBS 2001). In other words, economic development has been at the cost of reduced environmental quality and resources of later generations. Township enterprises have greatly contributed to rural urbanization and local peasants' income, but, at the same time, they created serious environmental problems. In 1995, there were 1.2 million pollution generators emitted by 16.9 per cent of the total number of township enterprises in China at the time (Chen et al. 1999).

Since 1978, China's economic growth has lifted many millions of Chinese out of poverty, but it has come at the cost of the country's environment, much of it already degraded since 1949 by concentrating on the development of heavy industry in urbanized areas. China's main environmental problems relate to:

- air pollution and acid rain;
- water supply;
- water and river pollution;
- the loss of forested areas, which contributes to soil erosion;
- the loss of fertile agricultural land for housing and industrial development and urbanization.

Energy and air pollution The poor quality of air in the majority of China's urban areas derives primarily from the burning of coal as a source of energy and the ever expanding motor vehicle ownership and usage (Bradsher 2007).

China relies on domestic coal for about two-thirds of its energy needs and as it has abundant supplies of generally low grade coal and limited supplies of other sources of power such as oil, it naturally uses its domestic coal as the prime source of energy to drive economic growth. At the same time the environmental consequences of this policy for economic growth are compounded by:

- a massive programme of building more residential properties which contain energy-consuming modern equipment;
- increasing car ownership and usage and a reliance on heavy trucks for transporting goods throughout the country;
- the use of coal-fired power plants and industrial furnaces which operate inefficiently and use inadequate pollution controls.

Whilst acknowledging the environmental consequences of this policy, the debates are justified on the grounds that: (1) the elimination of poverty through continuing economic growth has a much higher priority than the resolution of environmental problems; and (2) all industrialized nations have, as they developed, created environmental problems which they delayed addressing until they were richer.

The coal-based economy is firmly entrenched. Despite the damage being inflicted on the environment by the discharging of CO_2 into the atmosphere the dependency on coal is difficult to change in the near future although nuclear power is a possible answer – it is estimated that when fully developed nuclear power could provide 20 per cent of China's energy demands (Wells 2005). During the past five years, renewable energy has been a development tendency. Renewable energy is helping China complete its economic transformation and achieve 'energy security'. China has rapidly moved along the path of renewable energy development. About 17 per cent of China's electricity came from renewable sources in 2007, led by the world's largest number of hydroelectric generators. Total installations of hydropower reached 145,000 MW in 2007. China has set a target of 190,000 MW for 2010; in terms of wind resources, China has the largest wind resources in the world and three-quarters of them are offshore. At the end of 2008, wind power in China accounted for 12.2 GW of electricity generating capacity and China has identified wind power as a key growth component of the country's economy; with regard to solar energy, China produces 30 per cent of the world's solar photovoltaics.[8]

Pollution in China is having a major impact on human health. The results of the limited research show that environmental pollution has become a major cause of death in China. The research of Kahn and Yardley (2007) reports that according to a World Bank study of Chinese pollution published in 2007:

- 99 per cent of China's urban population of 560 million now breathes air that is 3.5 times more polluted than the level considered safe by the European Union;[9]
- emissions of sulphur dioxide from coal and fuel oil, which can cause respiratory and cardiovascular diseases as well as acid rain, are increasing even faster than China's economic growth;
- other major air pollutants, including ozone and smaller particulate matter, called PM 2.5, cause more chronic diseases of the lung and heart than the more widely watched PM 10.

---

8   http://en.wikipedia.org/wiki/Renewable_energy_in_China.

9   A major pollutant contributing to China's bad air is particulate matter, which includes concentrations of fine dust, soot and aerosol particles less than 10 microns in diameter (known as PM 10). The level of such particulates is measured in micrograms per cubic metre of air. The European Union stipulates that any reading above 40 micrograms is unsafe. The United States allows 50. In 2006, Beijing's average PM 10 level was 141, according to the Chinese National Bureau of Statistics.

Water supply and pollution Kahn and Yardley (2007) and Yardley (2007) argue that water quality is perhaps an even bigger problem than air pollution. Whilst southern China is relatively wet, the north, where a large proportion of China's population lives, has suffered from a water shortage, with every indication that desertification could well take place in the near future. For example:

- the coastal cities of China, where 75 per cent of the wealth is based, are short of water;
- the water table in the north has been remorselessly depleted to the point where some wells in Beijing and Hebei Province have to be sunk more than half a mile before they reach fresh water;
- industry and agriculture in the north use nearly all of the degraded flow of the Yellow River, before it reaches the Bohai Sea.

The government has responded to this situation by building a major network of canals, rivers and lakes to transport water from the Yangtze River system to the Yellow River system, but even if this project is successful it will still leave the north short of water.

Despite the scarcity of water there is no appropriate policy and method to conserve water. By world standards water remains inexpensive and, according to the World Bank, China uses between four and 10 times the amount of water per unit of production than the average in other industrialized nations (Kahn and Yardley 2007; Yardley 2007).

At the same time the dumping of waste into surface water by factories and farms is uncontrolled. The Ministry of Environmental Protection Administration is of the view that one-third of all river water, and vast sections of China's great lakes, the Tai, Chao and Dianchi, have water of the most degraded level, making it unfit for industrial or agricultural use as the following quotation so graphically puts it in describing conditions around Lake Tai:

> Toxic cyanobacteria, commonly referred to as pond scum, turned the big lake fluorescent green. The stench of decay choked anyone who came within a mile of its shores. At least two million people who live amid the canals, rice paddies and chemical plants around the lake had to stop drinking or cooking with their main source of water. (Kahn 2007)

Climate change and desertification Desert occupies 25 per cent of the land area in China. With the climate of Central Asia becoming drier as part of the worldwide climate change process, and with overgrazing and deforestation, the deserts are growing by about 950 square miles each year. With fewer trees to block their path, fierce sandstorms, which sweep eastwards each spring from the northern desert areas, have become a regular feature of China's weather systems pushing the edge of the sands ever closer to Beijing:

A choking blanket of particles coats houses, cars, and people, and the city's hospitals become flooded with patients suffering from respiratory ailments. The dust clogs machinery, shuts airports, and destroys crops, forcing thousands of rural Chinese off their lands. Clouds of it blow throughout Asia, carrying pollution and potentially infectious disease. (Ratcliff 2003)

To combat this threat the Chinese government is planting a network of forest belts some 4,480 km (2,800 miles) long across the northwest rim of China skirting the Gobi Desert, the so-called China's Great Green Wall. By 2050 it claims that much of the arid land will have been restored to a productive and sustainable state. The first phase of this afforestation programme has been completed. At the same time Beijing has its own desertification project, which aims to increase the tree coverage around Beijing from the current 13.4 per cent of land area to 27 per cent.

There are other potential problems arising from global warming not the least of which is the danger of increased flooding and encroachment of the seas on the river plains of China on which the majority of China's richest cities have been developed – Shanghai, Tianjin. Even today, Shanghai, which is built on a river delta area, is slowly sinking under the weight of the buildings that have been constructed on land which is vulnerable to flooding. There are plans for a massive dam to protect the city from rising sea levels which are expected to have a significant impact within the next 20 years. In addition, 11 satellite cities, each accommodating more than one million people, are proposed and are being built for Shanghai on land which is at risk from rising sea levels (Wells 2005).

The conflicting politics of pollution and growth Since 2002 the Chinese government has acknowledged that environmental degradation has reached intolerable levels and that the growth of the economy should be held at a more sustainable level. However, there have not been many changes so far. For example, in 2006 China burned 18 per cent more coal than it did in 2005, whilst in the second quarter of 2007 the economy expanded at 11.9 per cent, its fastest in a decade.

The difficulty facing the Chinese government is moving to a position which balances economic growth and environmental protection could mean that the environmental problems are systemic. Through the economic reforms introduced in 1978 and reaffirmed in 1992 officials at the provincial, municipal and county levels have been required to increase economic growth and were given greater freedom to implement projects and policies that achieved that growth. The objectives of this system were to stimulate the economy and create jobs. Today the system has a momentum of its own, with local governments and business colluding to ensure that pro-growth policies are successfully implemented and environmental issues ignored.

Conclusion

China is a vast and diverse country. It is vast in terms of areal extent and population. It is physically diverse – topographically and climatically. It is socially diverse – ethnically, with the Han Chinese dominant; linguistically; and religiously. The size and diversity of the country inevitably result in complexity.

This complexity is compounded by the long history of social, administrative and cultural development dating from the fifth century BC through to the demise of the Qing Dynasty in 1912, when it was displaced by the People's Republic of China. During this period the country was at times dominated by foreign invaders who were eventually assimilated into the Han Chinese population.

This long history resulted in the development of an ideology that accepted centralized rule, was based on an agrarian economy which was self-sufficient and a society focused around a family centred social security system, which to a certain extent still underpins the Chinese psyche today.

At the turn of the twentieth century China faced a crisis of uncertainty. It had been attacked by Japan, was unwilling to adapt to changes in the outside world and was faced with a growing foreign presence in China. The republican movement led by Sun Yat Sen successfully ended imperial rule and the ROC began to attempt to modernize China – industrially, administratively and educationally – and to secure its international boundaries. It flirted with Marxism following the Russian revolution which showed many parallels with China and in 1923 the Chinese Communist Party was established.

Political and economic uncertainty continued until 1978 with:

- the Kuomintang attempting to impose central control over the warlords and the major urban areas and to secure China's boundaries against international intervention (1927–8);
- the civil war between the Chinese Communist Party and the Kuomintang, which resulted in victory for the Chinese Communist Party lead by Mao Zedong in 1949;
- the introduction of communist rule with major periods of uncertainty such as the Great Leap Forward and the Cultural Revolution.

This rather chaotic political and economic uncertainty was replaced by open door and reform policies introduced by Deng Xiaoping in 1978. Whilst in many ways a form of certainty has been introduced with the economic reforms, inherent in this new system is the uncertainty and complexity that goes with globalization.

There are several development challenges confronting China. In China, in which the population is increasing by 10 million people a year, it is inevitable that demands on the infrastructure and the environment are intense. At the same time the form of governance is changing from a centrally planning system to a socialist market economy. In the process, disparities arise in the equity of development, for example, discrepancies arise between the coastal regions and the poorer interior provinces, between the demands to sustain a still increasing population and the capacity of the environment and natural resources to accommodate an estimated 1.5 billion people by 2030.

Overall access and coverage of a range of basic social services has improved during the last two decades with notable progress towards, for example, reduction of infant and under-five mortality. There is almost equal enrolment of boys and girls in primary school; a rapid increase in housing stock and living space for families; and increase in access to potable water resources. Some important initiatives are also under way with respect to the prevention of desertification and land degradation.

On the whole, there is more that is similar than dissimilar with other developing countries in the kinds of strategic challenges China faces in achieving various goals and implementing provisions of conventions. Some of these challenges include:

- how to accelerate achievement of more regional equity and reduce inequities;
- how to maintain economic growth and avoid destruction and pollution of the natural resource base;
- how to expand the progress made in increasing access to basic services into higher quality and more widespread coverage;
- how to bring about the consistent implementation of existing laws and filling gaps in present legislation;
- how to better match gains in development of technical expertise with greater progress in managerial and regulatory competencies.

At the same time, however, there are some more systemic and probably deeper challenges associated with the overall stresses incurred in China's social and economic transition from a centrally planned economy to a socialist market economy. There are a number of instances in which national policy changes are under way or could benefit from major improvements, for example national health and nutrition; environment and energy and social protection.

China is wedded to an overall purpose of increasing the well-being of its people. Changing economic and social conditions over the next decade, partly spurred by leaps in technological applications, will open many possibilities for improvement in both societal and personal well-being. However, there remain a host of important national goals, partly inspired by global initiatives, the sustainable achievement of which would bolster both national and international development efforts.

Against this complex background the Chinese planning system, ranging from the National Socio-Economic Development Plan at the top, through provincial/regional plans to land use plans at the municipal/city/town/local area level, provides a one-solution-to-all-problems approach rather than reflecting the complexity of the issues with which it is attempting to deal.

Chapter 2
Evolution of the Chinese City and Regional Planning System

Introduction

This chapter explores the impacts of changes in society, politics and the economy to city and regional planning and its function from a historical perspective in Chinese history. In ancient China, methods of town construction were specified by either the code of rituals or *Fengshui*. It is useful to review and assess these ideas in the history.

However, the development of the modern Chinese cities had been very slow in the first half of the twentieth century. The function of urban planning was poor. There were neither national policies nor guidelines for city plan formulation. Numerous plans were made for Chinese cities, most of which were the concessions occupied by the Western countries. Even if the central government at the time attempted to produce city plans for the capital city, it had not been delivered.

The existing city and regional planning system was initiated in the centrally planned economy after the founding of the People's Republic of China in 1949. The planning approach had been influenced by the centrally planned economy. However, with the delivery of the market oriented 'open door and reform' policies by Deng Xiaoping in the 1980s, China started its transition from the centrally planned economy to a market economy. This transition has been associated with devolution, globalization, marketization, and rapid urbanization and economic growth. This chapter identifies the impacts and challenges of the Chinese city and regional planning system under the changing social, political and economic context in various eras in China.

Ancient Planning Concepts for Settlement and City Development

Concept of Order and Feudal Hierarchy

From the layouts of Chang'an, Kaifeng and Beijing, it is possible to find clear characteristics in Chinese ancient cities to regard the city as an complete and total entirety. Chinese cities in the ancient time were organized in an order with clear priorities from their forms or contents and built according to a unitary plan. The basic idea could be found as early as the year 2000 BC. According to 'Gun Built City and its Outline' (*Gun Zuo Cheng Lang*) in the chapter of '*Shiben. Zuopian*' in 'The Book of History', there were 'city' and 'outer ring city' at that time. Regulations were set out in the book to build a city to protect the emperor, and to build an outer ring city to settle the populace. It also regulated in the book that 'the gentlemen (higher class) should stay in a city and the populace should stay in rural areas'. In ancient China, it was held that a city should locate to the higher place instead of the centre. The concepts of location selection of a city proposed by Guanzhon (725–645 BC) are explained in the chapter 'Chengma' of the book *Guanzi* written by Guanzhong and interpreted by Chi in his book *Guanzi Research* (Chi 2004). It was recommended by Guanzhong when building a city that it should not be located at the foot of great mountains, but should be near to a broad river. The site selection for a city should be a piece of high land. However, it should not be too high, so that the supply of water is not disrupted. Neither should it be too close to the water, so that flooding can be avoided.

Guangzhong's concept of city building was based on the functions of a city. This was different from the later codes of rituals. Since the code of rituals had been established, the recommended concepts of the code of rituals, which were a set of moral principles and regulations, were more popular with the governments' rulers in the feudal society.

The code of rituals has greatly influenced China for more than 3,000 years. It also had great influence on the ancient Chinese planning concepts, explained in Chapter Kaogong Ji in the book of "*Ethical Code of Zhou's Rituals*" in the Dynasty of Xizhou (1100 BC to 700 BC). Chen (1984) interprets the ancient Chinese language of the *Ethical Code of Zhou's Rituals* and explains the concept of how an ideal city for a king should be built measuring 81 square *li*.[1] it should be provided with three city gates on each side. In the city, there should be nine transversal roads and nine perpendiculars ones. The width of the transversal roads should be nine times the span between two wheels of a carriage. A temple of ancestors should be located on the left, whereas the temple of the state should be located on the right. A court is arranged in the front part of city, and a market place in the rear. The court and the market place should occupy a total area of 100 *mu* (6,600 square metres).

In addition to definition of location for various activities, roads have also been hierarchically specified. It was suggested by the *Ethical Code of Zhou's Rituals* through interpretation by Chen (1984) that the width of the transversal roads is nine times the span between two wheels of a carriage; the width of ring roads is nine times the span between the two wheels of a carriage; and the width of rural roads is five times the span between two wheels of a carriage.

It was also regulated that a sovereign's city can be as large as 81 square *li*, whereas the cities for a duke should not be larger than 49 square *li*. The city of a marquis and counts should be controlled within 25 square *li*, and that of a baron should not be more than 9 square *li* (Chen 1984).

---

1 One Chinese ancient *li* is about 415 metres (1 *li* = 415 m); 81 square *li* is about 13.95 square km.

The core principles of the rituals are patriarchal relationship, moral principles, cardinal guides and constant virtues. It was proposed in the *Ethical Code of Zhou's Rituals* (Chen 1984) that king should be allowed to be a king, the minister be a minister, the father a father and the son a son. Within a group, the group interests ranged first, and the individual second. The individual should obey the group interests. An individual is required to be loyal to family, to the throne and to the country. In the feudal city, the forbidden city, which was both a living and official space of a king or an employer, should be located at the centre and symmetrical to both sides. There would be an axis to express the superior position of the emperor. The subjects of the feudal ruler should surround and protect the emperor according to their hierarchical positions. There was clear zoning for the functions of market, workshops and tombs. The heights of the buildings, number of rooms, the building materials, the structure and the colours had also been clarified according to the hierarchy of their classes. Each person should be at his own position without transcending regulations.

The traditional courtyard building, as the element of city, is the result of the code of rituals. In the courtyard building, the parents lived in the principal room. The elder son lived in the east-wing room, and the second son lived in the west-wing room. The servants lived in the outer yard. This was expressed in the philosophy of 'difference in inside and outside' and 'priority in ranking'. The grid pattern street links the courtyard building to *Hutong*,[2] to workshop, to market, and to government office. It was required to consider the geographic characteristics, climate situation, local building materials, natural costumes and other conditions to form the city pattern with a basic unity but variation in styles and highlighting in an organic entity. The general layout of the courtyard building and a city has been restrained by the influential philosophical concepts of Confucianism, but broke through with the social development in the history. The concepts of Confucianism are the basic philosophy that has greatly influenced Chinese people's moral character, regulated activities of the family, managed the organization and events of the state and pacified the whole of China. Confucianism is thus deeply ingrained in Chinese people's minds.

The Concept of Integrated Urban and Rural Areas

The concept of region in ancient China was of integrating planning and coordination between urban and rural areas. The concept could be found as early as in the chapter of 'To Develop Rural Land While Defining and Managing City' in the *Ethical Code of Zhou's Rituals* (1100 BC to 700 BC). The term of 'city' refers to large city, small city and urban proper. The term of 'defining city' is about the arrangement of scale and hierarchy of cities. The term of 'to develop rural land' relates to an appropriate treatment of relationship between city and countryside. It

---

2 *Hutong* means 'well' in the Mongolian language, it means the public water supply in the neighbourhood.

was understood in ancient China that a city was unable to develop independently. It was necessary to have a plan for regional integration and construction.

It was Shang Yang (fifth to sixth centuries BC) who had a completed explanation of the relationship between the city and its hinterland areas. According to Zhang Jue's interpretation of the ancient Book of Shang written by Shang Yang in 'Laiming Chapter', in an area measuring 10,000 square *li*, hills and mountains should occupy no more than one-tenth of the land; lakes and ponds one-tenth; valleys and streams one-tenth; the city and the roads one-tenth; waste land two-tenths; and fertile land four-tenths. In such an area, 50,000 peasants are required to supply the city with sufficient foods and material resources. The city and the roads will provide the people with a liveable environment (Zhang 1993).

This expressed Shang Yang's idea of integration between a city and its rural hinterland in term of support from rural areas in water, food, raw material, and others at the time. It was suggested by Shang Yang that to support an area of 10,000 square *li* (Chinese ancient *li*), which is about 1,722.25 square kilometres, 50,000 peasants were required to provide enough agricultural foods.

This concept of the integration of urban and rural areas illustrated a particular city region. It also defined the relationship and the draft ratio of land use.

Concept of Harmony with Nature

In the ancient time, the productivity of human beings was very low. The natural forces had significant impact on the volume of produce. People had to obey and show reverence toward nature. The earlier natural concepts in China can be found in the books of *Laozi* and *Yijing* that is also called *The Book of Changes*, which was published in the years of 770–221 BC. The book of *Laozi*, written by Li Er, who is also called Laozi, explains the philosophic ideas of Laozi, who respected nature with a simple but dialectical idea. Yang (2007) interprets the ancient language of *Laozi* into modern language which explains the dialectical idea of Li Er that an axle of a carriage concentrates 30 scrolls, it is possible to find the function of the carriage with the utility of hole in the centre of axle; moulding clay into a vessel, the vessel is able to be used with the utility of its hollowness; making a door and windows for a home, a room is able to be used because of the utility of its empty space; therefore, the being of things is profitable. The one being of things is serviceable.

When talking about the natural environment, it was the concept of *Laozi* that human beings should adhere to the ways of the earth; the earth follows the rules of the universe; the universe obeys the governing laws; and the governing laws are in harmony with nature. This emphasizes the priority of the nature in the activities of human beings. Laozi proposed to return to nature. He was a naturalist.

Yijing, a classical book, which is also called the *Book of Changes*, is about divination. The concepts of *Yijing* are said to originate from *Fuyi Divining*, and *He Lu* (*Rivers Map and Book*). The theory of 'Yi' (Changing) is too abstruse to have a full understand that the study of which has never come to completion.

However, the core idea is to consider nature (heaven and earth) and human beings integrated as a whole. *Fengshui* was derived from the 'Theory of Changes (*Yi*)'. There are some principles based on sciences. It expressed the experience in an ingenious way. The major principle of *Fengshui* is to depend upon the natural mountains and waters, and patterns of bios to seek the best place for living and development of human being by an approach of harmony between people and nature. It suggests following nature to avoid disasters and to seek the benefits by copying the changes of nature. The selection of location by *Fengshui* should seek the nature embrace, and it is necessary to protect the greening and some other natural geographic pattern. It also proposes that city and buildings should also embrace nature. Regardless of whether it is a city, a place or a house, they need to have surroundings to form a yard. This means to establish the inter-links between heaven and earth.

These ancient Chinese planning and construction concepts had generally been obeyed until the mid-nineteenth century, when industrial capitalism emerged in China, especially after the Opium War in 1840, when the Western countries entered China. City planning and construction in China has made many changes since then.

Planning before 1949

The development of the modern Chinese cities in the first half of the twentieth century has been very slow, and the function of urban planning was poor. Before 1949, there were neither national policies nor guidelines for city plan formulation. Only a few cities, most of which were the concessions occupied by the Western countries, had prepared their city plans. Some cities, which were occupied by a single Western country, e.g. Qingdao (by Germany between 1898 and 1910), Dalian (by Russia and then Japan since 1900), Harbin (by Russia in 1900) and Changchun (by Japan in 1932), had produced general city plan (Faculty of Urban Planning of Tongji University 1985). The other cities, e.g. Shanghai, Tianjin, Wuhan, Jinan, Shenyang (ibid.), were the so-called common concessions which were shared among several Western countries. Their city plans were undertaken by different countries within their own leasing land. The philosophies of these city plans and construction were very different from traditional Chinese planning concepts. They mainly copied earlier planning and city construction concepts of the Western countries.

In Shanghai, Shanghai Municipal Council in the concession area was established in 1853. The council took responsibilities for roads and dock construction, infrastructure provision, and regulations formulation. It was not until 1927 that Shanghai Public Work Bureau was formed to charge the duty of urban construction outside of the concession area (Li et al. 2006).

After the establishment of Shanghai City Centre District Construction Committee in 1929, the Great Shanghai Plan was produced. However, the concession areas and existing dense urban areas in Shanghai were not included in the plan, but targeted at new city centre development instead. This plan proposed a new spatial structure to northeast Shanghai (Sun 2006). Nevertheless, the plan was not implemented because of the Japanese invasion and the Second World War. The planning and design scheme recommended by the Plan was copied from city patterns of Western countries.

After the Second World War, all the concessions had been repealed to China. The Chinese government at the time began to prepare city plans for a few cities. In Shanghai, a Technical Advising Commission was formed at the beginning of 1946 to study approaches of city plan formulation for Shanghai. In the same year, the Shanghai Urban Planning Commission was established for city plan production and implementation.

After the draft Shanghai City Plan was produced in 1946, it had been revised three times between 1946 and 1949. Planning concepts of satellite towns, neighbourhood, decentralization and arterial roads were introduced by Chinese planners who came back from the Western countries.

It was proposed in the Plan that the neighbourhood should be regarded as the basic unit in the urban area. A hierarchical road network should be established in the city from a comprehensive perspective. The planning policies of the Plan specified to resettle industrial zones to satellite towns in the suburban areas from the city centre. These satellite towns should be separated by green belt. Development of high speed roads was also suggested in the Plan (Li et al. 2006)

At the same period of time, some plans of other large cities were also produced. They include 'Chongqing 10-Year Construction Plan', 'Tianjin Expansion Plan', 'Hangzhou New Urban Area Plan', 'Chengdu City Centre Plan', 'Nanchang 5-Year Development Plan' and 'Changsha New Urban Area Plan'. However, none of them had been implemented (Faculty of Urban Planning of Tongji University 1985).

In 1927, the National Government established its capital at Nanjing. Nanjing was specified as a Special City in 1928. In January 1928, the Capital Construction Committee was established as responsible for the planning and construction of Nanjing, as the capital of China at the time, the Capital City Plan was produced in 1928. This was the first and the most comprehensive city plan produced by formally by the Chinese government. The concepts of the planning were learnt from the European countries and the United States.

The guiding policies for plan formulation were learning scientific doctrines from Europe and the United States but associating with merits of arts in China. The guiding policies specified to adopt the planning approach from European countries and the United States at the macro planning level, but to deliver Chinese traditional patterns at the micro level in the planning process.

The main contents of the Capital City Plan consisted of 28 items including population forecast; definition of city boundary, central political zone, and municipal administrative districts; selection of architectural styles; road network plan and proposal of suburban road construction; layout of park and boulevard; traffic management; electricity and road lights; research of public housing; location selection for school, industries and ports; urban design; zoning ordinance;

implementation procedure; and fundraising. However, the Capital City Plan had not been delivered because of political struggle (Dong 2012).

Planning in the PRC 1949–1980s (Mao's Era)

The Chinese planning system, which was initially established in the centrally planned economy, is still playing an important role in transition to the market economy. Its characteristic 'Plan-led' approach to development is still the main development policy followed by the Chinese government.

From the establishment of the People's Republic of China in 1949 to the beginning of the twenty-first century, modern Chinese urban planning has evolved through three different phases of plan preparation: from the initial phase – the Master Plan format, adopted from the former Soviet Union, to a comprehensive plan phase in the 1990s, influenced by the theories and practices of Western developed countries, to a new practice, the Urban Development Strategic Plan, which was introduced to cope with rapid development and the transition to a market economy. The changes in planning have been driven by social and economic developments, especially the move from a centrally planned economy to a market economy and entry to the WTO. Chinese urban planning is facing increasing uncertainties and pressures to increase the capability of competitiveness while more concerning the interests of multiple stakeholders in development compared with the former first two phases.

In this chapter, urban planning under the different social and economic contexts that have existed in China since the founding of the People's Republic of China in 1949 is analysed. In the last 50 years, Chinese urban planning has developed from a 'blueprint' type in the centrally planned economy to a rational-comprehensive approach, and then to a flexible and pragmatic urban strategic planning under the pressure of competition in marketization and globalization. It is evident that Chinese urban planning reflects and reacts to the social, political and economic contexts, in which it develops and operates. The urban planning system and planning concepts have endeavoured to follow in the steps of social and economic development.

Extension of the National Economic and Social Development Plans

Establishing planning system   The industries in China were very poor when the People's Republic of China was initially established. In 1949, 89.9 per cent of the Chinese population was in rural areas and only 10.1 per cent in urban areas (Hu and Ma 2012). The modern economy merely contributed 10.4 per cent of total GDP in the year of 1952 (ibid.). Urban development in China at the time mainly tried to promote industrial development, typically heavy industry. Cities were intended to guarantee the success of these industrial development projects. The Chinese government asked all the people to be 'self-dependent' to build the New China. It was a concept of Chinese politicians that the cities should no longer be concerned with a consumption function, which created capitalism and the bourgeoisie, rather they should be redeveloped as centres of secondary industry to create the proletariat, regarded as the basis of the New Socialist China. Industrial development was therefore the priority in cities, and political policy greatly influenced urbanization, city planning and development. During the initial industrialization of China in the First Five-Year Plan (1953–7), industrial investment was concentrated in large and medium-sized cities in order to maximize economic efficiency and inter-industry linkages. As a consequence, industrial cities grew extremely fast. In 1951, there were 22 large cities with a population of over half a million. These were the main industrial cities, e.g. Beijing, Baotou and Lanzhou. One benefit of this policy was its contribution to the establishment of Chinese industrialization in urban areas.

Associating with the First Five-Year National Economic and Social Development Plan ('Five-Year Plan'), the Chinese urban planning system was initiated in September 1952, when the State Commission of Finance and Economy organized the first National Workshop of Urban Construction to make the necessary preparation in policies and resources for large-scale economic development defined in the First Five-Year Plan from 1953 to 1957. It was decided in the Workshop that the development and construction of new cities and towns, or the renewal of existing cities, should follow the policies of the National Long-term Plan. Several decisions were made in the Workshop as government policy. It was decided to establish a hierarchical city construction administrative system. The Ministry of Construction was established in central government in March 1953. The Department of Urban Construction of Ministry of Construction was the executive authority in central government for urban construction and planning.

In each of the 39 major cities, e.g. Beijing, Shanghai, Baotou, a municipal urban construction commission chaired by the mayor of city was established. It was also specified by government policy that a city master plan should be produced by all the cities according to the guidelines of the government's ordinance, 'Urban Planning Formulation and Building Design Schedule of the People's Republic of China (Draft)'. This ordinance was written with the assistance of experts from the Soviet Union. A critical term, 'urban planning areas', was introduced. This special definition has greatly impacted urban development negatively later after the reform and open door policy when China was entering its rapid development era.

The concepts of the centrally planned economy from the former Soviet Union dominated Chinese social and economic development policies, and city and regional planning. At this period, economic development of the poor and socialist China had to be supported by the former socialist Soviet Union. The technologies and theories of development and industrialization were all introduced from the Soviet Union. The definition and understanding of the city master plan at the time was that urban planning should extend the Social and Economic Development Plan. The city master plan should allocate land necessary for the designated economic and industrial development projects. Giving those policies physical expression, the functions of the city master plan were to allocate the land and select the locations for the designated national major industrial development projects.

The first National Conference of Urban Construction was held in June 1954. An ambitious decision was made at the Conference that the new cities and major industrial developing cities must complete their city master plan by the end of the same year (Ren 2000). This was the beginning of urban plans formulation in China. In order to guide the production of urban plans, the Ministry of Construction issued an Ordinance of 'Temporary Measure of Urban Plans' Formulation' in 1956, the main principles of which were copied from the 'Measure of Urban Plans' Formulation' of the Soviet Union.

In the centrally planned economy, one important characteristic was that investment, production and consumption were all controlled and assigned by central government. Because the basic function of city plan was the assignment of economic development defined in the Five-Year Plan, the main objectives of the city master plan were to divide or zone the functioning areas of a city, to organize the linkages of various activities, and to propose the infrastructure facilities for development.

There was no requirement for communication among the citizens or different authorities, nor developers, as the developments were the responsibility of the government. Planners did not need to be concerned about uncertainties in the development – all the elements for the development, including resources, were allocated by the government. The urban plans were simple and easy to implement.

A Scapegoat of Ambitious Development Policy and Abolishing Planning

Planning has been often been criticized as being too slow to cope with social and economic development. However, when planning proposals were ambitious under the pressures of the government's social and economic development policy, if the policies failed, planning might then be the scapegoat being criticized as too ambitious. The experience of planning in China confronted such circumstances in the Second Five-Year Plan period (1958–65).

The successful social and economic development in the first Chinese Five-Year Plan stimulated the Chinese government to execute the ambitious development policies in the Second Five-Year Plan period. As a consequence of the over-ambitious 'Great Leap Forward' policy, it was decided by the government that Chinese cities should be fully modernized within 10–15 years. All city master plans were to be revised to meet the demands of urban expansion, to take into account the great amount of construction that would be needed as well as the necessary land allocation for ambitious development. The cities were then developing without appropriate control. Large amounts of land were allocated for city expansion without careful study and analysis of the affordability of proposals, which threatened cultivated land. The situation was a disaster. Some cities were too ambitious in their development objectives. For example, they proposed to reach a population of two million from 100,000 in a few years. During the national review of policy, the proposed over-ambitious urban development and the disorderly situation as a consequence were criticized as the failures of urban planning. It was decided at the Ninth National Planning Conference in November 1960 that urban planning should be suspended for at least three years (Tang 2000) because of its ambitious objectives.

From 1966 to 1976 during the 'Cultural Revolution', the policy of the central government mainly emphasized 'class struggle' instead of social, economic and urban development; and most of the organizations involved in urban planning in China were closed down. Many urban planners were transferred to the countryside. The 'Cultural Revolution' deteriorated the living quality of Chinese people. The housing and urban infrastructures were in great shortage. The average living space per capita was only 3.6 square metres in 1978, which was 0.9 square metres less than that of 1949, when the People's Republic of China had been founded (Wang 1999a).

Even though Chinese urban planning had become a scapegoat for the failure of the government's ambitious policies in economic development and suspended for more than 10 years, it had at least contributed to the land allocation and infrastructure provision for Chinese industrialization mimetically after establishment of the People's Republic of China. At that time, urban planning was functioning under the centrally planned economy, where urban development and urbanization were seen as auxiliary products of industrialization.

Planning after the 'Open Door and Reform' Policies since the Era of Deng Xiaoping

Pressures from New Political, Social and Economic Policies in the Open Door

After being suspended for more than 10 years since the 'Great Leap Forward' and then the 'Cultural Revolution', the urban planning system in China was re-established in 1978. It was decided at the National Conference of Urban Development organized by the State Council in March 1978 that all cities and designated towns in China should produce or revise city master plans and detailed plans (Wang 1999a). This marked the beginning of the re-establishment of urban planning in China. In 1979, the National City Construction General Bureau was established within the State Commission of Construction. The planning administrative authorities of the central, provincial and municipal governments were resumed, or established as a consequence.

After the execution of the 'open door and reform policy' in China in the late 1970s, several new economic policies were introduced. These new policies encouraged and greatly impacted on urban development. In this period, four Special Economic Zones and 14 Coast Open Door Cities were established to attract overseas investment and new technologies, as well as the experimental experiences of the economic reforms. The urban plan mainly attempted to guide urban development to achieve the objectives of the open door and reform policy, which were decided at the Third Plenary Session of the Eleventh Central

Committee of the Chinese Communist Party in December 1978. However, the principles of urban planning still followed the notions and practices of the centrally planned economy and the political ideas of the Cultural Revolution. It was not only urban planning, but the whole social and political environment, which found it difficult to depart from the notions of the centrally planned economy.

After the suspension of planning for more than 10 years and the lack of educational facilities to train planners, there was a great shortage of capable urban planners in China at the beginning of the 1980s. The planners at the time had graduated before the Cultural Revolution (1966–76) and were of professional backgrounds in architecture and engineering. Architect-planners viewed planning as urban design with an emphasis on the aesthetic character of a city (Tang 2000). They regarded their task as deciding the nature of the city and the proposed population size. They tried to apply national planning norms and standards regulated by central government to work out the spatial requirements and land allocation in the plan formulation. These plans and relevant norms and standards were regarded as the main consideration for the approval of applications of development planning permission.

In 1978, the central government promulgated the ordinance of 'The Opinions of Strengthening the City Construction Management' (MOC 1978). It was decided in this ordinance that city master plans should be subject to a strict examination and approval procedure, and all the master plans of metropolitan and capital cities of the provinces, and of those cities with a registered urban population of more than 500,000 had to be submitted to central government for examination and approval (ibid.).

However, with inward investment to the Special Economic Zones and Developing Cities, the 'blueprint' type of master plan with National Norms and Standards published by central government, was found to be inappropriate to deal with new issues and problems in development, especially after the beginning of the transition to the market economy.

To increase local governments' proactivity was an agenda of transition to a market economy from the centrally planned economy. Incremental devolution of relevant power to local municipal governments was a consideration. Consequently, at the National Conference of Urban Planning in October 1980, a decision was brought forward that the major task of mayors in local municipal governments should be to plan, build and manage cities. This was part of devolution, one of the main themes of reform in China. The achievement of economic and urban development was also regarded as one of the main tasks of the local governments and a criterion to assess their performance, especially that of mayors. At the same conference, the national policy of controlling the size of large cities, developing medium size cities in a reasonable way, and promoting greatly the development of small sized cities was decided (Wang 1999a). This policy has influenced the urbanization process and urban planning. It was the government which decided the size of cities instead of natural development in the market.

The policy was to restrain the development of large cities and metropolitan areas. One major concern in the urban plan approval process was to check the proposed scale of the city for the purpose of controlling the population in large cities. In the same year, 'Temporary Measures for Urban Planning Formulation and Approval Process' was issued as a Ministry Ordinance by the State Commission of Construction. For the first time it was suggested by the government in China that the city master plan should avoid the 'blueprint' approach, but adopt a kind of development strategy instead. Nonetheless, planning was still strictly controlled from the top. Urban planning was subservient to the policies of the government, which examined and approved plans through the provincial and central governments.

Techno-rationalism in Urban Planning

The city master plan in this period began to include social, economic and regional analysis. The influence of the region on the social and economic development of a city was considered for the first time. The consideration and discussion of introducing an integrated systematic plan of a city and towns within a city region context was initiated. It was the beginning of the departure from the 'blueprint' type of city master plan, though it was an incremental process. The city master plan became more comprehensive. One main feature of the Chinese master plan was its incorporation of several specific plans, e.g. historical protection plan, environmental protection plan, and rehabilitation plan for the city centre.

The principles of managing urban development were emphasized. The concept of an 'urban planning area' initiated in the first Five-Year Plan in the centrally planned economy was redefined in the 'Regulation of Urban Planning', a government ordinance, in 1984 (MOC 1984). An 'urban planning area', which consists of urban built-up districts, inner suburban districts, and the areas needing urban development and construction, is the only area controlled and managed by urban planning and is the area that is available for urban construction.

The definition of an 'urban planning area' was a significant aspect in the Chinese city master plan because all development and construction within the defined 'urban planning area' could not take place without planning permission. The urban planning areas were significant for three reasons:

- They are the only areas that could have urban construction and development, and be managed by urban planning.
- Land within the area defined as urban land is under state ownership. The areas outside the boundary of 'urban planning areas' are not permitted to have any urban development and construction. These lands are under so-called 'rural collective ownership'.
- All development and construction within the urban planning areas had to conform to the policies and proposals of the city master plan. This was further confirmed in the City Planning Act 1989.

It is since that time that urban planning has been regarded as techno-rationalism in China. Urban planning was treated as a type of science and technology and a technological process operated by urban planners who, as specially trained experts, were able to predict and to control urban development by producing urban plans in a rational manner. This establishes the term of 'techno-rationalism'.

Establishing Statutory Planning System in the Transition to the Market Economy

Influences of the Western Planning Theories

One of the key features of Chinese reforms and transition to the market economy has been operated under the process of marketization. To have a rational market system, it is necessary to have laws to regulate operations in a market. The establishment of a legislative framework of different laws and relevant acts to replace the former administrative system has been a priority target of Chinese government. This procedure has included the establishment of a statutory urban planning system.

This process was assisted by academic exchanges in urban planning between China and other countries. Chinese academics and government officials also introduced new urban planning ideas from the Western countries after their study tours. They published papers on the urban planning theories and practices of the Western countries in journals (Tang 1993; Wang 1993; Xue 1995; Yu 1994, 1995; Zeng 1992). Some Western planning concepts began to impact on Chinese urban planning practice.

It is evident that systematic and rational approaches to planning, which believe in order, central control, rationality and comprehensiveness, fit well to the philosophy in China in which the centrally planned economy still had great influence in social and economic development in the 1980s. These two planning approaches and their notions were accepted by the Chinese urban planning profession very quickly, especially as the Chinese urban planning profession had been seeking appropriate 'scientific' and 'objective' planning methods since the 1980s. The planning system established at the time was comprehensive and top-down as a consequence.

The statutory urban planning system began to be established in the late 1980s. The City Planning Act 1989 (NPC 1989a) was approved by the People's Congress in 1989 and was implemented on 1 April 1990. Several other relevant acts were approved by the People's Congress. These included the Land Administrative Act (NPC 1986), Environmental Protection Act (NPC 1989b) and others.

The City Planning Act 1989 is applicable to all the development and construction activities within the urban planning area defined by a City Comprehensive Plan. In central government, several regulations have been promulgated as supplementary to the City Planning Act 1989. Provincial and municipal Peoples' Congress promulgated executive regulations and by-laws as local statutory documents according to the City Planning Act 1989.

National and professional planning norms and standards were also formulated and published at the same time as a guidance and reference for plan preparation and approval. They included 'City Land Use Categories and Construction Land Use Standards', 'City Residential Area Design Norms', 'City Road's Design and Transport Planning Norms', and others.

Transition to Neo-liberalist Market-led Economy

After the Third Plenary Session of the Twelfth Central Committee of the Chinese Communist Party in October 1984, economic reform started in the urban areas. The transition to the socialist market economy was initiated. As the key policy, development has been regarded as an absolute priority in China. The major work of the government has been emphasizing the promotion of rapid and comprehensive growth. During his tour to South China in 1992, Deng Xiaoping (Deng 1992) warned not to throw obstacles in the way of areas that can grow fast. Areas with the potential for fast growth should be encouraged to develop as rapidly as possible. Even the debate on 'socialism' and 'capitalism' was abolished. The market-oriented transition was further encouraged and promoted. China has evolved to Chinese style neo-liberalism.

The Chinese neo-liberalism has greatly impacted cities, governmental administration and planning as a consequence. The function of local government has shifted from central plan executor to the operator of local economy. One main feature of Chinese reform was devolution (Harvey 2005; Wang et al. 1999; Wei 2000; Yu 2005). With devolution, increasing of economic capability and local entrepreneurialism for promotion of competitiveness has then been the priority task of local governments (Yu 2005).

One outcome as a consequence of the devolution is competition among cities. In order to attract the scarce resources of capital flows, municipal governments have to compete with each other both in China and in the outside world. This competition drives municipal governments towards local state entrepreneurialism.

Application of systematic and rational planning approaches in China satisfied the needs of local state's entrepreneurialism. A comprehensive statement of goals in rational planning is not only the purview of planners, but also local state entrepreneurialism. Rational planning suggests goals for ideal future development, which are also what local politicians need, since urban development and construction and attraction of inward investment have been specified as the main tasks of local municipal governments. Plans were used for place marketing purposes.

Marketization in China means less government investment in development, and more diverse types of fundraising. Chinese marketization encourages diverse ownership, especially increasing private ownership. China's transition to the market economy has made a great contribution of private business in economic and urban development during last three decades.

The former strict control of migration from rural to urban areas was relaxed, and a significant rural–urban migration developed. During the last three decades, it is estimated that 80 million rural migrants have moved to urban areas (Wang 2002a). This large migration creates great pressure on urban planning policies and provision of job opportunities, housing, transport and various public facilities in urban areas. Forecasting the future distribution of population is extremely difficult because of the migration which implies existing planning technology leaves some aspects to be desired.

At the same time, the sustainable development principles proposed at the Rio Conference, and in Agenda 21, which should be adhered to by nearly all the governments in the world, have raised new concepts for urban development and planning.

All these changes as the consequences of the neo-liberalist market-led economy brought new tasks and challenges for planning. Thus, the planning system initiated in the centrally planned economy has to adjust to cope with new social and economic conditions, as suggested by the Chief Planner of the Ministry of Construction that planning has to cope with the objective world, to meet the demands of the social and political system instead of the other way round (Chen 2000); otherwise it would have been inappropriate as a planning and management tool to perform appropriately in the transition to the market economy.

Innovations of Urban Planning to Cope with Political, Social and Economic Changes

A Trial-and-Error Learning Process

There used to be many debates and discussions on the Chinese planning system regarding its stable, inflexible as 'blueprint' plans in the literature (Duan 1999; Li 2003; Xu and Song 1998). However, recent research on detailed planning practices in China reaches a different view that Chinese planning in practice is flexible and pragmatic to promote development.

The terms of the City Planning Act 1989 and the Ministry's Ordinance 'Regulations for Urban Plan's Formulation', 'Implementation Measures for City Planning Act', and the planning by-laws promulgated by local governments are loosely worded, and defined in a way so as to allow for judgement and discretion to be used in both urban plan formulation and its implementation. The City Planning Act 1989 also encouraged the application and implementation of new technologies in urban planning (Article 8, NPC 1989a). Planners are able to undertake wider and complicated research beyond the specified contents and structures defined by the 'Regulations for Urban Plan's Formulation' while producing City Comprehensive Plans. It is also possible for them to freely apply some theories and methodologies in the planning process. It is only suggested that the plan should be in line with the specific condition of China (Article 5, NPC 1989a). The contents of the plan should include the designated function of a city, the goals of the development and its size. The plan should define the standards, norms and criteria for the main construction, the land use structure, function division, and the comprehensive arrangement for all types of constructions (Article 9, NPC 1989a). The 'Regulations for Urban Plan's Formulation' confirm that a City Comprehensive Plan should allocate reasonably the urban infrastructures, specify the long-term and short-term development profiles, and guide (or indicate) how the city should develop in a rational way (Article 14, NPC 1989a). There are no strict or detailed definitions and standards for City Comprehensive Plan formulation methods, only some defined contents and the structure to form the plans. Chinese urban planning is a trial-and-error learning process not only in practice but in the system too.

Within this context, during plan formulation, the applied methodologies will be dependent upon the capability and professional background of the urban planners who produce them. Some methods are 'blueprints' similar to the master plans in the 1950s or that of the early 1980s, others are more flexible with a rational-comprehensive planning approach, even pragmatist planning approach. Although the general structure is similar, the technologies and contents of the plans vary. Generally speaking, the urban macro plans, i.e. city comprehensive plan and city development strategic plan, consist of aesthetic principles to 'image' a city and this has been the main approach in planning practice. It is because of the differences in planning approaches that the performance of urban macro plans is very different.

Planning in a Multiple Stakeholders Society of the Market Economy

In the centrally planned economy, all investment, production, quantity of products and even their prices were decided by central government. There was no private investment. Planning might not have to deal with uncertainties and different interests of stakeholders. However, in the market economy, there are more stakeholders, particularly due to private investment in urban development instead of investment by government only in a centrally planned economy. The urban planning management approach initiated in the Chinese centrally planned economy has to update to deal with the complexity and conflicts of diverse interests. An innovation was needed.

The Control Detailed Plan, as a type of regulatory plan, was first innovated in Wenzhou, Zhejiang Province, where the elements of a market economy had often been very active. Since the 'reform and open door policy', and the transition to a market economy and the commercialized reform of urban land use, private development and private investment have prospered in Wenzhou. However, there was no regulation to deal with private development, either in the City Planning Act 1989 or in government ordinances.

Wenzhou urban planners introduced 'Zoning Ordinance' principles from practices in the United States, incorporating local situation and characteristics to form a new type of urban planning instrument in their city. They named it the 'Control Detailed Plan'. Through experiences and lessons of Wenzhou, in order to deal with pluralistic stakeholders in development in the transition to the market economy, the Ministry of Construction issued an ordinance to promote the Control Detailed Plan as a new planning tool for the whole country in 1992 (Chen 2000). The intention of the Control Detailed Plan is to guide urban development and construction on a specific plot of land according to detailed controlled regulations of plot ratio, height of buildings, building density, provision of utilities by following planning policies and schemes defined in a City Comprehensive Plan, and the relevant national or professional norms and standards in urban planning.

Since implementation of the 'reform and open door policy', particularly after endorsement of the market economy by the Chinese government in the Third Plenary Session of the Twelfth Central Committee of the Chinese Communist Party in October 1984, and further promotion of the market economy at the Fourteenth Communist Party Congress in October 1992, Chinese cities have had to produce their plans and implement plan proposals in a climate of uncertainty, conflict and competition. These uncertainties, conflicts and competitive context exist because of the need to attract inward investment; deal with rapid land supply to meet the requirement of economic development; make provision for infrastructure and other physical development to accommodate industrial investment; change the relationship between central and local government because of devolution of the administrative structure; and mediate among diverse interest groups, as well as promote the capability of cities for competition.

Nevertheless, the present urban planning system and the contents of City Comprehensive Plans, initiated under the centrally planned economy, are complicated and time consuming, and find it difficult to cope with rapid development. The pressures of competition and rapid social and economic change have driven the municipalities to undertake innovation from the bottom up at local governments, to seek an alternative planning approach to deal with the pressures and uncertainties.

Local governments require an analysis to compare the strengths, weaknesses, opportunities and threats of cities (Li 2003) when they are involved in competition. Local municipal governments hope that with support from urban planners through their research and planning, the city can expand rationally to be larger and stronger through a 'scientific' approach (Zhang 2002). The approach of the 'Urban Development Strategic Plan' was the consequence of such a requirement.

Guangzhou Urban Development Strategic Plan as a non-statutory plan, first emerged in 2000. This innovative planning approach was then introduced to many Chinese cities, especially most large cities, very quickly. The appearance and application of the 'Urban Development Strategic Plan' illustrated that the traditional comprehensive plan cannot cope with rapid changes (Li 2003). It is Li's argument (2003) that while making the City Comprehensive Plan, urban planners often seek the perfect written documents and plans instead of carrying out actual research into strategic issues. Plan formulation process has emphasized the control of the scale of the urban population and the protection of cultivated land. Even so these are the criteria for plan approval set by the central government.

The task of the Urban Development Strategic Plan is to provide a spatial structure with long-term, general and comprehensive features (Qiu 2003a; Zhang 2002; Zou 2003b). This is achieved by coordinating the city's social and economic development strategies. Economic development, the coordination between urban planning and social economic development and proposals to increase competition, are the main objectives of the strategic plan. The Urban Development Strategic Plan tries to focus on the role and function of a city from a much wider aspect compared with that of a City Comprehensive Plan. The plan would review the city within the context of the whole country, or the region (Dai and Duan 2003; Luo and Zhao 2003).

Pressure from competition and rapid changes means that the 'Urban Development Strategic Plan' in China is characterized by fastness with a short period of plan formulation, efficiency with an approach of problem orientation, and flexibility without any regulation and restraints in plan formulation (Luo and Zhao 2003).

The innovative 'Control Detailed Plan' and 'Urban Development Strategic Plan' are all bottom-up from the local municipality, the opposite of the top-down planning system from central to local government. It is an approach of urban planning to 'keep pace with the times' (The Sixteenth Central Committee of the Chinese Communist Party).

Urbanization in China

A simple definition of urbanization is that it is a process of transition from agriculture to urban land use and population. It is not only migration from rural to urban areas, nor the expansion of cities, but a changing process of living styles and industrial patterns. It is the process of change from agriculture, or primary industry, to secondary and then tertiary industries. China has been under this process of transition during last 100 years. However, this process was very slow before the late 1970s.

Before 1949, wars continued in China. There was neither any intention for economic development nor for urbanization from government policies. After 1949, the process of urbanization has been fluctuating, but mainly at a slow speed until the late 1970s. Urbanization in the period between 1949 and 1979 had only increased by 10 per cent (Zhou 2002). It was because Chinese urbanization had been greatly influenced by political and economic policies, three main policies of which had been to slow down Chinese urbanization. The first main reason was the poor economic capacity as the consequence of continuing wars since the mid-nineteenth century. There had not been a period of peace, which should be the basic requirement for economic development, for at least 10 years. The second influenced policy was the strict migrants' control policy from countryside to city, even city to city in the era between the 1950s and the 1980s. This strict control policy restrained the urbanization process. The third impact came from the central planned policies and centrally planned economy.

Bachmann and Leung (2008) summarize the way in which Chinese urbanization has developed as follows. In the period 1950 to 1970s there was minor rural urban migration and government policy was concerned to:

- use economic surpluses from agriculture to subsidize the development of heavy industry in the large cities;
- absorb any redundant agricultural population in the countryside; and
- restrict the mobility of the working population through the '*hukou*' registration system.

The former Chief Planner of the Ministry of Construction, Chen Xiaoli (2000), suggests three main reasons for control of urbanization in China by government in the central control economy between 1949 and the late 1970s:

- China was a country with a large population but a low level of industrialization with poor economic capability at the time. In order to increase the national capability and peoples' quality of life, industrialization became the priority of policies.
- Under the central economic planning system, development, building and expansion of cities were the responsibility of the government. In addition, the government provided subsidies to the urban citizens. One more citizen would mean more financial subsidy. Because of the shortage of finance and poor capacity in the economy, the state had to control urbanization, and the numbers of urban citizens.
- As a mainly agricultural country for thousands of years, there was a need to keep people on the land in order to provide enough food to feed one-fifth of the population.

Kirkby (1985) suggested that urbanization in China has in fact been the process of 'industry/strategy-oriented urbanisation'. He (ibid.) argued that it was the emphasis on the targets of industrialization, which greatly affected the speed of urbanization in China. In order to reach the industrial target, the Chinese government ignored the construction of urban infrastructure and an increase in agricultural production.

Since the open door and reform policy beginning in the late 1970s, there have been several forces to stimulate urbanization in China. These driving forces have included plan-led investment by government in urban areas, township development and rural urbanization by local non-governmental investment, i.e. peasant, private business persons (Wang 2000), and the impulse of FDI, mainly locating to the most rapidly developing areas on the coastline. The modes of rural urbanization can be further divided into two types i.e. 'Sunan Type' and 'Wenzhou Type'.[3]

It is the view of Bachmann and Leung (2008) that from 1978 to the 1990s, following the reforms of Deng Xiaoping, rural trade and labour mobility were liberalized, with the agricultural population being allowed to produce and sell their products in towns and rural areas. Subsequent reforms in the 1980s led to a modernization of rural enterprises and linked them to the nationally promoted export strategy. These reforms, allied with the policy of controlling migration to the large cities, led to the development of both the secondary and tertiary sectors in the smaller towns and cities, which began to absorb excess rural labour. Thus at this time the policy towards urbanization was to:

- strictly control the development of large cities;
- rationally develop medium sized cities;
- vigorously promote the development of small cities and towns.

In the 1990s urbanization policy became a major political issue with opponents to the role of towns and small cities in economic and urban development claiming that:

- they were not capable of generating the type or volume of industrial production needed to develop the national economy;
- inadequate urban and economic management in towns and small cities was giving rise to uncontrolled development and poor living conditions;
- larger cities offered economies of scale and greater productivity and should therefore be the focus of industrial and urban development.

One important feature of Chinese urbanization is that it has been hastened because of the rapid growth in the national economy and social development after the reform and open door policy. The number of cities has increased to 663 from 192 in 1978. According to the definition in China, a metropolitan area should have an urban population of more than one million; a large city has an urban population of between 500,000 and one million; a medium sized city encloses an urban population between 200,000 to 500,000; and there are less than 200,000 people in designated small sized cities. Table 2.1 illustrates the increase in the number of cities in all categories. In addition to the increasing number of cities, within the same period, the number of designated towns increased to about 20,000 from 2,173. The urban population reached 456 million (36.1 per cent of the total population) from 170 million (17.9 per cent of the total population).

---

3 Sunan refers to Southern part of Jiansu province in China; Wenzhou is the name of a Chinese city. These two types of urbanization will be explained later in the chapter.

Table 2.1 Number of cities in different categories and their percentage, between 1949 and 1999

Year	Total no. of cities	Met' areas No. of cities	%	Large cities No. of cities	%	Medium cities No. of cities	%	Small cities No. of cities	%
1949	136	5	3.7	8	5.9	17	12.5	106	77.9
1958	176	11	6.3	19	10.8	36	20.4	110	62.5
1963	174	15	8.6	18	10.4	54	31	87	56.5
1970	176	11	6.3	21	11.9	47	26.7	97	55.2
1978	192	13	6.8	27	14.1	60	31.2	92	49.4
1983	289	19	6.6	29	10	73	25.3	168	51.8
1988	434	28	6.5	30	6.9	110	25.3	266	58.6
1992	517	32	6.2	31	6	141	27.3	313	60.5
1999	667	37	5.5	49	7.4	216	32.4	365	54.7

Source: China Statistics Year Book 2000.

Jiabin Lin of the Development Research Centre of the State Council of the People's Republic of China claims (2005) that:

- urbanization has accelerated in China since 1990 with (a) a growth rate of 1.1 per cent per annum between 1995 and 2003; and (b) 15 million rural–urban migrants per annum between 1990 and 2003;
- in 2002 there were 48 mega-cities;
- by 2010 45 per cent of China's population will live in what he calls mega-cities and that by 2020 this percentage will increase to 52 per cent and by 2030 to 60 per cent;
- the key issues for the management of these mega-cities are security of water supply; urban transportation; air quality; disaster prevention and management and cultural conservation.

In expanding on these issues Jiabin Lin (ibid.) states that:

- China's water resource is 2,250 m^3 per person on average which is only 25 per cent of the world's average, whilst the per capita water resource of some severe water shortage cities, for example, Beijing, Shanghai and Tianjin is 300 m^3, 200 m^3 and 153 m^3 respectively. With the increasing consumption of both industrial and household water the situation can only worsen especially as the water table has continuously declined in most Chinese cities over the years.
- The disposal rate of sewage in Chinese cities is estimated to be 30 per cent. To improve this disposal rate to 50 per cent by 2010 and 80 per cent by

2030, and thus improve the water environment, it is estimated that a total investment of 1.2 trillion yuan (US$144 trillion) is needed.
- The drainage capacity in mega-cities is reduced as a consequence of continued urban expansion with a resultant increase in run-off, a reduction in natural drainage systems and frequent flooding.
- Motor vehicle ownership and usage has increased dramatically in recent years, with for example an additional 120,000 cars being registered for use in Beijing in the first half of 2002 and China's mega-cities generally lagging behind in the development of rail and mass rapid transport systems. Congestion is a major problem with an average driving speed of 12 kilometres per hour in the downtown area.
- Air quality is a major problem with a high concentration of suspended substances, sulphur dioxide and nitrides with industrial smoke emissions and motor vehicles being the major causes of air pollution.
- The disposal of domestic waste is a major problem, with landfill accounting for 80 per cent of the total treatment, although a programme of constructing incinerators is being implemented.
- Chinese cities are less well prepared than their Western counterparts in the prevention and management of disasters deriving from earthquakes, fires, wind storms, floods and geological hazards.
- There is a general disregard of the need to protect and conserve the cultural environment in Chinese cities in the race for economic development.

In a global context, and in contrast with Lin's definition of 'mega-cities', Hall (2005) defines mega-cities as high-density metropolises with a population of more than 10 million people which are hubs of trade, culture, information and industry and are key in the process of globalization which can be subdivided into three categories:

- cities coping with informal hyper growth;
- cities managing dynamic growth;
- mature cities coping with aging. (Hall and Pfeiffer 2000)

Three mega-cities are identified in China – Shanghai, Beijing and Tianjin, all falling within the category 'cities coping with dynamic growth' which are characterized by high-quality housing, where the municipal governments have intervened in the housing process through slum clearance programmes and the provision of housing for rent and more recently for sale and the development of high-quality public transport systems, metro and bus, to serve the high-density residential and commercial developments, as well as the private motor vehicle where they are developing 'Los Angeles' style motorway networks.

Following a subsequent in-depth analysis of urban development trends throughout the world Hall (2005) identified a new phenomenon – the global Mega City Region (MCR) – which he sees as a solution to the problems facing mega-cities through a rebalancing of homes, jobs and transport. These MCRs develop through a complex process of simultaneous decentralization at a regional scale stretching over distances of 150 square kilometres, and a recentralization of employment centres at a more local scale. These employment centres are surrounded by overlapping commuter fields, served primarily by the motor car and are thus increasingly polycentric. Whilst the MCRs are seen as a solution to the problems of mega-cities there are also difficulties in that they suffer from fragmented governance which mitigates against the effective planning and action that these regions need if they are to work effectively.

Hall identifies two such MCRs in China – the Pearl River and Yangtze River Deltas, with the Pearl River Delta stretching from Hong Kong through Shenzhen to Guangzhou and with an estimated population of 120 million by 2020 and the Yangtze River Delta stretching from Shanghai-Suzhou-Hangzhou-Nanjing and with a population of 83 million in 2020 (Hall 2005). In a global context he concludes by stating that:

> the overwhelming need in all these great metropolitan areas is for effective metropolitan governance across the entire mega-city-region. Such regions are the new reality of urban existence in the 21st century. They are ... both the solution and the emerging problem. They are a Solution because the offer the prospect of re-equilibrating homes, jobs and transport across a new and vast spatial scale. But they are also the Problem because this demands effective planning, powers and action across a very wide Metropolitan scale. Unless this opportunity can be grasped, the evident risk is that such regions will be characterised by a deepening economic and social imbalance and polarisation, between rich central cities and marginalised poor peripheries. The signs are already evident. There is some time to grasp the problem and resolve it – but less than we might think. (Hall 2005)

Thus as well as coping with the urbanization problems deriving from rapid economic growth and rural–urban migration, China has also to address the more 'modern' issues associated with the evolution of MCRs.

The other feature of urbanization in China is that it has been closely associated with the Chinese transition to the market economy. This transition has been an important impulse for Chinese urbanization. However, the urbanization stimulated by the forces of the market economy creates a main feature which is the unbalanced development and different speeds and approaches of the urbanization process in China. The growth rate of both the economy and urbanization in the eastern coastal area has been much faster than that of the central and western areas. There are several reasons that can explain this phenomenon.

Unlike the former Soviet Union, it is argued that China has never been a 'pure' socialist planned economy (Wei 2000). State-owned enterprises, collective enterprises and a few small private businesses have usually been mixed together, even during the period of tight central control. The non-state-owned sectors were mainly located in the eastern coastal areas. This helped economic development and urbanization to progress rapidly, particularly in the process of the transition to the market economy in the 1980s. The non-state sectors have made a significant contribution to development and urbanization. The economic development context, typically the favourable policies promoted by central government to attract FDI, traditional linkages with overseas in business and trade, higher education standards, and a convenient transport network have provided an appropriate environment for the rapid urbanization process in east China.

During the process of development and urbanization in more than three decades, the major and most important co-urban areas in China were establishing along the eastern coastal area. Among the nine major co-urban areas in China, six of them, i.e. the Yangtze River Delta, Pearl River Delta, Beijing-Tianjin-Hebei Province, Liaoning Middle and South Area (Liaoning Peninsula), Shandong Peninsula, and Fujian Coast (South) area are located along the coast areas (Figure 2.1). Of these six co-urban areas, the Yangtze River Delta (with Shanghai as its central city), the Pearl River Delta (with Hong Kong and Guangzhou as its central cities), and Hebei Province (with Beijing and Tianjin as cores, it is also called the 'Greater Beijing Area') are the most important ones in social, economic and political perspectives. These three co-urban areas have been functioned as the development poles until now after the Chinese open door and reform policy. The Pearl River Delta area with Shenzhen and Guangzhou as the core cities was the development pole in the 1980s;

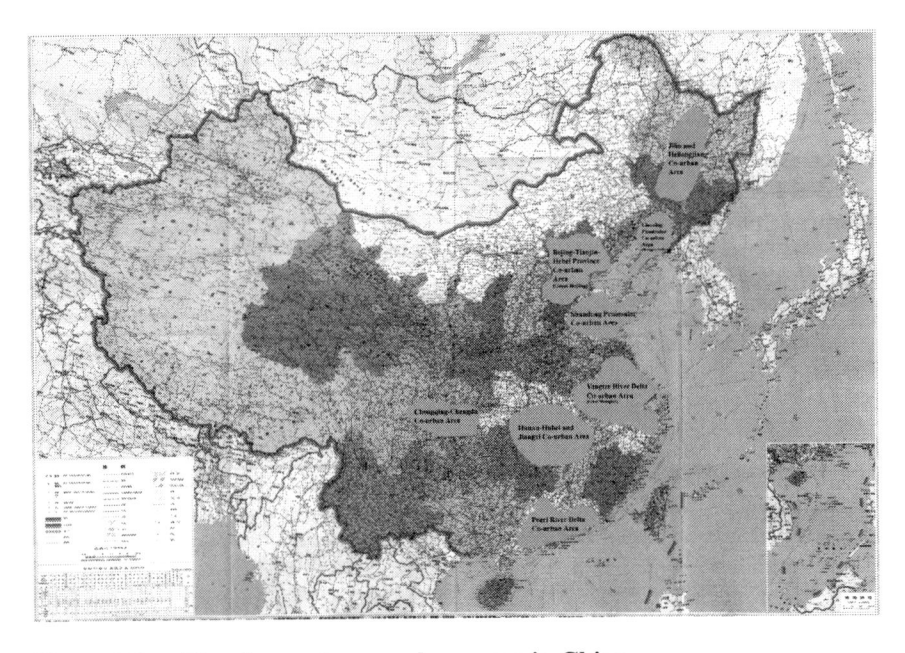

Figure 2.1   The nine major co-urban areas in China
Source: Adapted by the author.

the Yangtze River Delta with Shanghai as the core city has played a significant role since the 1990s; and the Beijing-Tianjin-Hebei Province, with Beijing as national political centre and Binhai Tianjin as major development pole since the twenty-first century, is the key player in Chinese economy. According to the presentation given by the former Chinese Construction Minister Mr Wang Guangtao on 15 October 2002, these three co-urban areas are occupied by 9.92 per cent of the total population and 2.04 per cent of China's total land area. Nevertheless, they contribute 31 per cent of total GPD and attract about 65 per cent of FDI (Table 2.2).

Table 2.2 Index of the three main co-urban areas in 2000

Co-urban areas	Population (thousands)	Area (km²)	GDP (100 million yuan)	GDP per capita (yuan/person)	FDI (100 million US dollars)
Beijing-Tianjin-Hebei Province	30,970	48,629	5,402	17,442	53.87
Yangtze River Delta	69,050	92,937	14,800	21,433	104.01
Pearl River Delta	25,640	54,747	7,522	29,336	104.67
Percentage of the whole country	9.92	2.04	31		64.67

Source: Quoted from former Minister Wang Guangtao's presentation, on 15 October 2002.

Chinese economic development and urbanization has created an affluent and rapidly expanding Chinese middle class in the cities and towns who are concerned to provide a better standard of living for themselves and their families by purchasing second housing with more space for the children to play in a less polluted atmosphere.

The middle class is generally established by entrepreneurs, small business owners, and professionals. It is not difficult for them to have a standard mortgage involving typically a 25 per cent deposit. They usually acquire a town house of 200–400 square metres with no more than four bedrooms, and a small garden. The emergent middle class and their demands of better living standards have greatly impacted the Chinese city and regional planning system.

Three Main Strategies for Urbanization

Chinese urbanization has several challenges. The first is the shortage of available land. The average cultivated land per capita is only about 7.6 hectares; however the total cultivated land in China as a whole has been confronted by the risk of reduction because of urbanization. There are around 15 million rural to urban migrants every year. The other challenge of Chinese urbanization comes from the severe shortage of water and its unbalanced distribution. For example, the population in north China concentrates one-third of the total Chinese population, however, the water resources only occupies 6 per cent.

To have an appropriate and rational strategy for Chinese urbanization is then a key issue. However, it has been a major debate for the strategy of Chinese urbanization promotion for many years. The most important strategy has been defined as the following: 'To keep strict control over the size of large cities and to have a rationally development of medium-sized cities and to promote the development of small cities, so as to have a rational structure of productivity and population' (Article 4, NPC 1989a). There have been debates in Chinese academic world whether the development of large cities is more efficient than small sized cities. However, it is obviously inappropriate to make a unitary policy when dealing with urbanization in China since China is too large and too complicated, and there are such diverse local situations and development levels, that a unitary policy could not work well.

It was not until the Fourth Session of the Ninth National People's Congress in 1999 that the policy of integration and diversity of development of large, medium and small cities according to local features was decided in the document of the Tenth National Social and Economic Development Plan.

Promotion of development strategies for small cities and towns   The Chinese open door and reform policies were first implemented in rural areas. With successful reform of agriculture in the rural area, rural productivity had been significantly increased. However, surplus labour in the rural areas began to become a social and political problem. As a result, in the 1980s, surplus labour from the rural area was dispersed through local job transfers to industry or services, i.e. the so-called process of 'leaving agricultural work without migration; entering factories without migrating to cities' (Fei 1984). It was in fact a process of rural urbanization, or 'promoting the development of small towns'. There were two main types of models. One was called the 'Wenzhou Type', which was driven by private business, e.g. family or multiple family businesses. The other one was the 'Sunan Type' (South Jiangsu Province), which was a collective economy depending upon township and village enterprises (TVEs). These two types of model have contributed to urbanization and employment of surplus rural labour, as well as local economic development.

However, the application of this strategy has created many problems, typically the development of small towns based on the development of township and village enterprises (TVEs). These enterprises enlarge the costs and damages to the environment, natural and economic resources, and risks to sustainability.

Except for some small towns in east China, the development of majority small towns in China is very difficult. The function of small towns in connecting large and medium size cities and carrying forward services to hinterland rural areas is very weak. The services in retail, finance and banking, intermediary agents, are too poor to provide appropriate services to rural areas. Shortage of fiscal and land resources, because of the existing Chinese fiscal and land management system, are the main reason that restrained the development of small towns. These are the priority issues that have to be addressed in Chinese urbanization and small town development (Yu 2010).

It is significant that there will be multiple 100 million migrants from rural areas to urban areas. It is impossible and impractical for large cities to accommodate such large number of migrants from rural areas. It is evident that small cities and towns should play a significant role and function in absorbing rural surplus labour, and provide job opportunities for the peasants who lose their land because of urbanization, industrialization and concentration of agricultural production. The small cities and towns are the places for accommodating large amounts of the migrants. It is also necessary to realize that even when China has completed its urbanization, in order to feed more than 1.3 billion people, or 21 per cent of the world population, the rural areas have to be maintained and further developed. The rural population may decline but the rural areas will not disappear. Small towns will still function as a hub for service provision and interaction between large and medium cities to rural hinterland areas (Qiu 2006).

Promotion of development in large cities The other model emphasizes the development of large cities. This is mainly an academic argument. It is suggested that valuable resources should be allocated to the large cities to establish a co-urban area, or mega-city, in Chinese urbanization. From the view of social, economic, environmental and infrastructure development, the scale efficiencies of large cities are higher than that of small cities and towns (Rao 1989). The other critical argument is that China is a country with several shortages in natural resources, typically in available land for development. It is calculated by some researchers that China is feeding and supporting 21 per cent of the world's population and its urbanization and industrialization with 7 per cent of the world's cultivated land, 7 per cent of its fresh water resources, 4 per cent of its coal and other mineral resources, and 4 per cent of its petrol and 2 per cent of its gas (Qiu 2010).

From the angle of opportunities, typically when there is a great gap between regions, and between large cities and small cities, the large cities are more attractive for the migrants from rural areas from the perspective of incomes and jobs. During last three decades, it has been the tendency that major migrants from rural areas are moving to eastern China and the large cities, typically those large cities and metropolises on the eastern coast, e.g. Shanghai, Shenzhen, Guangzhou, Xiamen, and others.

Some researchers (such as Zhang 1993) argue that in order to have an efficient development in terms of cost and benefit, the alternative way for Chinese urbanization is the development of large cities. Wang (2002) contended that the problem of land shortage and a large population in China could be solved by the development of the large city as a model for urbanization.

However, Chinese urbanization and industrialization is challenged by other critical issues of global warming and pressure to reduce carbon emissions at the same time. The large cities are the main sources of carbon emissions. If migrants continue to move to large cities and make the large cities even bigger, the reduction of carbon emission in China will be more difficult.

Integration and diversity of development of large, medium and small cities according to local features A further alternative and existing model for Chinese urbanization is of diversity for development patterns of large, medium and small sized cities simultaneously. This model suggests that large, medium and small sized cities should be developed simultaneously, with diversity in the regions of the east, the centre and the west. It proposes to consider the diversities of local characteristics, typically differences in the level of production, economic development, and the foundation of urban development. Since China is such a large country, it is inappropriate to have only one unitary policy and defined model for urbanization. It should be carried out according to local features.

A report by the National Development and Reform Commission, P.R. China, entitled 'The Understanding of the Urbanisation in China and the Basic Idea for Promoting the Urbanisation Process in the Ninth Five Year Plan' (1995) indicated that urbanization in the east of China has achieved a high level. The strong economic capability of the area has been the impulse of the establishment of the co-urban areas. The report suggested the decentralization of the functions of large cities and promotion of the development of small cities and towns. The development target is to establish co-urban areas with integrated coordination among mega-cities, large, medium and small cities in east China. The central area of China is still at the middle stage of its urbanization process. Although the structure of its city and town system has already been established with reasonable spatial layouts, there are not enough prioritized central cities with suitable major urban functioning activities in the area. It was suggested in the report that development of large and medium sized cities should be emphasized to achieve more concentrated urbanization. In western China, urbanization is only at its beginning stage. The primary industry is still the major contributor to the economic structure. There are few central and major cities with comprehensive urban functions in the region, especially in economic development. Investment is needed in those small cities with high-quality infrastructure facilities and convenience for communication. Small and medium sized cities with potential capacity should be further developed to be promoted as medium or large size cities through plan-led approach. These large and medium size cities are then able to function as central and major cities to provide services in their catchment areas. Medium and small sized cities with single economic functioning activities, e.g. raw material resources only, should be developed as multiple-functioning cities. It is also recommended to improve the capacity of the existing large cities to play significant role and function in the urbanization in the area.

Chinese Transition to the Market Economy

Political and economic reform in China is different to that in East European, former socialist countries, where the system was drastically changed into a capitalist system through 'shock therapy'. Reform in China towards a market economy has been happening in a gradual way.

The Chinese transition from the centrally planned economy to a market economy is associated with devolution, globalization and marketization. This includes the operation of the economy transferred to the market from the government. Within the government system itself, more powers have been decentralized to local governments.

Devolution

The major innovation in devolution is the emergence of municipalities as the key players in city and regional development in the economy, with the objective of promoting an increase in GDP and urban construction. The function of local government has been shifted from executors of decided plans made by the central government to operators of local economic and social development. To support this change, enforcement of the capabilities of fundraising for development at the local level has been promoted.

The reform of the fiscal investment system has been the most important devolution. In the fiscal system, tax revenue sources are separated and shared between the central government and local governments. The tax is divided into local taxes, central taxes and shared taxes. The division of taxes is complicated. They are collected by the Local Tax Bureau and the State Tax Bureau separately. This is summarized by Wang et al. (1999) and explained in Table 2.3.

Table 2.3   The tax categories

Type of tax	Elements of tax category
State tax	1. consumption tax; 2. value-added tax and consumption tax collected by Customs Office; 3. income tax on state-owned enterprises; 4. income tax on banks and non-banking financial enterprises; 5. business tax, income tax and urban maintenance and construction tax levied on banks and insurance companies; 6. tariffs.
Local tax	1. business tax and urban construction and maintenance tax; 2. income tax on local enterprises; 3. individual income tax; 4. urban land tax; 5. cultivated land occupation tax; 6. fixed asset investment direction adjustment tax; 7. vehicle use and licence tax; 8. real estate tax, contract tax; 9. stamp duty tax; 10. animal slaughter tax, agricultural tax and animal husbandry tax.
Shared tax	1. value-added tax, of which central government's share is 75 per cent; 2. natural resource tax. The resource taxes on offshore oil belong to the central government while the resource taxes associated with land belong to local government.

Source: Wang et al. 1999.

Devolution in the fiscal system has given local governments much financial power. Because of the division of the state tax, the local tax and the shared tax, the tax collection activities of local governments have been encouraged. The local fiscal capacity has increased so rapidly that local finance is even higher than that of the state government. In fact, the state income has declined. The share of state income in total national income dropped to 16.3 per cent in 1990 from 32.8 per cent in 1978. (Wang et al. 1999). Moreover, the capability of budget control by central government has been reduced dramatically.

Associated with the fiscal system reform has been the reform in investment. After 1984, when the urban reform started, investment decentralization policies were implemented (Wei 2000). Except for some large investment projects, or those projects which are significant to national social economic development, local governments have authority to make decisions in investment, particularly in urban infrastructure development and construction.

With further implementation of the fiscal reforms, central government has, however, found development sometimes out of control, or 'over-heated'. The central government has therefore tried to re-centralize financial resources and administrative power to implement its policies (Wei 2000) or to control resources.

With devolution, especially in the tax and investment systems, local governments have begun to seek economic objectives and increase their fiscal capacity. Increasing economic capability and local fiscal raising has therefore become the priority of local governments. Especially since reform in the urban areas, which started in the 1980s as a part of the devolution process, the achievement of economic and urban development has also been regarded as the main criterion with which to evaluate the local government capacity, especially that of mayors. Achievements in economic and urban development are attained by the growth of GDP and the expansion of a city or new urban construction. Progress in achievements will keep politicians in higher levels of government happy while satisfying local citizens, who will improve their quality of life. In order to accomplish these objectives, local governments have to attract inward investment, and seek various financial resources for urban development to provide a better quality of life while increasing the GDP.

Research by Logan (2002) shows that a large percentage of local government revenues is now drawn from land leasing and real estate development projects in which the municipality (or its district governments) is a partner. However, it has been realized recently that some aspects of devolution further worsen the problem of inequality and disparity in the regions and among the cities. Some less developing regions or cities, particularly those in the central and western areas, are able neither to generate more local fiscal resources nor to attract inward investment because of their poor or less developed capacity. The more developed regions and cities will find more opportunities in fiscal resources and inward investment. Their development will be further hastened. Successful cities have been growing so rapidly that there are some changes every year. There are places where everything is changing before one's eyes, and a visitor can return after a year's absence and be surprised at the transformation (Logan 2002). The situation is then that the rich areas become richer, but the poor areas become poorer.

Globalization

Globalization and the open door policy have stimulated the growth of foreign investment in China (Wei 2000), which has made crucial contributions to economic growth and urbanization. Foreign investment enterprises (FIEs) also provide many job opportunities. Many factories funded by Hong Kong and Taiwan investors are mainly labour intensive processing and assembling industries. They do not need skilled workers. The majority of their workers are from the rural areas. Some studies have reported that there are at least 150 million migrants from rural to urban areas (Shao 2001; Wang 2002c). These migrants have to do the work without skills or with lower skill requirements. They normally stay in dormitories provided by the factories. This situation is very common in Guangdong and Fujian Provinces.

Foreign investment provides capital for economic development, promotes employment, and stimulates urbanization and economic growth. Figure 2.2 illustrates the continuing increase in value of foreign investment until 2007 in China. Although FDI is a fluctuating process, generally it is attaining an increasing direction in China. Wei's (2000) research shows that foreign investment has become an important source of financing development in China, as evidenced by the increasing proportion of fixed investment provided by FDI, from 3.6 per cent in 1985, to 6.25 per cent in 1990, and 11.5 per cent in 1995. He suggested that the contribution of FDI to capital formation is the primary purpose of China's open door policy. The involvement of foreign enterprises also helps to transfer technologies from overseas to China and to increase the capability of adopting more advanced technology. With the coming of foreign investment enterprises to China, the integration of the Chinese economy with the global economy is promoted. At the beginning of the open door and reform policy, the first foreign investment enterprises entering China were 'processing or assembling industries' with high labour-intensive products, such as clothes, shoes and watches. These enterprises produced the goods according to samples designed outside China and assembling parts supplied by investors or clients from overseas. Materials for the products were imported from abroad, and the products were exported, too. The advantage of the low labour cost in Mainland China was used. While driving the Chinese economy to the global economy by these means, these enterprises also provided many job opportunities for migrants from rural areas.

Globalization in China can also be illustrated by changes in the three major indexes in international trade, foreign investment and international tourism. Table 2.4 shows the changes in these three aspects.

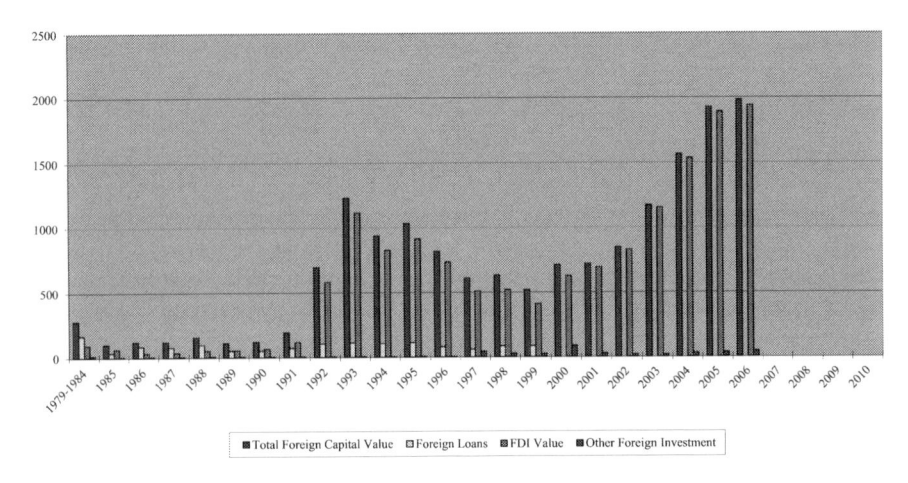

Figure 2.2 Usage of foreign capital
Source: China Statistics Year Book 2011.

Table 2.4 The major index for globalization (100 million US dollars)

	1985	1990	1995	1998	1999
Foreign direct investment	46.5	102.9	481.3	585.6	526.6
Total import and exports	696	1,154.4	2,808.6	3,240.5	3,606.3
International tourists' income	12.5	22.2	87.3	126	141

Source: China Statistics Year Book 2000.

Marketization and Urban Entrepreneurialism

Marketization started from the rural areas. Economic reform in the late 1970s and the beginning of the 1980s gave peasants the freedom to operate their businesses depending upon the market. Rural productivity increased rapidly. Township and rural enterprises contributed to rural economic development and urbanization. The success of the rural economic reform encouraged reform in the urban areas. In 1984, the government decided to shift reform from rural to urban areas. From 1984 to the beginning of the 1990s, the reform of state-owned enterprises saw separation between the ownership and business operation. State-owned enterprises began to enjoy more rights to undertake their business operations according to the market. The guiding ordinances of the operations of government enterprises were reduced. Price reform was released to the market but macro-controlled by the government at first. By the beginning of the 1990s, the prices of industrial consumption goods, agricultural products and raw materials were all decided by the market instead of government. Nevertheless, the government still had a monopoly on some prices and products.

The term 'socialist market economy' formally appeared in the Third Plenary Session of the Twelfth Central Committee of the Chinese Communist Party in October 1984. In 1992, the market-oriented transition was further promoted after the tour to south China by Deng Xiaoping. The debate on investment and development belonging to 'socialism' and 'capitalism' was stopped. The reform further expanded to investment and banking to establish a more appropriate administrative system for marketization. The market has generally played a basic function in resource allocation since then.

Marketization in China has been under an incremental and graduated process, which can be divided into three stages. The first stage was the marketization of products and their prices. Decisions on production, products and their prices were transferred to the market from decisions by the government. This stage of marketization has been completed. The second stage is market-oriented allocation and distribution of resources of finance, land, labour and technologies. Marketization of resources greatly affects the social and economic well-being of society. It is in fact marketization of capital. Marketization at this level is under way in China. Allocation of resources is no longer decided by government but by the market instead. Even some government resources, i.e. land, public buildings or infrastructure projects, and recruitment, go through a tendering or auctioneering process. The third stage of marketization is the ownership of assets and operation. This marketization started before the completion of the second stage. The third stage consists of the process of 'privatization', or the operation of multiple types of public utilities and services, including the involvement of the private sector.

Marketization is also expressed in the diversity of ownership enterprises, especially the increase in privately owned businesses. During the last two decades, private businesses have made an increasing contribution to economic development and urban construction. This can be illustrated by the percentage of investment in fixed assets by sources other than government (Table 2.5). It is clear that state investment has declined and is less significant when compared with other resources, typically fundraising in the market locally.

Nevertheless, the process of marketization has not been smooth. Several problems, not created by marketization, but the by-products of this process, have occurred. During the last 30 to 40 years, state-owned enterprises (SOEs) have played an important role not only in the economy, but also in the private lives of ordinary people. SOEs have provided all the welfare benefits, i.e. health treatment, housing, pension, and even primary and secondary education, to employees and their families. With reform and marketization, these benefits will also be marketized. Those employees who have relied nearly all their life on the SOEs and received low level of wages but reasonable benefits, face the problems of losing all the benefits they should enjoy. Even more serious for these employees is the situation where the SOEs they have worked for encounter problems of financial solvency or bankruptcy. The second problem comes from the operation

Table 2.5   Structure of funded sources for total investment in fixed assets (%)

Year	State budget	Domestic loans	Foreign investment	Self-raising and others
1981	28.1	12.7	3.8	55.4
1982	22.7	14.3	4.9	58.1
1983	23.8	12.3	4.7	59.2
1984	23.0	14.1	3.9	59.0
1985	16.0	20.1	3.6	60.3
1986	14.6	21.1	4.4	59.9
1987	13.1	23.0	4.8	59.1
1988	9.3	21.0	5.9	63.8
1989	8.3	17.3	6.6	67.8
1990	8.7	19.6	6.3	65.4
1991	6.8	23.5	5.7	64.0
1992	4.3	27.4	5.8	62.5
1993	3.7	23.5	7.3	65.5
1994	3.0	22.4	9.9	64.7
1995	3.0	20.5	11.2	65.3
1996	2.7	19.6	11.8	66.0
1997	2.8	18.9	10.6	67.7
1998	4.2	19.3	9.1	67.4
1999	6.2	19.2	6.7	67.8
2000	6.4	20.3	5.1	68.2
2001	6.7	19.1	4.6	69.6
2002	7.0	19.7	4.6	68.7
2003	4.6	20.5	4.4	70.5
2004	4.4	18.5	4.4	72.7
2005	4.4	17.3	4.2	74.1
2006	3.9	16.5	3.6	76.0
2007	3.9	15.3	3.4	77.4
2008	4.3	14.5	2.9	78.3
2009	5.1	15.7	1.8	77.4
2010	4.7	15.2	1.6	78.5

Source: China Statistics Year Book 2011.

of government. Some government authorities attempt to marketize their power to create profits, or to generate profits for their private interests from the 'public goods' under their control. This kind of inappropriate 'rent seeking' by government authorities seriously impacts on the proper working of the market operation.

Marketization and the importance of foreign direct investment (FDI) in revenue and income generation have stimulated competition among local governments. To attract FDI, local governments have offered more preferential policies by providing more land, improving infrastructure, protecting the interests of foreign investors and forming local–global alliances. Many developing zones have been established and huge areas of agricultural land have been converted to urban and industrial use, leading to the rapid decline in cultivable land (Wei 2000).

To attract the scarce resources of investment, Chinese local governments have to compete with other cities both in China and even in the world. The importance of FDI from inside and the inter-city competition for FDI from outside have strongly driven the local governments towards entrepreneurialism (Fu 2002).

Under such a situation, city and regional planning will strive to allocate the necessary resources to meet the proposed targets of economic and urban development by local government. One of the major planning objectives should seek to attract investment and to encourage economic development by granting planning permission rather than seeking to deter it by refusal. It is the general concern of planners that no reason should arise to drive away investment to the competing cities.

According to research by Hubbard and Hall (1998), urban entrepreneurialism was first introduced in the United States in the 1970s, and then implemented in Europe. Urban entrepreneurialism is the result of city competition. In order to deal with the pressures and to improve their competitive capacity, municipal governments have to implement a more active strategy to establish a more efficient, dynamic and productive spatial economy, and a new type of urban governance mechanism. This is the initial concept of urban entrepreneurialism.

With economic globalization, the competition among cities has become more serious. Even world mega-cities, e.g. London, Tokyo, New York, have adopted principles of entrepreneurial policy, to promote their status as the command centres of the global economy (Hubbard and Hall 1998). Urban entrepreneurialism is not universal; the concepts have been widely taken up by Chinese local government (Yu and Zhu 2009; Wu 2003).

One of the important measures of urban entrepreneurialism is city marketing. City marketing struggles to form the city image to be attractive to more investors and developers. During recent years, marketing has been a priority for major cities around the world as they attempt to promote their competitive capability.

The image of a city is the most important element in the marketing process. Investors and developers will make decisions to invest in a certain city according to their understanding of the city. The establishment of a city image will help to increase the confidence of potential investors and developers.

Healey (1998) referred to several criteria for deciding where to select the place to live or to invest. People want to know if they will be comfortable with the neighbourhood; whether the living environment is suitable for accomplishing the challenges of their daily life. Education facilities are also important for the decision as well as the kind of social world the children may encounter. People seek out places which can provide particular lifestyles. City and regional plans and the relevant development policies can provide some information to assist in their decision-making. From this perspective, Chinese city and regional plans have been functioning as one of the important means for city marketing to attract inward investment.

As a result of people's increasing concerns about conserving the natural environment and preserving local culture, living quality has come to be viewed as a most important issue by human beings. The greening of the environment and local culture have become important elements for marketing (Short and Kim 1998).

City and regional planning in China under the process of local urban entrepreneurialism has played a critical role in this aspect. The plans define the size and economic orientation and structure of a city and explain the course of action in the future. It is typically the statutory City Comprehensive Plan that recommends the urban economic development orientation through comprehensively analysing the background of a city. Moreover, the plan tries to protect the environment and local characteristics for human beings and the opportunity of attracting investment.

However, a city is a social, historical, cultural and economic complex. Implementation of urban entrepreneurialism in all aspects of a city will result in non-sustainable development. Of course, urban entrepreneurialism can be a main methodology to promote local economic development, particularly within the process of transition to a market economy and serious competition in China. The critical issue is to seek the most appropriate level of intensity, in other words, there needs to be an eclectic methodology in Chinese entrepreneurialism.

Planning in the Twenty-first Century

After 35 years of rapid economic development, China has been transformed from a poor developing country into one of the world's leading economic players. Three decades of rapid social and economic development, closely associated with rapid urbanization and significant rural–urban migration has created great challenges in terms of environment, social equity, regional polarization and sustainable development. Chinese planning is confronting new challenges and seeking approaches to cope with political, social and economic changes.

As a developing country, development is still regarded as a priority in China. China has benefited greatly from economic globalization. The changes in the economic and industrial structure as a consequence of globalization require urban planning to be flexible in order to consider economic concepts in the market economy. Until now, Chinese urban planning has been continuing to help promote competitive capacity and to encourage more inward investment. However, since China has become a key player in the global economy, typically after the world economic recession, and the significant impacts of the global weather changes, the international market, the supply of natural resources, typically petroleum, and capital flows in the world are no longer maintained as they were one decade ago. Chinese economic structure is under the pressure of becoming more self-sufficient, more energy efficiency and environmentally friendly. It was decided in the fifth plenary session of the Sixteenth Communist Party Central Committee to pursue a 'harmonious society' (Xinhua News 2003). The President of P.R. China, Hu Jingtao (2007) proposed the 'scientific concept of development' and 'building a moderately prosperous Society in all respects' in his report to the Seventeenth Communist Party Central Committee. The policy requirements are sustainable development through an energy-efficient and environmentally friendly approach, and to establish a harmonious society with an increase in employment, social security and poverty reduction. This indicates a shift of the government's development philosophy and planning policies. From the perspective of urban and regional development and its planning system, it is required to establish a self-adjustment mechanism to deal with the changes and uncertainties.

Public participation in city and regional planning and development in China has become a task in the planning and development process in the twenty-first century. This is the result of the incrementally increasing income of the normal residents, and the awareness and understanding of legislation and democracy.

In *Circular No. 13: Notice of Strengthening City and Countryside Planning Control* issued by the State Council (2002), it is specified to improve and enforce the supervision and monitoring from the society and the media. The function of the media and the public hearing should be promoted in the decision-making process to encourage public participation and to increase awareness of public monitoring. This has encouraged public participation in city and regional development policy-making in China. A important case is 'Xiamen PX (Paraxylene) Event'. A Paraxylene development project was approved by the State Committee of Development and Planning in February 2004. The project started its construction in November 2006 and was proposed to operate in 2008. It is expected that the annual industrial production volume can reach 80 billion yuan. However, it was because of the concerns of the impact on the local environment of the project that Xiamen citizens had strongly struggled against the project. Xiamen citizens express their opinions through the Internet. Public opinion has forced the local government to postpone the project while undertaking a new environmental assessment. A public hearing, on which 99 local citizens presented, was organized. It was because of the opposition to the project. A few days later, Fujian Provincial Government and Xiamen Municipal Government stopped the construction of PX factory in Xiamen.

The new policy requires planning policy-making to consider the interests of local people, the public and sustainability, while listening to the demands of developers. This has forced the existing development and planning concepts to change, which emphasize too much the market-oriented and economic-oriented positions, while ignoring the public interests, especially those of the disadvantaged (Wang 2000).

In the twenty-first century, sustainable development which seeks diverse objectives but focuses on delivering those objectives now and for the future has become a mainstream in development. Economic target is only one, but an important objective, as it will secure a better living quality. However, ecological environment protection has also been regarded as a priority in policy-making. The economic development should not be at the cost of increasing global warming and enlarging disparity and inequality. The objectives of economic development have tried to target poverty reduction, and improvement of the existing environment, while protecting the non-renewable resources.

According to the nationwide survey on Chinese cities between November 2010 and February 2011 by the Chinese Society of Urban Studies, guided by the author, 97.6 per cent of prefectural cities and above, including 15 sub-provincial cities and four metropolises under direct jurisdiction of the State Council have introduced 'eco-city' or 'low carbon eco-city' or 'low carbon city' as their development strategy. From the survey, it was found that 53 per cent of these cities are already under their process of different scale and types of ecological urban development; 28 per cent of them invite planning consultants to produce eco-city plans; and 19 per cent initiate their strategic policy for eco-city development (see Figure 2.3).

The data from the survey illustrates that low carbon eco-city development strategy has been an agenda of many local governments. The objective of low carbon eco-city is to have an integrated social, economic and environment development by reducing usage of natural resources, carbon emissions, applying of green technology, changing the existing development pattern of high natural resources consumption, reducing the income disparity between different groups of people.

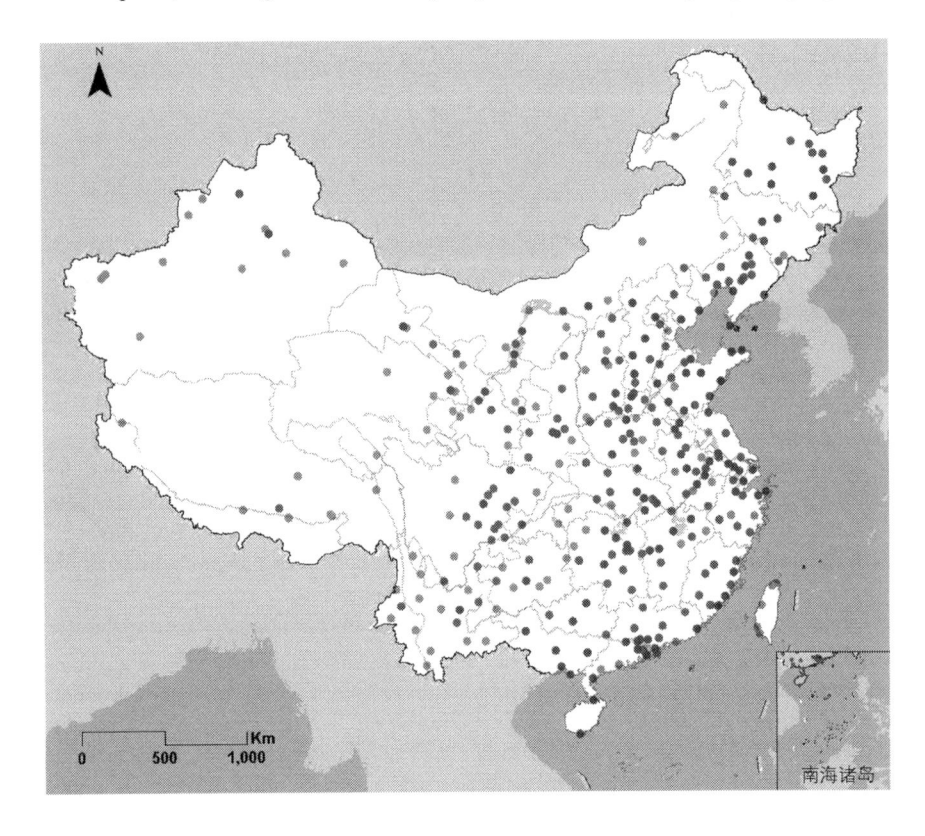

Figure 2.3   Location of low carbon eco-cities in China

Conclusion

China has more or less completed its transition from a centrally planned economy to a market economy and maintained its economic development at an average annual rate of 8 per cent during last two decades. In this chapter Chinese planning that was initially established in the centrally planned economy has been evaluated. From the evaluation, it shows that planning has inevitably played an important role in the process and its characteristic 'Plan-led' approach to development.

From the analysis, it is possible to conclude that the Chinese planning system has tried to cope with, and been impacted by, the political, social and economic changes in China during last six decades. After the founding of People's Republic of China in 1949, the existing Chinese planning system was established. This system has been in place through several eras of political, social and economic changes. In the centrally planned economy, because of central control of resources for development, urban planning extended and incarnated the social and economic development plans. Urban planning made a contribution to the allocation of resources and infrastructure to promote industrialization in China. Urban plans were easier to produce and implement as they were only required to service designated and approved projects, which were decided, invested and built by the government. The urban plan allocated the land rationally by considering the coordination needed between industrial production and the living conditions of residents. Planning norms and standards were needed to produce a reasonable spatial layout.

During the 'Great Leap Forward' and then the 'Cultural Revolution', urban planning, as a victim of the ambitious development objective defined by the government, was suspended for one decade. However, after the open door and reform policy, the urban planning system in China was re-established. Urban planning at the time began to be influenced by Western planning theories and practice, but the main concepts of urban planning still came from the centrally planned economy approach. Nevertheless, it was realized that there was a problem with the 'blueprint' city master plan.

The City Planning Act 1989 illustrated the beginning of the statutory planning system. In the 1990s, the transition to a market-oriented economy was further promoted. Economic development and urbanization were hastened. Some planning concepts and approaches from Western planning theories have been introduced to China. The most acceptable and practicable planning approaches were the systematic and rational planning approaches. The notions of these two types of planning have been adapted to the Chinese situation, probably because there is still some influence from the centrally planned economy during the transition.

The Chinese transition to a market economy has three main characters of devolution, globalization and marketization. It is because of devolution that local government has more authority and responsibility for economic and urban development. They have become the key players in development. The function of local governments has been shifted from executors of decided plans to operators of local economic and social development.

With economic globalization and competition, and the delivery of marketization, the local governments have had to introduce an entrepreneurial approach to encourage local economic development.

Although development is still regarded as a priority, Chinese development is confronting the challenges from its restructuring to be more self-sufficient, more energy efficient and environmentally friendly.

At the same time, as a result of the increasing income of its citizens, a Chinese middle class has appeared. They require more democracy in the policy-making process. The new emerging phenomenon has greatly impact Chinese city and regional planning.

Until now, it is possible to conclude that Chinese city and regional planning has always been linked closely with political, social and economic changes; and tried to reflect and react to the changes. It is important that the City Planning Act 1989 encourages the application of new techniques. It is significant to find the possibility for urban planners to apply a new planning approach with discretion in practice. Further transition to the market economy has created complex challenges for urban planning. The transition has brought about a new social, economic and political context, which creates new tasks and challenges for urban planning. It forces the planning innovations from bottom-up in local municipalities to meet the demands and changes of the market economy. The applications of the Urban Development Strategic Plan and Control Detailed Plan are such examples. Chinese urban planning has tried to 'keep pace with the times'.

Chapter 3
The Government System and Administrative Framework for Planning and Development

Introduction

The Chinese government structure was initially established according to the system of the former Soviet Union in the centrally planned economy. In addition to comprehensive economic administrative organizations, the ministries were established depending upon the different industrial sectors for the convenience of managing various economic development within the centrally planned economy. Since the transition to the market economy, the Chinese government has restructured its framework and amended its administrative system several times, typically at the central government level to cope with the social and economic changes in the transition. Some industrial ministries have been repealed. The functions and responsibilities of these ministries have been concentrated to the National Development and Reform Commission and the Ministry of Trades.

In local government, its structure has to follow the administrative framework of the central government in order to follow up policies from different ministries since China is still a hierarchical top-down administrative system.

The purpose of government's restructuring was trying to reduce the institutional cost while enforcing the management capability of the central government's institution. Nevertheless, the impacts of changing its administrative system has not seriously influenced that of urban development and planning. The urban development and planning has been regarded as public policies and a tool of the government to provide 'public goods' for local citizens and to intervene in the market by addressing its failures (Blower 1994; Pigou 1932; Simmie 1993). Since Chinese reform and resumption of the planning system in the late 1970s, the planning and construction has always been the task of the Ministry of Construction and its administrative hierarchical system from top down. However, city and regional development is a comprehensive process. A city, as the major human settlement, consists of most social, economic and cultural activities that either impact, or are participated in by, people. There are several departments of government with different responsibilities and functions relating to urban development. This chapter explores and analyses the ministries in the central government and its hierarchical departments at provincial and municipal levels which have important influence on people's daily life in Chinese cities.

The Government and Administration of the People's Republic of China

The Chinese Communist Party

The structure of China's government is hierarchical, and involves a complex set of inter-relationships between the Chinese Communist Party (CCP), which has ruled the country since 1949, and the instruments of government at central, provincial and lower levels.

The organizational structure adopted by the Chinese Communist Party is replicated throughout the country at the different hierarchical levels of government and administration. The most senior body of the Chinese Communist Party is the National Chinese Communist Party Congress which meets once every five years and brings together more than 2,000 delegates from party organizations across the country. Its main function is to 'elect' a central committee of about 200 members There have been 17 such Congresses since the establishment of the Chinese Communist Party in 1921, the last of which was in 2002 with the adoption of a resolution on the amendment to the China People Congress Constitution to enshrine the 'scientific outlook on development', which includes several items of 'taking development as its essence, putting people first as its core, comprehensive, balanced and sustainable development as its basic requirement, and overall consideration as its fundamental approach' (Xinhua News 2003).

The Chinese Communist Party Central Committee, which in theory exercises the powers of the National Chinese Communist Party Congress until its next meeting, meets in plenary session from time to time. Currently the Eighteenth Central Committee, which was elected by the Eighteenth Party Congress, has 205 full members and 171 alternative members. The first meeting of the Central Committee is held immediately after its election by the Party Congress where it elects a new Politburo (25 members) and its smaller Standing Committee (seven members), which are responsible for progressing the work of the Central Committee when the latter is not in session. The real decision-making powers lie with these groups. The Politburo and its Standing Committee contain the most influential and powerful people in China and the present membership of the Standing Committee from the second plenary session of the Eighteenth Central Committee of the Chinese Communist Party is Xi Jinping, Li Keqiang, Zhang Dejiang, Yu Zhengsheng, Liu Yunshan, Wang Qishan and Zhang Gaoli. Beneath the Standing Committee is a secretariat that services the whole operation.

The operating principle of the Chinese Communist Party is democratic centralism, whereby the party members contribute to debate on issues before the central leadership (i.e. the Politburo Standing Committee) decides on its position. At this point the decision becomes policy that all must accept. Central to the idea of democratic centralism is the maintenance of the Chinese Communist Party's leadership, i.e. there is no question of the Chinese Communist Party sharing power.

Government and Local Administration in the People's Republic of China

Central government   The central government in the People's Republic of China is required to follow the leadership of the Chinese Communist Party and to implement party policy.

The most important national body is in theory the National People's Congress (NPC). Its membership of nearly 3,000 delegates is elected by China's provinces, autonomous regions, municipalities and the armed forces for a period of five years, and is required to meet once a year, although this did not happen during the Cultural Revolution period.

The NPC's functions and powers include:

- adopting the People's Republic of China constitution or amendments to that constitution;[1]
- electing a president of the People's Republic of China;
- enacting or amending laws;
- examining and approving the state budget;
- approving/rejecting decisions of the Standing Committee;
- establishing provinces, autonomous regions, municipalities and special administrative regions;
- deciding on questions relating to war and peace;
- electing a president of the Supreme Court.

The NPC has its own permanent standing committee of 150 members elected from the membership of Congress, which generally meets once every two months and carries out the work of the NPC when it is not in session.

The State Council is the highest unit of state administration: 'it is the cabinet which oversees China's vast government machine ... [and] sits at the top of a complex bureaucracy of commissions and ministries and is responsible for making sure party policy gets implemented from the national to the local level' (BBC News 2008).

The premier is chairman and its membership consists of vice premiers, state councillors and ministers. It generally meets once a month and its most important roles are to draft and manage the national economic plan and the central government's circulars. It is assisted in its work by its Standing Committee of the State Council which consists of the premier, four vice premiers, state councillors and the secretary general who are appointed by the NPC. It generally meets twice a week. The premier, who is responsible for the day-to-day work of government, can be considered to be the equivalent of the prime minister in the United Kingdom. The NPC chooses the premier although his nomination and formal appointment are made by the president of the People's Republic of China. In 2013 the twelfth

---

1   There have been four constitutions – in 1954, 1975, 1978 and 1982 with revisions in 1988, 1993, 1999 and 2004.

NPC reduced the number of ministries and ministry level state commissions to 25 as follows:

- Ministry of Foreign Affairs
- Ministry of National Defence
- State Development Planning and Reform Commission
- Ministry of Education
- Ministry of Science and Technology
- Ministry of Industry and information
- State Ethnic Affairs Commission
- Ministry of Public Security
- Ministry of State Security
- Ministry of Supervision
- Ministry of Civil Affairs
- Ministry of Justice
- Ministry of Finance
- Ministry of Human Resources and Social Security
- Ministry of Land and Natural Resources
- Ministry of Environmental Protection
- Ministry of Housing and Urban-Rural Development
- Ministry of Transport
- Ministry of Water Resources
- Ministry of Agriculture
- Ministry of Commerce
- Ministry of Culture
- National Health and Family Planning Commission
- People's Bank of China
- National Audit Office.

The links between the Chinese Communist Party and the military (the People's Liberation Army, PLA) are an important element in maintaining the Chinese Communist Party's control of the government. Mao Zedong always insisted that the 'army should be under the leadership of the Communist Party'. During the early years of communist rule, most of the country's leaders owed their positions to the military positions they had held during the civil war, and links between them and the military were close.

As a consequence, the party has attempted successfully to control the armed forces through the Central Military Affairs Commission – an 11-man commission chaired by Communist Party Secretary General (or Communist Party Chair). This commission has the final say on all decisions relating to the PLA, including senior appointments, troop deployments and arms spending. The commission also controls the paramilitary People's Armed Police, who guard key government buildings.

After the founding of the People's Republic of China, the chairmanship of the Central Military Affairs Commission was held by Mao Zedong, Deng Xiaoping, Jiang Zemin, Hu Jintao and Xi Jinping.

Local government and administration At the next level down in the governmental structure (the provincial level) China is governed through 22 provinces (excluding Taiwan), five autonomous regions, four municipalities and two special administrative regions. The people responsible for running these bodies are all appointed by the Chinese Communist Party, and their ability to deviate from the party line is limited because they know their career development depends on continued party support. However, in recent years the centre has lost some control to the regions, especially in the field of economic development because of devolution of development power.

Power and decisions come down from the provincial level through an intermediate level of counties and cities down to the local level townships. At each level the party and the government structures sit side by side but with the party's representative always being more powerful. Although the 1982 constitution of the People's Republic of China provides for three 'de jure' levels of local government (province, municipal and country), in practice there are five levels of government as follows:

- province level (provinces, autonomous regions, municipalities, special administrative regions);
- semi-provincial level cities (15);
- prefecture level (prefectures, autonomous prefectures, prefecture-level cities, leagues);
- county level (counties, autonomous counties, county-level cities, districts, banners, autonomous banners, forestry areas, special districts);
- township level (townships, ethnic townships, towns, sub-districts);
- village level (neighbourhood committees, village committees or village groups).

Each of the different levels of local government has its own People's Congress and associated standing committees except two levels below county level. The heads of autonomous regions, autonomous prefectures and autonomous counties must be members of the ethnic minority represented. In 2005 the People's Republic of China administered 33 province-level regions, 333 prefecture-level regions, 2,862 county-level regions, and 41,636 township-level regions. For each of these local government levels there is a corresponding level in the civil service of the People's Republic of China.

With the exception of the provinces in the northeast, most of the provinces have boundaries which were established during the historical Yuan, Ming and Qing dynasties and they serve an important cultural role in China where people tend to be identified in terms of their native provinces. The most recent administrative changes to the provincial structure include the promotion of Hainan (1988) and Chongqing (1997) to provincial level status and the establishment of Hong Kong (1997) and Macau (1999) as special administrative regions. Taiwan, controlled by the government of the Republic of China, is regarded as the twenty-third province of the People's Republic of China.

The five autonomous regions are province-level units of local government, each with a designated ethnic minority which is guaranteed more rights under the constitution than the 22 provinces. For example, they have a chairman who must be from the ethnic minority group associated with the autonomous region as opposed to a governor in the regular provinces. The autonomous regions were established in 1949 after the People's Republic of China became the ruling government, and are modelled on the then Soviet approach for dealing with ethnic minorities.

Metropolises are large cities that have the same status as provinces. In general, the metropolitan area of a municipality is a small part of its total area with the rest of the municipality consisting of towns and farmland. There are four metropolises in China relating to the most important Chinese cities – Beijing, Tianjin, Shanghai and Chongqing.

Beijing is the capital of the People's Republic of China, with an urban population of 7.34 million. It is the country's political, cultural, scientific and educational centre, and a major transportation hub. Situated on the northern edge of the North China Plain, it is sheltered by mountains to the west, north and east, whilst to the southeast is an important fertile plain. Beijing's temperate continental climate produces four clearly contrasted seasons: a short spring, rainy and humid summer, long and cold winter, and a very pleasant autumn.

Shanghai is China's largest city, with an urban population of 9.54 million. It is located halfway down China's mainland coastline, where the Yangtze River reaches the sea; has an important and comprehensive industrial base and possesses a large harbour. It has a highly developed commerce, banking and shipping industry in addition to major industries including metallurgy, machine-building, shipbuilding, chemicals, electronics, instruments, textiles and other light industries. Shanghai is central to the development of the national economy. The Pudong area separated from the old city by the Huangpu River, is being developed as a modern, multifunctional, export-oriented district, which is intended to transform Shanghai into an international economic, banking and trade centre and a modern international city.

Tianjin is a major industrial and commercial city in north China, with an urban population of 9.5 million. It is 120 km (approx.) from Beijing, and is an important port for international trade, and ocean-going shipping. Its traditional industries include iron and steel, shipbuilding, chemicals, power generation, textiles and foodstuffs, whilst it has successfully attracted modern industries such as motor vehicle manufacturing, petroleum exploitation and processing, and the production of watches, TVs and cameras. Binhai New District is recently treated as an economic development pole in China with great support from the Chinese central government. The Sino-Singapore Eco-city is also located in the Binhai New District.

Chongqing is the largest industrial and commercial centre in southwest China with a total population of 30 million and a major transportation hub for land and water transport. It is situated in the upper Yangtze valley at the confluence of the Yangtze and Jialing rivers; has an urban population of 6.14 million; is a comprehensive industrial city, with advanced iron and steel, chemical, power generation, motor vehicle, textiles, foodstuffs and pharmaceuticals industries. It is becoming increasingly important in: (1) China's strategy for the development of western China; and (2) in the opening-up and development of areas along the Yangtze River.

The two special administrative regions (SARs) were established when the former European colonies of Hong Kong and Macau were returned from the United Kingdom and Portugal to the People's Republic of China in 1997 and 1999 respectively. They were granted SAR status to provide them with a high degree of autonomy under the 'one country, two systems' arrangement introduced by Deng Xiaoping. Both SARs are small, and neither has adopted the administrative structure of Mainland China. For example, Hong Kong is divided into 18 districts, each with a consultative district council, whilst Macau is administered as a unitary body by the SAR government, with no further subdivisions. Both SARs are directly responsible to the Central People's Government, as provided for in Article 12 of the acts establishing the two SARs, whilst Article 30 of the 1982 Constitution for People's Republic of China provides for the establishment of special administrative regions.

The prefecture level does not exist at either the metropolis or SAR levels of local government. However, in the provinces and the autonomous regions it is the second level of the local administrative structure. In December 2005 there were 333 prefecture units consisting of prefecture-level cities (283); prefectures (17); autonomous prefectures (which are prefectures with one or more designated ethnic minorities, 30); and leagues (three in Inner Mongolia only).

Prefectures were formerly the dominant second-level unit of local government. However, between 1983 and the 1990s they were gradually replaced by prefecture-level cities. Today, 17 prefectures remain in the ARs of Xinjiang and Tibet, and the provinces of Yunnan, Guizhou and Qinghai.

Leagues are effectively the same as prefectures, but they exist only in Inner Mongolia. Similar to prefectures, leagues have mostly been replaced with prefecture-level cities. The name is a relic from earlier forms of administration in Mongolia.

Prefecture-level cities thus make up the majority of prefecture-level units and they generally consist of a city core with its associated hinterland of agricultural land, townships and villages.

The 15 largest and important prefecture-level cities, most of which are the capital city of a province, or special economic zones, or important industrial cities, known as sub-provincial cities, are given a level of power higher than a prefecture, but still lower than a province. Although these cities still belong to provinces, their special status gives them a high degree of autonomy within their respective provinces.

A similar case exists with some county-level cities known as sub-prefecture-level cities that are also given more autonomy. These cities are given a level of power higher than a county, but still lower than a prefecture. Such sub-prefecture-level cities are often not put into any prefecture but are directly administered by their province, e.g. the Pudong District of Shanghai and Binhai New District of Tianjin. Although its status as a district of a direct-controlled municipality should define its prefecture-level, the district head of Pudong and Binhai New District is given sub-provincial powers.

At the next level down – the county level – there is a range of different administrative units all possessing roughly the same powers, including the:

- counties (1,464) areas with towns, villages and farmland;
- autonomous counties (117) areas with towns, villages and farmland and one or more designated ethnic minorities;
- districts (862) which were formerly the subdivisions of built-up urban areas and today incorporate towns, villages and farmland;
- county-level cities (374) which are large administrative regions that cover both urban and rural areas;
- banners (49) and autonomous banners (3) which operate in Inner Mongolia;
- two special districts and one forestry district in Mainland China.

In general, urban areas are divided into districts, while rural areas are divided into towns, townships and ethnic townships. Sumu and ethnic Sumu are the same as townships and ethnic townships, but are unique to Inner Mongolia.

Rural areas are organized into village committees in a central administrative village, which is a bureaucratic entity or a natural village for systems such as the census and the mail system but is not really significant in terms of political power. In urban areas, the village level unit exists as a sub-district of a district of a city, called '*Jiedao*', and administers many communities or neighbourhoods. Each has a neighbourhood committee to attend to the needs of the occupants of that neighbourhood or community.

It is clear that the structure and operation of the local government administrative system is complex and there is ongoing speculation of an impending reform to its structure to reduce the different levels of administration from five to three in rural areas (province, county, village). There have also been calls to:

- abolish the prefecture level, with some provinces having transferred some of the power prefectures currently hold to the counties they govern;
- reduce the size of the provinces.

The idea is to separate counties from prefecture-level cities and make them under direct jurisdiction of a provincial government. This is what China's local government administrative system used to be before the promotion of urbanization.

In China, an indirect election system is adopted to elect the local administrative officers whereby the local voters elect representatives to the People's Congresses who then have the right to nominate the officers to the different departments within the different units of local government. In rural areas, villagers can vote directly to elect the head of a central administrative village, which may consist of several nature villages.

The Legal System

There are three elements to the Chinese legal system as follows:

- public security, which is responsible for investigating crime and detaining suspects;
- people's procuratorate, which approves arrests, establishes an a priori case against suspects and initiates prosecutions;
- people's courts, responsible for the passing of judgments.

The vast majority who appear before the people's courts are convicted. The Supreme People's Court reported that in the period 1992–7 only 0.43 per cent of criminal cases resulted in a not guilty verdict (Mackerras 2001: 100–101).

The Supreme People's Court and the Supreme People's Procuratorate are responsible to the NPC and its standing committee and they exist at different levels in a hierarchical relationship which moves from higher to intermediate to basic and special people's courts.

The independence of the legal system from state control was removed during the Cultural Revolution. Since the introduction of the reform era, there has been a growing recognition that the law should be independent of the party and the state and today there is a consolidated legal framework which deals with most aspects of social life.

In 1995, the Compensation Law came into effect which in theory protects and compensates those whose legal rights have been violated by the state. In 2007, the Property Right Law came into effect which intends to provide private properties.

The Ministries with Responsibilities for Urban Development in Central Government

The National Development and Reform Commission (NDRC)

The National Development and Reform Commission used to be the most important ministry in the central government in the era of the centrally planned economy. It still plays a crucial function in social and economic development today. The National Development and Reform Commission is the ministry that is responsible for the formulation of the National Social and Economic Development Plan (Five-Year Plan). It is also responsible for the allocation of industrial development projects with national, regional and economic significance; coordination, assessment and approval of local large infrastructure development projects applied by local governments. The tasks of the NDRC cover very wide areas in terms of national social and economic development.[2]

In addition to formulating and delivering national economic and social development (Five-Year Plan), the tasks of the NDRC also include coordinating and balancing nationwide economic and social development; undertaking research and analysis on domestic and international economics; decision on development targets and policies relating to the national economy, and regulating overall price level and optimization of major economic structures.

The National Development and Reform Commission should monitor macroeconomic and social development trends, providing forecasts and information guidance; and coordinate and address major issues in social and economic operation. It should also adjust economic performance and coordinating the transport of important goods and materials.

In fiscal and financial aspects, coordinating with the Ministry of Finance and People's Bank of China, the National Development and Reform Commission should participate in the formulation of fiscal and monetary policies, and supervise and inspect the implementation of price policies. The prices of important commodities with national and important tariff significance should be adjusted and approved by the NDRC.

In terms of major infrastructure projects development, the NDRC has responsibility for approving, authorizing, reviewing and allocating central government's budget to major development and infrastructure projects with national and regional social and economic significance.

The National Development and Reform Commission has also been involved actively in regional planning work. During last two decades, the NDRC has initiated a type of regional plan, the plan of development functioning zones that cover the whole country, to clarify different regional functions of development.

Since climate change and global warming are becoming a major challenge for human beings, the NDRC has been defined as an important policy-maker in terms of sustainable development, energy efficiency, reduction of carbon emissions, promotion of recycling economy, and development of environmentally friendly industries and clean production.

The Ministry of Transport (MoT)

The functions and tasks of the Ministry of Transport include planning, development and administration of water transport, sea ports, motorways, national highways, and urban and inter-city freight and passenger transport.

---

2 http://en.ndrc.gov.cn/mfndrc/default.htm, accessed 11 June 2009.

The Ministry of Transport should take responsibility for the formulation of development strategy, plans, policies and regulation of nationwide motorway, highway and seaway transport. It should also be responsible for the delivery and construction of different hierarchic roads. It is the responsibility of the Ministry of Transport to charge toll fees on highways. The Ministry of Transport undertakes the policies, norms and standards of both urban and inter-city road construction.

There used to be an overlap between the tasks of the Ministry of Transport and the Ministry of Housing and Urban–Rural Development. As regarding urban transport, the local transport authority is responsible for urban transport management and road construction. However, it is the responsibility of the local city planning authority to produce urban transport development plans and traffic management policies. It is then generating a problem of coordinating between the transport development plans and road network development. In the Twelfth National People's Congress, the former Ministry of Railways was merged into the Ministry of Transport.

The Chinese railways have been controlled and managed by this ministry for 64 years since 1949. The functions and tasks of the Ministry of Railways included planning, construction and administration of railways for the whole of China. The partnership of railway construction and operation is unusual but is only operated in a few regions. Every railway station in China was also under the direct jurisdiction of this ministry. Railways have been a highly monopolized industry and operation.

It is because of this monopoly that contradictions between the railway authority and planning authority in location decisions of railways and stations are very common. Some decision of locations for railways and stations are totally different from city and regional development planning, but there is not any compromised solution from the railway authority. 'To be or not to be' is the only solution from the railway authority since it is the only investor and decision-maker in all issues of railway development in China. This monopoly and totally independent hierarchical management system of railways also create the phenomenon of inefficient use of rail resources. The conventional railway is unable to be used as a means of local public transport.

The Ministry of Housing and Urban–Rural Development (MoHURD)

The Ministry of Housing and Urban–Rural Development is responsible for policies of housing development, urban and rural planning and their development and construction. The ministry provides national guidance for urban water supply, gas and district heating, public utilities and sanitation. MoHURD also has responsibilities in development, use and protection of underground water in urban areas, supervision of urban flood prevention and drainage, and approval of city comprehensive plans of designated cities. National and professional planning and construction norms and standards, including urban transport, e.g. bus and taxis, and urban rail transport are made and promulgated by the Ministry of Housing and Urban-Rural Development.

In terms of housing policies and provision, the Ministry of Housing and Urban-Rural Development has been authorized more powers in decision-making. In April 2010, the central government's circular of 'Strictly Control Rapid Increasing of Housing Price in Some Cities' (State Council 2010) empowered the Ministry of Housing and Urban-Rural Development to appraise, monitor, investigate and inquire after the accountability of provincial governments in their task of housing price control and affordable housing provision. In order to provide appropriate land to meet the demand for housing, the ministry is able to interfere directly in the planning and development of projects for housing. In order to address the housing problem for lower income households, MoHURD has also been authorized to allocate housing subsidies from the central government by working with the National Development and Reform Commission and the Ministry of Finance, the organization that used to be responsible for national fiscal allocation at the central government. Moreover, the Ministry of Housing and Urban-Rural Development has the responsibility to establish a nationwide housing information system for any resident and family.

The Ministry of Land and Resources (MoLR)

The Ministry of Land and Resources is responsible for planning, administration, protection and rational use of natural resources, e.g. land, mineral and marine resources. It is regulated by the State Council of the People's Republic of China that the Ministry of Land and Resources should enact and promulgate relevant ordinances and regulations in governing the management of land, mineral and marine resources except marine fishery resources managed by the Ministry of Agriculture. The ministry should be responsible for the formulation of technical criteria, rules, standards and measures for the planning and management of land, mineral and marine resources.

It is the task of the MoLR to compile and implement the national general plan for land and resources and to participate as a member of the steering committee for examination and verification of city comprehensive plans submitted for approval by the State Council. It is the responsibility of the ministry to organize surveys and evaluation of mineral and marine resources, to produce plans for the protection of mineral and marine resources for the prevention and mitigation of geological disasters.

There are two main tasks of the Ministry of Land and Resources that significantly impact city and countryside development. One task is to develop and deliver regulations for assignment, lease, evaluation, transfer, transaction and purchase of land use rights on behalf of the state. Without availability of land allocation for defined urban development by MoLR, urban planning and its delivery are meaningless. The other task is to make policies and regulations to protect cultivated land and to enforce control over different purposes of land used for agriculture. This task is closely associated with the first task. The Chinese central government's policies have been continually to protect cultivated land as a priority. Land allocation for urban development should not reduce the total volume of cultivated land. However, the great pressures of rapid urbanization, industrialization and economic development create contradictions between urban development and cultivated land protection, between the central and local governments, and between the Ministry of Land and Resources and the Ministry of Housing and Urban-Rural Development.

Urbanization, industrialization and economic development require more input of land. Nearly 50 per cent of the Chinese population is still in rural areas. China is still under its process of urbanization and industrialization. 300 to 500 millions rural populations will move to urban areas. These migrants require space and land for the provision of housing, infrastructure and public facilities, and jobs. However, China is a country with a population of more than 1.3 billion. The provision of food to its people has been regarded as a national security issue. Protection of 12,000 hectares of basic cultivated land of the whole of China is defined as a key national policy, which is defined as an untouchable 'red line'. However, the resources, typically finance of local government for development, mainly come from land leasing. It inevitably creates conflict between the central government and local government.

As regarding to the contradictory tasks between the Ministry of Land and Resources (MoLR) and the Ministry of Housing and Urban-Rural Development (MoHURD), MoHURD should promote urban and rural planning, development and construction. It requires available land with increasing of the urban population because of migrants from the rural area. However, the task of MoLR is to protect cultivated land and allocate land for urban and rural development. The limitation of land supply may impact the achievements of MoHURD. The contradictions of different interests between the two ministries are unavailable.

The Ministry of Environmental Protection (MoEP)

As specified by the State Council, the Ministry of Environmental Protection should take responsibility for the establishment of an environmental protection mechanism, i.e. the administrative work of formulating and delivering national environmental protection policies, plans, regulations and acts. The Ministry of Environmental Protection should formulate and promulgate national and professional norms and standards of environmental protection, and plans for the functional division of environmentally protected areas. The ministry should also take responsibility for the management and monitoring of environmental protection for major development areas, river catchment areas, and water resources in China. MoEP is functioning as a leading body to investigate and supervise local government's actions on emergent treatment of environmental disasters, and playing an active role as a coordinator in inter-regional environmental dispute.

MoEP is a key player in achieving national targets of reducing CO_2 emission. This task includes defining the total emission control level of major polluters and licensing mechanism of polluters' discharge; and supervising and reviewing achievement of pollution emission reductions by local governments.

When the central government is preparing its financial budget, the Ministry of Environmental Protection has the power to propose and to allocate national fiscal budget in the field of environmental protection, especially in the aspect of addressing climate change and promoting recycle economy.

Entrusted by the State Council, the Ministry of Environmental Protection is responsible for the organization and coordination of environmental impact assessments to those major economic and technological development, or regional major development, projects. It is also the responsibility of the Ministry of Environmental Protection to monitor environmental protection and to promulgate relevant treatment regulations and mechanisms for water, air, land, noise, light, solid waste, chemistry products and automobiles.

This ministry should be involved in the formulation of ecological environmental protection plans, assessment of natural resources and their development and use, which may impact the ecological environment; protection of wildlife and biodiversity in natural protection zones, natural beauty areas, forests and wetlands.

Nuclear power has been an alternative to address demands of energy supply. It is the responsibility of the Ministry of Environmental Protection to supervise and to execute the security of nuclear power stations and radiation. The task includes relevant policy, plans and standards making, and participating in treatment of emergent events. The security of nuclear facilities, radiation resources and application of nuclear technologies and protection of nuclear pollution in their usage and development of nuclear mining resources are supervised by the Ministry of Environmental Protection.

Shortage of Coordination among Government Departments with Responsibilities for Development

The Chinese government framework has adapted the structure of the former Soviet Union. Ministries were established according to different industrial sectors. This structure worked soundly to manage the social, economic and political operations within the centrally planned economy. However, in the market economy, this structure may have to change especially in economic development spheres as most resources to promote economic development depend upon the market. In the urban development perspectives, it is more complicated as urban development is a comprehensive process. A city, as major human settlement, is a platform attracting all activities of human beings. It is also a centre to promote economic development. The characteristics of the Chinese government's framework have inevitably brought about multiple departments with different responsibilities for various urban development tasks. These government departments, especially the ministries of the central government, have different resources and programmes as well as different interests in urban development. It is a crucial issue to have an appropriate coordination mechanism among various responsibilities and tasks of the government's departments to avoid conflicts and overlaps. Otherwise, from the administrative arrangement of the Chinese government, the operation cost would be very high.

In land use aspect, the tasks of the two ministries are closely related to this issue in the central government – the Ministry of Housing and Urban-Rural Development and the Ministry of Land and Resources. The Ministry of Housing and Urban-Rural Development is responsible for the production of the policies and regulations related to city and countryside planning and development, and for the approval of city comprehensive plans of designated cities. The Ministry of Land and Resources is responsible for preparing the policies and regulations of general land use plans to control land resources. In order to coordinate these two types of plan, several acts and government's ordinances have to clarify the relationship between them.

It is regulated by the Land Administrative Act that:

> the City comprehensive plan, town plan and village plan should be connected to general land use plan. The land use defined by city comprehensive plan, town plan and village plan should not scale out the land use for the city, town and village defined by general land use plan. Within the areas defined by urban planning, town planning and village planning, development of land should follow urban, town and village plans. (Article 22)

Article Five of City and Countryside Planning Act 2008 specifies that 'City comprehensive plan, town plan and village plan should be produced according to national economic and social development plan, and be connected to general land use plan'. The No. 11 Circular of 1997 by the State Council, *Further Strengthening Land Management to Protect Cultivate Land Resources*, also specifies the relationship between the two types of plan: 'urban development plan should be connected to general land use plan, the land use defined by urban plan should not be scale out the volume of land use defined by general land use plan'.

In addition to the overlap and contradiction aforementioned among various government departments, the overlap and duplication of responsibilities also exist in water administration between the Ministry of Housing and Urban-Rural Development and the Ministry of Water Resources. The former is responsible for the water supply within urban areas and underground water resources in urban areas, while the latter is responsible for water supply outside urban areas, and general water resources management.

Hierarchical and Sectoral Administration

The Chinese government administrative structure was established vertically top-down in the centrally planned economy. This system has been continued even as

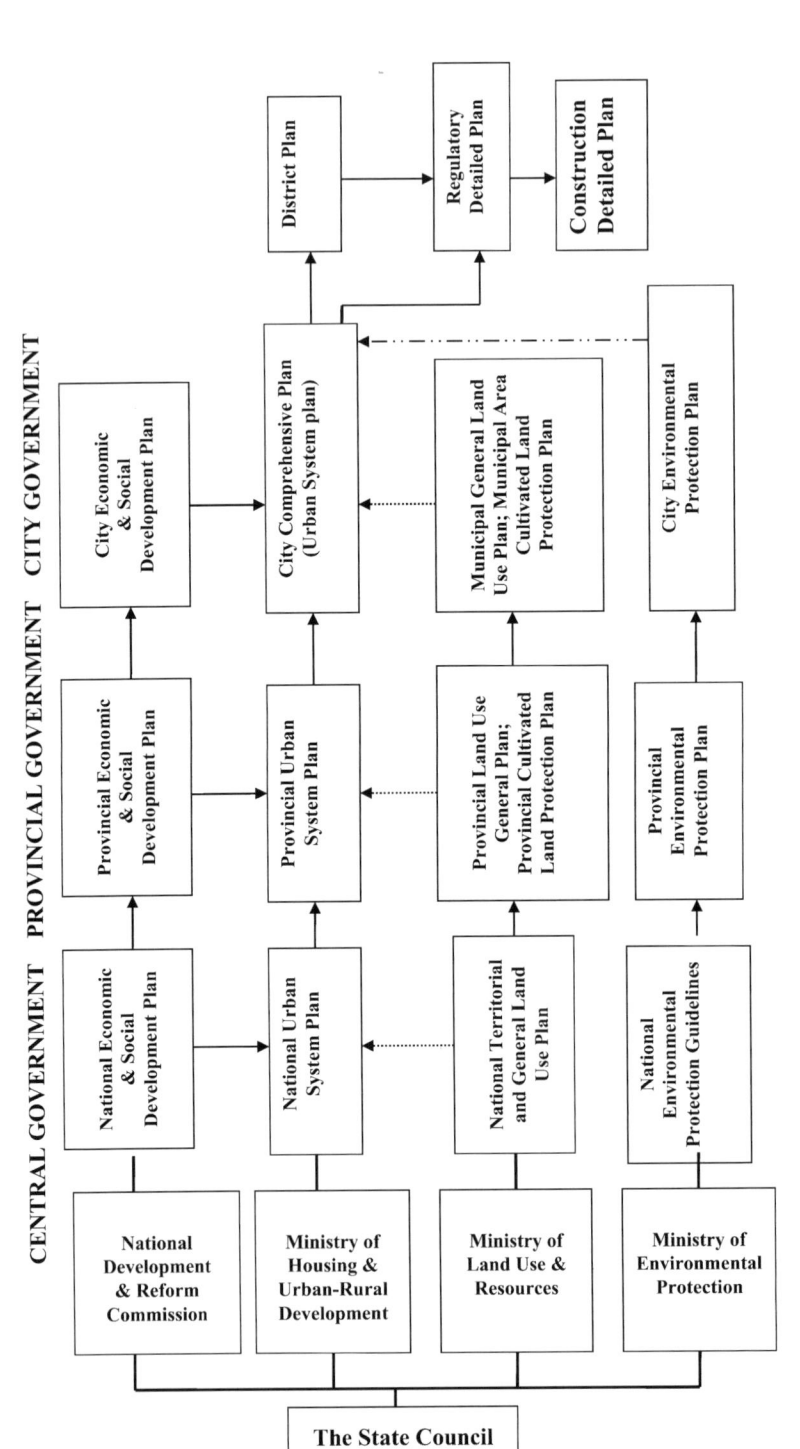

Figure 3.1 Hierarchical planning system in China

Source: Created by the author.

China started its process of transition to the market economy. The planning system is associated with the government's administrative structure from the top-down centrally planned economy. As argued by Khakee (1996), a key aspect of reform in decentralization is sectoral bias. Local governments have to negotiate and bargain with different sectoral ministries in the central government. Establishment of corresponding sectoral departments at provincial and municipal levels provides a pre-condition for negotiation with central government. Figure 3.1 summarizes the hierarchic government structure and its planning system, which provides the whole framework for regional, urban and rural planning and development.

The main characteristics of Chinese government structure and its planning system is a sectoral framework in addition to its top-down approach. The whole government structure is formed by various sectors, or ministries, that produce and implement its own plan. The departments of local governments within the same sector should produce and implement its sectoral plan that should also follow the policies of higher hierarchical plans.

This structure was formed before the reform and open door policy when local government was not an autonomous authority. Its leverage ability on local economic development was very weak as was the incentive to facilitate economic development. In the centrally planned economy, the Chinese local governments did not have independent revenue. Allocation of resources was centrally controlled from the top down. All local revenues were collected by the central government and distributed according to the national economic and social development plan. The departments of the local governments were characterized by sectoral dominance or sectoral separation. These departments, such as the urban planning bureau, construction bureau, environment protection bureau and transport bureau were both responsible to their higher hierarchical sectoral departments and local city party chief and mayor. Within this system, the central and local governments had a common interest (Zhang and Meng 2007).

These characteristics have created great uncertainties and inefficiencies in policy-making and delivery because of different individual interests of government departments. The contradictions and conflicts not only appear between different policies promulgated by various government departments in central government, but also in the detailed operation and management in local government.

Features of Local Government

Chinese local governments can be divided into two tiers of provincial and municipal governments. There used to be prefectural commissioner's offices which were representing agents of provincial governments charging rural areas. However, this type of government agent was cancelled in order to promote the urbanization process. Its territory was merged with municipal governments in the late 1980s and beginning of the 1990s.

Generally speaking, the governmental administrative structure of provinces and municipalities would be similar to that of the central government, especially at the provincial government level. Before the reform and open door policy, since central government controlled and allocated all development resources, one department in local governments both at provincial and city levels had to correspond to one ministry (sector) of the central government. Otherwise, it was impossible for the department of local government to seek funding and to follow sectoral policies from the central government.

After the reform and open door policy, major changes took place at the municipal government level. The structure of municipal government had to be amended. However, even with some pressures on provincial government, especially in economic development, the structure of this tier of government has not changed significantly.

The provincial government's administrative framework mostly follows that of the central government. However, at the municipal government level, especially since the 1980s when Chinese reform had been delivered in urban areas, municipal government has to take main responsibility for economic development, and urban and rural planning and development which were devolved to municipal governments as their main task. The structure of municipal government had to be amended to cope with these changes. This government structural change was more significant in urban and rural planning. The tasks of region, city and rural planning and development, which are operated by one individual government's department in either the central government, the Ministry of Housing and Urban-Rural Development or provincial governments, the Department of Housing and Urban-Rural Development, has been broken up into several specific municipal bureaus of planning, contraction, sanitation, greening and parks and others (see Figure 3.2).

Since urban development and construction was defined as the main task of municipal government in the beginning of the 1980s, urban planning has then been regarded as the 'hub' for development and construction. The urban planning administration is formed as an independent bureau under direct jurisdiction of a municipal government. In some cities, the land administration is amalgamated to urban planning administration to reduce the contradictions between them.

In many cities, especially in large cities, a city planning commission chaired directly by local mayor is established for coordination and approval of development projects. The city planning commission should prepare city and city-regional development policies and make decisions for development projects. A planning administrative bureau as an executive agency of the commission is responsible for daily planning administration.

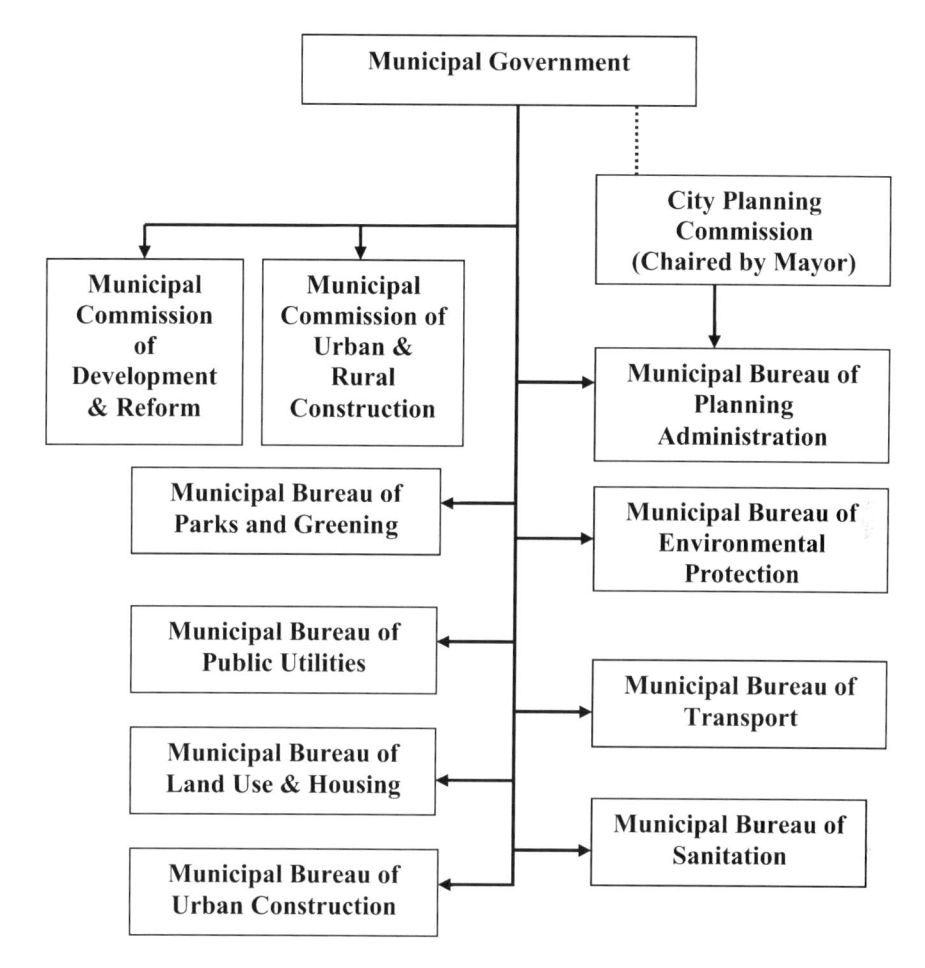

```
                    ┌─────────────────────┐
                    │ Municipal Government│
                    └─────────────────────┘
                              │
         ┌────────────────────┼────────────────┐
         ▼                    ▼                ▼
  ┌──────────────┐   ┌──────────────┐   ┌─────────────────┐
  │  Municipal   │   │  Municipal   │   │  City Planning  │
  │ Commission   │   │ Commission of│   │   Commission    │
  │      of      │   │   Urban &    │   │(Chaired by Mayor)│
  │ Development  │   │    Rural     │   └─────────────────┘
  │   & Reform   │   │ Construction │            │
  └──────────────┘   └──────────────┘            ▼
                                         ┌─────────────────┐
         ┌──────────────────────┐        │ Municipal Bureau│
         │  Municipal Bureau of │ ◄───   │   of Planning   │
         │   Parks and Greening │        │  Administration │
         └──────────────────────┘        └─────────────────┘

                                         ┌─────────────────┐
                                         │ Municipal Bureau│
                                         │        of       │
                                         │   Environmental │
                                         │    Protection   │
                                         └─────────────────┘
         ┌──────────────────────┐
         │  Municipal Bureau of │
         │    Public Utilities  │        ┌─────────────────┐
         └──────────────────────┘        │ Municipal Bureau│
                                         │   of Transport  │
         ┌──────────────────────┐        └─────────────────┘
         │  Municipal Bureau of │
         │  Land Use & Housing  │
         └──────────────────────┘        ┌─────────────────┐
                                         │ Municipal Bureau│
         ┌──────────────────────┐        │   of Sanitation │
         │  Municipal Bureau of │        └─────────────────┘
         │  Urban Construction  │
         └──────────────────────┘
```

Figure 3.2   Organizations responsible for development and construction in municipal government

Source: Created by the author.

Local Municipal Government in Transition

The Devolution of Fiscal and Tax System and its Impact on Local Government

The three trends of devolution, marketization and globalization in Chinese reform and transition to the market economy have significantly impacted the operation and performance of municipal governments. Most local municipal governments have to introduce and operate entrepreneurialism to promote local economic development.

Since 1978 when China started its open door and reform policy, the targets of reform had been concentrated on promoting economic development instead of former political struggle against capitalism and capitalists, as well as participating in the political movement of the Cultural Revolution. The reform was to release the centralized planning and operation of the social, economic and administrative system which was initially established in the centrally planned economy. Devolution stimulates and increases the interest and capabilities of local government in economic development. Devolution can be divided and operated in four steps and stages.

The first is the devolution of the state-owned enterprises. In the centrally planned economy, most of the state-owned enterprises were directly controlled by the central government. By 1994, the state-owned enterprises directly controlled by the central government were reduced to 117 according to the list of State Owned Asset Supervision and Administration Commission (2011). The second but more critical devolution is the decision for investment. In October 1984, the decision was made by the Chinese State Council that the investment in industrial production projects should be made depending upon the scale of investment from then on. A project with investment of over 30 million yuan should be approved by the National Development and Planning Commission; the investment with less than 30 million yuan for a single project would be delegated to local provincial or municipal governments. In 1987, it was further decided by the Chinese central government that only the projects of energy, transport, industries with investment above 50 million yuan should be submitted for the approval of the National Planning and Development Commission (the predecessor of the National Development and Reform Commission). However, the governments of Guangdong Province, Fujian Province, Hainan Province and Shanghai have been authorized to make investment decisions in the development projects of energy, transport and industry with maximum investment of 200 million yuan. These few provincial governments enjoyed a large power in decision-making for the investment in their territories. These were the experimental areas for foreign direct investment. All the special economic zones that were experiencing the market economy at the beginning of Chinese reform are all located in these provinces. The delegation of large power in investment was to empower and stimulate local governments in economic activities.

The third stage of devolution is the tax and fiscal system. The devolution of the tax and fiscal system has been operated through several steps to gradually form a complicated system. The first step was happened in 1980 when the Chinese central government changed the former totally controlled approach of tax collection by applying a system of so-called 'division of revenues and expenses between the central and local governments; and hierarchical contracting for tax collection' (*Hua fen shou zhi, fen ji bao gan*). In order to promote the proactivity of local governments in tax collection, the approach of 'Fiscal Contracting System', as the second step, was introduced and implemented in 1988. It was only after the delivery of this system for six years, a new fiscal and tax system of so-called

'Different Tax Categories Division', as the third step, was applied to replace the former ones in 1994.

The main reason for delivering this new tax system was the decline of fiscal capability of the central government after implementation of the 'Fiscal Contracting System' in 1988. In order to promote the fiscal capability of the central government, the 'Different Tax Categories Division' system was introduced in 1994 and is still operating now. However, with the application of this new tax system, the percentage of fiscal income of Chinese local governments declined rapidly to 44.3 per cent in the year of 1994 from 78 per cent in the year of 1993 (Dong 2008). It is more critical that the percentage of local governments' expenses has been kept at 70 per cent of total expenses of the whole country since implementation of the tax system in 1994, and further increased to 75.5 per cent in the year of 2006 (ibid.). This means that there is a deficit of 30 per cent between local governments' fiscal income and expenses.

At the same time, in the aspect of finance, banks in China were authorized to make the decision on loans according to the market. It means that power of government on the decision of loan provision has been banned. This policy or reform of decision of the bank loan has reduced the former power of local government to influence banks, and restrained the fiscal capability of local governments. Local governments have to seek other alternatives of financial incomes. However, the tasks and responsibilities of local governments have been increased with further delegation from the central government but without increasing their resource-raising capacity at the same time. The alternative approach is seeking income from land leasing which has then become a major financial resource for municipal governments. It is the opinion of Logan (2002) that most revenue incomes for local government are from real estate projects. The research by Dai Shuanxing (2009) provides evidence that the percentage of income from land leasing of Chinese local governments within their total incomes has increased to 50.7 per cent in the year of 2007 from 16.6 per cent in the year of 2001. This is the consequence of the tax system design of 1994. According to the tax division specified in the existing tax and fiscal system, the majority of income from land lease, e.g. urban land use fee, real estate tax, cultivated land occupation tax, land added value tax, is controlled by the local governments (ibid.). It is inevitable for Chinese local governments to expand their city size since their major incomes have to be generated by urban extension and occupying more agricultural land for industrial and real estate development. It is a common phenomenon that Chinese local governments are very keen to hasten the process of industrialization and urbanization. The income from land leasing for the industrialization and urbanization provides the main financial source for the operation of local governments and their investment.

It is also because those Chinese local governments have to depend upon income from land lease. The land price has been increasing tremulously during the last two decades because the winner of a land leasing tender should offer the highest price. As a consequence, housing prices have also been increasing rapidly. The affordability of housing has become a severe problem in many Chinese cities.

The fourth stage of devolution is the power for fundraising. It was a decision of the Chinese central government that the local governments of Shanghai, Tianjin and Hainan Province were able to make their decision on raising international loans at a maximum of 30 million US dollars, 10 million US dollars by the local governments of Beijing, Liaoning, Dalian and Guangzhou, and five million US dollars by other local governments for industrial development projects. In order to attract investment, the competition among local governments has become severe. Approaches in competition are very diverse, such as restraining consumption of products from other cities, which have been prohibited by central government after operation for a couple of years; provision of favourable policies in perspectives of taxes, land prices and other incentive packages.

An important consequence of devolution in China is that local governments are now controlling social capital resources. As a stakeholder in development, the Chinese local governments have to bargain and negotiate with the central government for revenues, policies and investment. This has been illustrated in the phenomena that the central government's policies have been sometimes ignored by the local governments in the process of delivery.

The Relationship between the State-owned Enterprises and the Local Governments after Devolution

Management and control of the state-owned enterprises has been changed in the devolution process. This has changed the relationship between local government and the state-owned enterprises. The state-owned enterprises have become independent economic players in the market since the changes of its management approach. In order to achieve economic development targets, the local governments have to collaborate with the enterprises. Enterprises may also need support from the local governments in land leasing, taxes, and other possible favourable policies. It becomes a common phenomenon that local government often associates with enterprises, typically those major state-owned enterprises that control resources to form a partnership with local economic development.

The reform and the changing economic status of local governments, typically the requirement of GDP increase and the pressures of local social and economic development, have forced local governments to introduce and apply entrepreneurialism in their delivery of local economic development objectives. From this aspect, local government has similar targets with entrepreneurs. They are the stakeholders in economic development, marketing and resources attraction. It is also in the interests of enterprises to work closely with Chinese local government, particularly in the process of transition to the market economy. It is because of an imperfect market and inappropriate information provision that the cost of enterprises is inevitably high. Involvement of government in information collection and provision as well as other activities can reduce the high cost of enterprises. For local governments, they have their own interests and motivation to work closely with enterprises. The interest of local government mainly lies in increase of GDP. It is common that local government regularly organizes trade events or fairs that provide opportunities for cooperation with the local enterprises and attract inward investment. From this perspective, local government has been functioning as a business 'broker'. The other important task of local government of establishing a good investment climate, including the provision of infrastructure and local incentive packages, is targeting to build an appropriate platform for entrepreneurialism.

The Relationship between the Central Government and the Local Governments

The new status and the function of local government in economic development have greatly impacted the performance of local governments and their development and planning policies. It has not been an unusual phenomenon that the local policies and development actions of Chinese local governments have not fully followed that of the central government. The adjustment and re-interpretation of the central government's policies in the local area have been common in many Chinese cities. Generally speaking the policies of the central government are macro control ones, such as definition of control index of loans, or scale of fixed capital investment. These indexes can be interpreted and allocated according to the local business and development features and their performance. Local government will decide what types of business and industries should be supported, and others should be controlled or left in the market. It provides great potential discretionary opportunities for local governments.

Conclusion

This chapter explains and analyses the Chinese government administrative framework in city and regional planning and the impacts to this framework after the reform and open door policy, especially in the process of transition to the market economy.

Devolution and marketization as the most important elements of Chinese reform and open door policy have been significant to the functions of local governments. Since devolution, typically after the reform of the tax and fiscal systems, pursuit of local economic development expressing by GDP increases of local governments has been stimulated (Liu 2008). With improvement of economic capacity, local governments are able to increase their financial incomes by increasing tax collection. The other benefit will be the potential career promotion of the local senior leaders. There are benefits both in economics and in politics (ibid.). The dual functions of local governments divide their role into two aspects. From one side, the local governments should have the responsibility to deliver policies of the central government in regional and local social and economic development; they should guide and govern the market in operation, and provide necessary infrastructure and other public goods to foster and improve market system (Yin et al. 2006). At the same time, it is because of the promotion of local economic development being an aspiration of all Chinese local governments that the priority of policies and service of local government is to increase GDP. Whenever there is any decline in GDP, local governments are involved actively and directly in economic activities, which should be operated in the market economy instead of the work of government. It is because of this function of local government in economic development that the Chinese local governments have enjoyed a special role with both functions as a local economic rules maker and one of stakeholders in economic development.

In order to promote economic development, illustrated by increasing GDP, local governments are very keen to invest in infrastructure, and rehabilitation of old towns and historical city centres. This phenomenon inevitably creates problems of low efficient and duplicated development of major infrastructures, e.g. seaports, airports and power stations in most cities because of competitiveness among them. The other critical problem of this type of economic development is that the achievement in GDP has been at the cost of the environment, inefficient use of natural resources, and social inequality and injustice.

However, the public services in China are much lagging behind economic development. This is illustrated in many perspectives, e.g. inequality of development, disparities between different regions and between urban and rural areas. Wang et al. (1999) comment that it is because of placing rapid economic development as a priority by Chinese policy-makers that they have great intension to sacrifice public services and quality for the goals of economic development. Policy-makers were ready to stand for a certain degree of inequality or widened disparity. It is their view that:

> A so-called gradient theory dominated the thinking of Chinese policy-makers for much of the 1980s. The theory divided China into three large geographic regions – eastern (coastal), central and western – and likened them to rungs on a ladder. According to the theory, the government should capitalise on the advantages of the coast first. Only after the coast had become sufficiently developed should attention be turned to the central region. (Wang et al. 1999: 175)

As a consequence of the policies, the Gini coefficient in China has already reached around 0.5 at a very dangerous standard. Although the general income of most Chinese people has been increasing since the open door and reform policy in the late 1970s, the absolute poor group still occupies about 11 per cent of the total population, the population of which may achieve 140 million. This creates a great risk for social stability. In order to address this problem, the Fourth Plenary Session of the Sixteenth Central Committee of the Communist Party of China, held between 16 and 19 September 2004 set up an important target to build a well-off society of a higher standard in an all-round way to the benefit of China's 1.3 billion population. However, it will not be an easy objective, as there is a long way to go.

Chapter 4
Regional Planning and Regional Governance Innovation

Introduction

This chapter discusses regional planning and its mechanism in China. Different from city plans or the Economic and Social Development Plan (the so-called Five-Year Plan), which have been formulated for several decades, the Chinese statutory regional planning is still at the earlier learning stage. However, this type of plan has attracted great interest and attention from different ministries of the central government that intend to influence policy-making at regional or provincial level. It is typically when China is still under its process of transition in terms of economy and urbanization that regional development planning is a significant tool of government in intervention to development. The consequent result is that the Ministry of Housing and Urban-Rural Development (MoHURD) and National Development and Reform Commission (NDRC) are actively involved in regional and provincial plans making. Two types of regional plan, i.e. the Provincial Integrated Urban System Plan which is produced by MoHURD, and the regional plan for the Development Priority Zones that is produced by NDRC, are formulated to guide regional development. The contents, relevant main principles of these two types of regional plans, and the tasks of provincial government in development will be discussed in the following sections of the chapter. A new regional mechanism of coordination – Cities Alliance – is also explored.

Task of Provincial Government in Planning

The responsibilities of a provincial government in city and regional development consist of producing regional plans, e.g. the Provincial Integrated Urban System Plan, the Plan for Development Priority Zones, and addressing development contradictions and promoting coordination among cities and towns under the jurisdiction of a provincial government; and to protect regional natural and environmental resources. It is also the responsibility of a provincial government to promote mutual benefits and to share regional infrastructures and public utilities within the territory; and to increase entire regional competitiveness of a province.

The Provincial Integrated Urban System Plan expresses general development strategic policies of a whole province, or a region. It should follow principles of the National Integrated Urban System Plan and other national policies, and provide development guidance and policies to the lower hierarchical plans, e.g. City Comprehensive Plan, but without directly intervening in policy-making by local governments.

As regarding plan implementation, it is specified in the City and Countryside Planning Act 2008 (NPC 2008) that a provincial government and its Department of Housing and Urban-Rural Development should be responsible for establishing an appropriate mechanism to deliver the plan, and to monitor and supervise plans formulation and implementation by local (municipal or town) governments and their departments of planning administration.

There is not any act to specify the formulation of Development Priority Zone plan formulation, but by the Circular No. 21 of the State Council (2007), the Provincial Development Priority Zones Plan is produced by the provincial government with the Provincial Department of Development and Reform as the executive agency and involved in various city and town governments under direct leadership of the provincial government.

Provincial Agencies Responsible for Planning

The Provincial Department of Housing and Urban-Rural Development is the organization responsible for city and regional planning tasks in a province. This department is also the executive agency for the formulation and delivery of a Provincial Integrated Urban System Plan under the leadership of the provincial government.

When the Provincial Integrated Urban System Plan is decided for formulation, the provincial government usually establishes a steering committee chaired directly by a provincial governor. The members of the committee consist of deputy governors and directors general of varies provincial departments and the mayors of local municipal governments within the province. A provincial urban and rural planning committee and an experts committee should also be established for plans production. The tasks of these two committees are to monitor, coordinate and provide technological assessment and support during the plan formulation and later implementation.

Objectives of Provincial Integrated Urban System Plan

A Provincial Integrated Urban System Plan is a statutory regional spatial plan. It functions as an important tool of the provincial government to rationalize regional development and urbanization. It was first defined by the City Planning Act 1989, and then further specified in the City and Countryside Planning Act 2008 and the 'Notices of Strengthening and Improving City and Countryside Planning' issued by General Office of the State Council (2000), as well as Circular No. 25 of *Notice of Strengthening City and Countryside Planning Control* promulgated by the State Council (2002). Since being defined in the 1980s, the policies of integrated urban system plans have been adjusted to cope with development policies in different periods. It is the view of Wang Kai (2007) that the integrated urban system plan was mainly to control the size of large cities, to promote reasonably medium size cities and to actively encourage development of small size cities in order to have a rational urban system, while establishing Chinese cities as open, multifunctional and social economic centres in the 1980s. In the 1990s, integrated urban system planning policies were to guide the movement of population and spatial allocation of urban system in the process of rapid urbanization with the precondition of protecting ecological environmental sensitive areas. It was the first time to propose different approaches to Chinese urbanization and urban development policies. In the twenty-first century, integrated urban system planning policy is focusing on sustainable development by emphasizing integrated development among population, natural resources and environment, and coordinating development among large, medium and small size cities (ibid.).

The main objectives of the Provincial Integrated Urban System Plan are to address contradictions and to coordinate development among cities and towns within the provincial territory by allocating regional spatial resources. The Provincial Integrated Urban System Plan should target inter-regional spatial resources protection and development coordination, typically major industrial and infrastructure projects that cross administrative boundaries of cities and counties (Liu et al. 2008). It should coordinate the interests and benefits of different stakeholders, including different government departments, private businesses and local communities in the market (Li et al. 2006). It is the opinion of Li and Xu (2004) that in the market economy, the task of a Provincial Integrated Urban System Plan is to organize the government's resources to achieve national and provincial development strategies and to address the failures of the market to provide 'public goods'. The plan should contribute to the well-being of all the people and to form a harmonious society.

Nevertheless, in the process of regional coordination, it is inevitable for a few cities or regions to obtain large benefits compared to those of others. In order to achieve maximum benefits for local entrepreneurial cities, the losers in the process may withdraw from the cooperation mechanism (Chen and Wang, 2006). The compensation mechanism to reduce the potential negative impacts of the loser in the cooperation should be established in the cooperation. This will be further analysed when discussing the Cities Alliance later in this chapter.

The Production Process of a Provincial Integrated Urban System Plan

When the formulation of the plan is formally started, the provincial government usually promulgates a circular to local municipal and county governments. In the circular, it normally explains the objectives and purposes of the Provincial Integrated Urban System Plan, and provides information on the members of the plan formulation steering committee. The circular often requires local governments to identify a government department as the coordinator in the process of plan making. The Provincial Department of Housing and Urban–Rural Development, on behalf of provincial planning steering committee, usually invites a planning institute to produce the plan. When the first draft plan is ready, consultation with of all the provincial departments and local municipal governments should be organized. The Ministry of Housing and Urban-Rural Development from the central government is sometimes involved in provincial plan production by organizing special workshops of experts' assessments of the plan.

The planning team should amend the planning policies according to the comments, suggestions and opinions from related seminars and workshops. After general consulting to provincial departments and local municipal governments, the Provincial Department of Housing and Urban-Rural Development should report the results of consulting to the steering committee, and provincial governor, as well as the Standing Committee of Provincial People's Congress. It is common that the aforementioned decision makers may raise different opinions and require amending planning policies and objectives. The plan will be amended again according to the proposed opinions and suggestions from the decision-makers, who are usually the local leaders at provincial and municipal levels.

Generally speaking, there is no public participation in the process of producing a Provincial Integrated Urban System Plan but notices in the media indicate that the plan is under production or has been approved by the central government.

The plan should be agreed and approved first by the Provincial People's Congress and then submitted for approval by the State Council. While approving a Provincial Integrated Urban System Plan, the State Council should consult the nearby provinces of the planning province and relevant ministries and commissions of the State Council through the mechanism of 'Inter-Ministries Allied Meetings'.

Contents of a Provincial Integrated Urban System Plan

The content of a Provincial Integrated Urban System Plan is defined by the City and Countryside Planning Act 2008 (Article 13) to decide the spatial layout of cities and towns and major infrastructures; to decide the size of each city and town; and to specify strictly control zones of ecological environment and natural resources protection.

One of the principle tasks of a Provincial Integrated Urban System Plan is to define the strategy of urbanization and urban development within the provincial territory. The plan should define the small towns that are able to provide services to their rural hinterlands as a priority development policy. According to the central government's circular (MoHURD 2010), several regulatory items for development should be clarified in the plan. These regulatory items are:

- Specification of controlling and banned development zones within the province. These include natural ecological conservation area, forest areas, large lakes and water resources and their catchments, flood discharging areas, and other ecological sensitive areas.

- Spatial allocation for regional major infrastructure development, which consists of motorways, major distributor roads, railways, sea ports and airports, regional electricity stations and high-voltage fences, major gas stations and pipelines, regional flood prevention projects, regional water and irrigation project, and regional water draw engineering projects.
- Spatial location of major inter-city infrastructures, such as access points of urban water resources, urban sewage discharging, and solid waste treatment plants, inter-city transport facilities, regional logistics centres.

Several special regional plans, which include regional greening plan, regional water supply plan, regional sewage treatment plan and inter-city rail network plan should be attached to a Provincial Integrated Urban System Plan.

It is regulated in the circulars (Circular No. 13 of 2002 by the State Council, and Circular No. 3 of 2010 issued by the Ministry of Housing and Urban-Rural Development) that a Provincial Integrated Urban System Plan should consist of three mandatory items of controlling development areas; major regional infrastructure development projects, which are organized by a provincial government; and major infrastructure allocation that requires the coordination with nearby provinces or metropolises.

Impact of Provincial Planning Policies in Local City's Plans Formulation and their Implementation

After completion of a Provincial Integrated Urban System Plan, the provincial government would then promulgate a circular to guide and require local municipal governments to follow the development policies, especially the policies of sub-regional planning guidance when preparing the local city comprehensive plans.

Delivery of a Provincial Integrated Urban System Plan is usually led by a planning coordination committee chaired by the governor (or a deputy governor) of a province. The members of the committee come from various provincial departments of social and economic planning, transport administration, finance, urban–rural development, environmental protection and others. This committee takes responsibility for decision-making on approval of development, infrastructure project, natural resources and environmental protection.

The head of a Provincial Department of Housing and Urban-Rural Development is required to report to the Provincial People's Congress and provincial government, and submits a report that records the process and achievement of the Provincial Integrated Urban System Plan to the Ministry of Housing and Urban-Rural Development.

Although it has been regulated by the Ministry of Housing and Urban-Rural Development that a Provincial Integrated Urban System Plan should be used as an important material for consideration during planning management (development control), a Provincial Integrated Urban System Plan had neither great influence on the lower hierarchical plans, nor being able to adjust or coordinate local development in practice. It is the view of some researchers (Liu et al. 2008) that the difficulties in implementing a Provincial Integrated Urban System Plan are because its planning concepts have been greatly impacted by the legacy of the centrally planned economy. Contents of a Provincial Integrated Urban System Plan are mainly prescriptive without appropriate flexibility. It is difficult to be followed by the lower hierarchical plans.

However, it is the view of Li and Men (2004) that the difficulties in implementing the policies of a Provincial Integrated Urban System Plan in the coordination and adjustment are because of lacking appropriate administrative delivering mechanisms. Jiafu Liu and his colleagues (2008) argue that in the market economy the provincial department of planning administration has neither administrative power, nor financial ability to influence or supervise the development decisions at local municipal or county bureau of planning administration.

In practice, the intervention of a provincial government in planning and development decision-making has predominantly been focused and based on the policies and contents of a City Comprehensive Plan. It is only at this stage when local municipal governments are submitting their City Comprehensive Plan for approval that the provincial governments are able to insure the policies defined by the Provincial Integrated Urban System Plan to be fulfilled in the producing process of City Comprehensive Plans. The power to approve a City Comprehensive Plan by a provincial government is the possible channel for a provincial government to influence city planning and development policies of local governments. It is required by City Planning Act 1989 and updated by City and Countryside Planning Act 2008 that a City Comprehensive Plan should be submitted for approval by a provincial government, even some designated cities whose City Comprehensive Plans should be submitted to the State Council for approval, they should be agreed first by the provincial governments.

Nevertheless, after the approval of a City Comprehensive Plan or in the process of its implementation, it is difficult for a provincial government to have any influence on the planning and development policy of a city, even if local government does not follow either a Provincial Integrated Urban System Plan or even the City Comprehensive Plan produced by themselves.

In order to address this problem, it is regulated by the City and Countryside Planning Act 2008 that a provincial government or its department of planning administration should supervise and monitor plans making and implementation in local governments (Article 51, NPC 2008), and to instruct local planning authorities to provide compulsory penalty notice to illegal development if they fail to do so (Article 52, ibid.), and to instruct local bureau of planning administration or to cancel directly the planning permission if it violates regulation defined by City and Countryside Planning Act 2008 (Article 53, ibid.). However, in practice, it is difficult to operate these supervisory functions by a provincial government for two reasons. First, although it is regulated that the provincial department of housing and urban-rural development should supervise and monitor plan making and implementation, there are not enough staff or any mechanism, e.g. inspectors or any special division, which has been defined by the planning acts or any of central government's circulars, to do the work. There is no appealing system in Chinese planning.

Second, implementation of local City Comprehensive Plan is through the system of project base planning application and permission, which is the responsibility of the local planning authority. In fact, both central and provincial governments have not any information how a City Comprehensive Plan is implemented. They are not able to have better information than a local normal citizen. Asymmetric information has been a major problem in the Chinese city and regional planning system.

Li Xiaolong and Men Xiaoying, the senior planning officers of the Department of Planning Management of the Ministry of Housing and Urban-Rural Development, argue that the shortage of supervision before and after the delivery of the plans is a critical problem of plans for implementation. It is their view that because of lacking necessary effective supervision at higher hierarchical planning authorities to the lower authorities, local planning official is able to change policies, and relevant controlling indexes or ordinances that have been decided in statutory regulatory detailed plans under the force or orders from local leaders (Li and Men 2004). It is typically when the local bureau of planning administration is one of local government's departments that should be under its jurisdiction and responsible to the local government (Zhou and Huang 2003).

Development Priority Zones in the Chinese Regional Planning

In order to effectively deliver the strategic ideas of 'Scientific Outlook on Development'[1] and 'building a moderately prosperous society in all respects', to plan national population and economic structure, to have an integrated coordination and development in population, economy and natural resources, and to reduce regional differences in income and development level (State Council 2007), the Circular No. 21 of *Decisions on Preparation of National Development Priority Zones* was promulgated in July 2007.

The whole of China has been divided into four types of zones for a typical different development function, i.e. optimizing development zone, prioritizing development zone, controlling development zone and prohibiting development zone, with different regional policies.

There may be different administrative regions, e.g. provinces, or cities, within a defined Development Priority Zone. They should decide their industrial and economic development policies according to the planning policy framework established by the Development Priority Zone Plan (Deng and Du 2006).

---

1 This term is raised by Chinese central government to mean putting people first and aiming at comprehensive, coordinated and sustainable development to establish a harmonious society.

When making planning policies of Development Priority Zones, it is necessary to consider the resources and environmental capacity, and existing development density and potential tendency of a region, as well as general layout of national population, economy land use and urbanization structure (Han et al. 2011).

The Development Priority Zones of China are divided into two tiers at national level and provincial level. National Development Priority Zones and its plan were produced and defined under the leadership of a steering committee, which consists of numerous ministries or commissions of the State Council with the State Development and Reform Commission as the executive agency. The planning period is targeting the year of 2020. The plan should be implemented through a rolling process and adjustment for monitoring and reviewing. Provincial Development Priority Zones and their plans are formulated by a provincial government with participation in numerous city and town governments under direct leadership of the provincial government. The planning period is also targeting the year of 2020.

The reason to identify the different Development Priority Zones is mainly based on the fact that there are significant differences between regions in China. This concept should also be applicable at the provincial level. Development Priority Zones will either promote, or control, or prohibit physical development according to the defined planning policies of each zone. Regional planning depending upon the defined function of each zone is trying to balance and coordinate between natural ecological protection and development through rational spatial allocation of resources, including natural material, population, urbanization, major infrastructures, and public services within the whole of China, or within a province. The tasks of a Development Priority Zone are to (Deng and Du 2006):

- promote rationally regional division, cooperation and sustainable development;
- encourage coordination between economic development and resources protection while increasing the capacity of the natural environment;
- encourage connections between industrial development policies and regional development policies;
- reduce differences between regional development and income.

It is the view of Jun Li and Wangbin Ren (2011) that planning for National Development Priority Zones is in fact planning for national general layout of productivity. It can be regarded as a type of national economic development strategic plan. Development Priority Zones have two tiers of plans at either national level or provincial level. National Development Priority Zones should define urbanization structure, agricultural development structure, ecological security structure and maritime strategic structure for the entire country. When producing the national plan, it is required to leave certain spaces for planning for Provincial Development Priority Zones to define their spatial structure. Provincial major Development Priority Zone is specified as the responsibility of a provincial government. The policies of the Provincial Development Priority Zone plan should follow Development Priority Zones planning framework.

Definition of Four Types of the Major Function Zones

Optimizing Development Priority Zone

Optimizing Development Priority Zone refers to the regions that have already been in the process of a high density development, but are at the beginning stage of the decline of natural and environmental capacity and resources. Within this type of Development Priority Zone, the existing development pattern, which is to achieve rapid economic development by occupying large amounts of cultivated land, consuming large amounts of natural resources and discharging large amounts of pollution, must be changed. The quality of improvement and increase of development efficiency should be regarded as a priority in the development. However, Optimizing Development Priority Zone is still functioning as leading regions in social and economic development in China. They are still functioning as Chinese key players in the global economy. From the view of China as a whole, this type of priority zone refers to the Pearl River Delta area, the Yangtze River Delta area, the area of Beijing-Tianjin-Hebei Province, and high density super megalopolis areas in central, western and northeast China, and transitional cities with the natural resources drained.

In this type of Development Priority Zone, development is interpreted as optimizing the allocation of resources, increasing capability for efficient and effective use of natural and environmental resources, and regional energy consolidation for industrial transition and upgrading.

Prioritizing Development Priority Zone

Prioritizing Development Priority Zones are the regions with the merits of potential capability in environment, and able to absorb and concentrate both economic development and population from other regions. Within this type of Development Priority Zone, development should achieve the following targets:

- to improve existing infrastructures and investment environment;
- to promote industrial cluster development;
- to strengthen capacity and scale of economy;
- to speed up urbanization and industrialization;
- to adopt industries that transfer from optimizing Major Function Zones; and
- to accommodate population moving from both controlling and banned Major Function Zones.

Prioritizing Development Priority Zones are designated as major carriers for national economic development and the concentration of population. This type of zone is defined to include major cities in central, west and northeast China with resources that are capable for large development, or the areas along major national transport corridors, and areas in east China that are less developed at present but with potential development.

In Prioritizing Development Priority Zones, the defined development includes:

- large scale industrialization and urbanization through rational allocation and resources use;
- the great promotion of the tertiary industry that can promote achievements of regional economic development; and
- the concentration of population by absorbing migration from other regions.

Differentiating from the Optimizing Development Priority Zone, Prioritizing Development Priority Zones should be encouraged to provide more job opportunities, to improve development conditions for enterprises and businesses, to increase capacity of capital accumulation, and to stimulate urbanization. Labour intensive industries are regarded as the major industrial development model. The key policy of urbanization in the zone is to increase the capacity of the urban area to absorb rural to urban migrants.

Controlling Development Priority Zone

Controlling Development Priority Zones are the regions with poor capacity of natural and environmental resources and inappropriate conditions for the concentration of large amounts of population and economic activities, or areas with sensitive ecological environments. In such a zone, conservation of the ecological environment is a priority over development. This type of zone should be gradually become a national or regional significant ecological functioning zone by improving the local ecological environment and moving people out of the zone to be resettled in Prioritizing Development Priority Zones. However, reasonable and small size development, or special industries of local materials but the impact of environment is controllable, are possible.

This type of Development Priority Zone is ordinary referred to the upper reach of major river catchment areas, important forests, main grain production areas, or other major agricultural areas.

Prohibiting Development Priority Zone

Prohibiting Development Priority Zones usually are different types of natural and ecological protection areas that should be strictly protected and forbidden for any development according to laws. These areas include state natural protection parks, world natural or cultural heritage protection sites, national scenic spots, national forest parks and national geological parks.

Principles for Establishment of the Development Priority Zones

The underpinning principles for the establishment of Development Priority Zones are to build a harmonious society, to reduce differences between regional development, and to form a rational economic and population spatial layout in the whole country from the perspective of 'scientific outlooks on development'. From this perspective, the objectives of Development Priority Zones should address the unequal development and the pattern at the cost of environment and high consumption of energy and natural resources, which is the major problem in China during the last three decades. Unequal development has existed among different regions and provinces, and between urban and rural areas. Development Priority Zones encourage consolidated development both in industries and population, mainly in mega-cities but with supplementary supporting by other cities and towns in different hierarchical urban system for the objective of effective and efficient uses of land, water and resources.

An important underpinning principle of Development Priority Zones is obviously the realization of the risk of and non-sustainability of existing economic development and urbanization patterns that are not only at the cost to the environment and natural resources, but also the cost of agriculture and rural areas. In order to change the development pattern and to deliver the approach of Development Priority Zones, it is specially emphasized in the State Council's circular (State Council 2007) that development should obey nature and regard protection of natural resources and ecological environment as a precondition. The development should fully consider the capacity of the environment by ensuring ecological safety and promoting the quality of the environment. All development should benefit integration between urban and rural areas, and to avoid the former approach of corrosive (negative) impacts to rural economic development and the interests of rural villagers. Development should provide necessary space and conditions for movement of rural migrants to urban areas.

Planning Concepts for the Development Priority Zones

As a developing country, China has adopted the concept of 'development is the absolute principle' to promote its economic development and increase the quality of life of its citizens. However, the development approach in the last three decades illustrates some severe problems, typically in environmental natural resources protection and huge energy consumption.

At the same time, although there have been notable changes in social and economic development in China, income and development polarization between regions and cities, and between urban and rural areas, has become a severe problem. The difference between a rich province, e.g. Shanghai (a metropolis at provincial level), and a poor one, e.g. Guizhou Province, is about nine times in GDP per capita. At city level, the difference between the top one, e.g. Kelamayi, Xinjiang, and the bottom one, Guyuan, Ningxia, is 36 times GDP (Li and Xu 2004).

These problems have been realized by the Chinese government and people that the existing development approach is unsustainable. It is also a great challenge and intention for Chinese central and local governments to address the polarization in economic development and to have a more balanced or harmonious society while fully considering the local differences in means of regional planning.

The application of Development Priority Zones is hoping to seek a rational, appropriate and ordered development, and to reduce the differences in regional development and incomes. When producing a plan for Development Priority Zones, there are several issues that need to be considered. The first one is to balance the relationship between local interests and overall interests. This should include balancing the interests of local cities or towns and the interests of a region (or a province); it should also balance the interests of a province and that of the whole country. Arbitration and compromising mechanisms should be established for the purpose of balancing different interests among stakeholders even though it is required by the central government that local interests should fulfil overall interests, or vice versa, the overall interests should consider the local interests (State Council 2007). Planning for Development Priority Zones is inevitably required to find an applicable solution to stimulate the market activities. Planning for Development Priority Zones should also express the government's policies in spatial development and general layout. However, development of a zone (region) depends upon the input from the private sector, or the market. It is crucial for governments to produce reasonable regional policies and other public resources, e.g. infrastructures, public facilities, to lead input from the market to achieve development objectives.

Within a Development Priority Zone, regardless of the fact that the main function has been defined, it is able have other subordinate functions. Otherwise, the approach of a Development Priority Zone may not be successful. For example, in a Prioritizing Development Priority Zone, besides industrial development, it may also need to have an ecological protection area, agricultural area and tourist attracting areas to meet the different demands of local residents. Within the Controlling or Prohibiting Development Priority Zones, special products with local characteristics are allowed and encouraged to be produced for economic development and local job promotion. These issues should be considered in the planning process and planning policy-making.

A Development Priority Zone can cover one province, or several provinces at the national level; or a city or several cities at a provincial level. It then creates a challenge of coordination among different administrative boundaries in planning. The most obvious challenge is to break down administrative boundaries and to produce planning policies at inter-provincial, or inter-city level, while establishing a practicable mechanism to deliver regional planning and development policies.

Impacts to Regional Planning because of the Development Priority Zones

Establishment of Development Priority Zones and delivery of its plans provides clear development targets for different areas in China. The establishment of the zones and their planning policies are seeking to combine both regulatory and discretionary management approaches in the process of policy delivery. In Controlling and Prohibiting Development Priority Zones, a regulatory approach that targets for ecological protection and environmental improvement and restoration should be operated. While in Optimizing and Prioritizing Development Priority Zones, there is discretion for development control, especially in terms of industrial guidance, land allocation and population increase.

It is obvious that the capacity of environment of a region has to be regarded as a priority in development and planning policy-making when defining various types of Development Priority Zones (Deng and Du 2006; Han et al. 2011; Zhang and Wu 2009). Environment capacity has greatly impacted the decision to designate the functions, development tendency and phasing of a zone.

The planning approach of the Development Priority Zones has impacted traditional Chinese regional planning. In terms of spatial structure, the new approach is considering environmental and social capability instead of former industrial or economic capability, which used to be the traditional regional planning and development approach in China for several decades since the 1950s. After the definition of a type of Development Priority Zone for a region, spatial structure and phases of development will be decided according to its main function through spatial allocation of the economy, population, public services and provision of major infrastructure.

However, there are some problems that need to be addressed in the process of delivering Development Priority Zones, typically the contradictions and overlapping responsibilities between different ministries in the central government and the organization in the lower hierarchical governments.

It is clarified by the State Council (2007) that establishment of the Development Priority Zones and formulation of their plans is the responsibility of the National Development and Reform Commission but associating with different ministries in the State Council, e.g. the Ministry of Housing and Urban-Rural Development, the Ministry of Territory and Land Resources, the Ministry of Civic Affairs and others. Nevertheless, it is due to the different national and regional planning responsibilities by these ministries, there are many contradictions and overlapping responsibilities among them.

To address the problem, some researchers (Liu and Lu 2012) suggest that Development Priority Zone Plans can be treated as one of the main materials for consideration when producing city plans. City planning should be changed to fit the demands of the Development Priority Zone planning. In principle, there should be some interactive relations between the Development Priority Zone Plan and city plans. It is the view of Yulong Shi (2008) that planning for Development Priority Zones is to form a macro planning framework. However, this type of plan does not have any tool to deliver or control its spatial development proposals. The Development Priority Zone Plan has to rely on city and countryside plan to deliver its regional development proposals and intentions.

In terms of reducing regional difference and polarization, which is defined as one of the major tasks of the Development Priority Zones, it is not promised either. Some researchers (Bo et al. 2011) have found that the Development Priority Zones are unable to decrease differences in regional development. On the contrary, if the areas within the defined Controlling Development Priority Zones or Prohibiting Development Priority Zones strictly follow the defined policies while the compensation policies or compromising mechanism have not been appropriately established, the differences in development and income between the regions will be further enlarged instead of reducing. It is because that the defined Controlling Development Priority Zones or the Prohibiting Development Priority Zones are mainly located in the less developing and poor areas, the regional development policies, including investment, industry, taxes and finance and ecological environmental protection are of great different comparing to other Zones. The defined Controlling Development Priority Zones or Prohibiting Development Priority Zones are vulnerable regions (Bo et al. 2011).

Innovation in Regional Governance Mechanism – Establishment of Cities Alliance in South Fujian Metropolitan Region

Establishment of Cities Alliance in the South of Fujian Province

Since Chinese open door and reform policy in the late 1980s, Chinese eastern coastal areas have operated experimental practices in the attraction of foreign direct investment and the delivery of the market economy by establishing either Special Economic Development Zones or Coastal Economic Development Zones. For more than two decades, Chinese eastern coastal areas have been developing as the most prosperous areas in China as the consequence of rapid economic growth, especially in the Yangtze River Delta area and the Pearl River Delta area. Fujian Province that locates between these two Chinese development poles is confronting serious pressure to catch up with the rapid development of the two Deltas and avoiding lagging behind. In fact, in the twenty-first century, the GDP of the Yangtze River Delta accounts for 20 per cent of the whole country; while that of the Pearl River Delta accounts for 28 per cent; but South Fujian Province only contributes 5 per cent. Fujian Province is challenged by the limitation of its market scale and consumption capacity, and the risk of regional competition within the province.

As explored in Chapter 2, one characteristic of Chinese reform is its devolution. Severe competition among cities has been stimulated by the devolution of development responsibility to local governments, the existing political mechanism of the appointment system of senior local government leaders for a term of four years, and the existing achievement assessment system to local government. In order to achieve better performance in economic development, typically a GDP increase, competition has been the main theme for cities and local governments. However, it has been gradually realized by local governments that the severe competition among them may create a 'loss–loss' situation. The cooperation is critical to promote capacity of regional competitiveness as a whole in the global economy.

In the course of interviews, local politicians and urban planners of the three cities held similar views as to cooperation. In 2006, when the author interviewed the vice-mayor of Zhangzhou responsible for planning and development, it was his view that 'The South Fujian Region should coordinate for economic development. Xiamen should play a leading role. There are boundless opportunities of mutual benefits and support of each other among the three cities through cooperation. Each city has its own advantages and disadvantages'. In order to have a strategic development to the west part of the Taiwan Strait and to increase economic vitality and coordination among Xiamen, Quanzhou and Zhangzhou in the south of Fujian Province, the Fujian provincial government decided to establish a Cities Alliance among Xiamen, Zhangzhou and Quanzhou municipal governments as a regional governance and coordination mechanism in South Fujian Province in 2004. The principles of a Cities Alliance are to have unitary planning for general spatial layout, to build infrastructures together by sharing resources, and to complement in each city's benefits for a coordinating development.

One of the objectives of this Cities Alliance in South Fujian is to form a megapolis region with a population of 10 million that will increase its capability of competitiveness through increasing local consumption capacity and inward investment attraction. The Cities Alliance is also targeting to have an appropriate coordination development within the area through mutually:

- supporting in industries and urban development;
- encouraging reasonable allocation for regional resources and productivity;
- attracting inward investment for industrial and economic development in the region.

The promotion of competitiveness in the region is the most critical objective for the establishment of the Cities Alliance.

Coordination of the Cities Alliance

The Cities Alliance establishes a mechanism of regular meetings which include meeting among the mayors of the three cities once a year, and several meetings depending on the needs among the heads of different municipal bureaus from the three cities. A secretariat was formed as executive body under this mechanism. The secretariat is associated with the city on duty, the mayor of the city on duty is the chair of the Cities Alliance. The city on duty is changed every two years.

The Cities Alliance of South Fujian Province has restructured political and administrative boundaries, and formed an internal social and economic institution and relationship within a megalopolis region. It is expected that a 'joint force' is able to be established to increase the regional capacity in competitiveness. From this perspective, the Cities Alliance of Xiamen, Zhangzhou and Quanzhou can be understood as a Chinese innovative politics-space governance model for effectiveness and efficiency of a regional politics-management-planning structure.

Since its establishment in 2004, the Cities Alliance of Xiamen–Zhangzhou–Quanzhou has operated and executed to attain some achievements in planning and development. The coordinated development is operated through a plan-led approach. Bureaus of planning administration of the three cities have jointly prepared several spatial coordination plans for effective and efficient regional infrastructure and economic development. Important jointly produced plans include a regional strategic development plan for the three cities to integrate City Comprehensive Plans and economic cooperation, 'Xiamen, Quanzhou and Zhangzhou Economic Development Corridors Plan', and several 'Spatial Coordination Plans for Contiguous Areas among Cities'.

Spatial coordination plans for contiguous areas between cities, i.e. the area between Haican of Xiamen to Longhai of Zhangzhou; and Nanan of Quanzhou to Xiangan and Zhangzhou, were formulated to address the economic cooperation at the boundary of adjacent two cities. The plans specified development priorities of each city to avoid the contradictions within the region.

Under the planning guidance of the coordination plans, i.e. 'Xiamen-Quanzhou-Zhangzhou Cities Alliances Coordination Development Plan', 'Coastal and Seaports Development Plan', 'Urban System Development Plan along Railway' and 'Action Plan for Major Infrastructure Projects', regional major economic development and infrastructure projects have been identified. The developing phases of these plans have also been confirmed according to the principles of building infrastructures together and sharing resources among cities. The related coordinating development policies can be illustrated by proposals for establishing inter-city light rail mass transit system, inter-city bus operation, and joint infrastructure development projects in water supply, sewage treatment and solid waste treatment between cities. Some development spaces have been preserved for future generations in terms of sustainable development.

Regional transport infrastructure is an important facility that can be used to promote regional cooperation. The Cities Alliance emphasizes on coordination of regional transport facilities construction. Based on three railways of Fuzhou to Xiamen, Xiamen to Shenzhen, and Longhai to Xiamen, a railway network in South Fujian Province has been established. The railway network provides the opportunity for development of logistics in the region. Small towns along the railways have also benefited to promote their economic development. Their capacity of serving rural hinterland areas has been increased as a consequence of the rail network. The market catchment area of South Fujian Province is then extended.

These three cities have a common feature. They are all seaport cities. Cooperation among the seaport authorities was formed by a unitary development plan to provide rational division of different functions of each port. Zhangzhou port authority was merged with Xiamen port authority. Maritime resources and the coastal lines of three cities were planned as a whole for development and conservation.

In the tourist development, a regional tourism coordination mechanism was established to form a unitary and coordinating regional market. Numerous strategies and marketing events were operated to have a boundary-free regional tourist market. These strategies and marketing events include restructuring tourist resources within the region, marketing jointly among the cities, and establishing a unitary regional tourism information platform.

Significances and Problems of the Cities Alliance

The Cities Alliance of South Fujian Province has played a significant and active role in regional coordination and governance. Considering the importance of a region in the global and national competition, this Cities Alliance has been developed for more efficient and effective resource exchanges and coordination at varies levels under the guidance of the provincial government.

The Cites Alliance of South Fujian Province is a new mechanism for regional cooperation and an experimental practice in local governments. This mechanism was formed as a top-down process in a province. As a new mechanism, the Cities Alliance is still at its initial and trial stage. Some problems and its limitations are inevitable.

Within the alliance of the three cities, Xiamen and Zhangzhou are more active in the Cities Alliance because these two cities are able to have more complements in resources. However, the relationship between Xiamen and Quanzhou is of competitive consciousness instead of coordination.

Xiamen is the most important city on account of its reputation, attraction and high quality in terms of social, economic, environmental and infrastructure development. Since the year of 1980, Xiamen has been defined as a Special Economic Zone. The city has enjoyed an incentive package, especially in income and corporate taxes for inward investment. Because of its role as a Special Economic Zone in China, it developed much earlier and quicker than the other two cities. Xiamen had been more mature in operation and service for inward investment attraction, infrastructure support and industrial development. It is a more prosperous city.

However, because of its size of territory and population, Quanzhou has become the city with the largest economic scale in term of total GDP in Fujian Province besides the capital city of Fuzhou. The city is well known for its market oriented and private business development. The majority of its enterprises are labour intensive with products of clothes, shoes and other daily essentials. Zhangzhou is a less developing city but its development rate has been increasing during the last few years. Zhangzhou and Quanzhou, especially Quanzhou, are rather concerned that Xiamen could attract most inward investment in the region after improvement of regional infrastructures owing to its advanced position.

The concerns and worry of the 'Matthew Effect' by the other two cities, especially after the improvement of the regional internal transport infrastructure, has resulted in Quanzhou hesitating to be involved in the coordination. As a Special Economic Zone, Xiamen enjoyed the lower rate of 15 per cent of corporate and income tax while enterprises in other cities have to pay 33 per cent. It was found from the survey of the author when interviewing some entrepreneurs in Quanzhou that some of enterprises are considering resettling their companies and factories to Xiamen while continuing to live in Quanzhou after the completion of the regional express railway. Despite the fact that the cities in the region might have realized the importance of coordination, competition is still a main phenomenon among them. Each municipal government tries to keep the existing enterprises in their territories and attracts more inward investment to their jurisdiction. Local governments maintain sharp vigilance and try to avoid the de-investment to them after improvement of regional transport facilitates. Xiamen municipal government even tried to attract enterprises from Quanzhou to resettle in Xiamen by providing its strengths of incentive package. This restrains Quanzhou from being proactive in the Cities Alliance.

It is impossible to have a successful Cities Alliance without mutual trust between partners. However, mutual trust should have to depend upon mutual benefits and mutual interdependence. When the partners realize that it is possible for them to achieve their benefits through coordination, they will then be able to strengthen their mutual trust. However, a critical issue has to be considered even within the partners of mutual interests. This requires establishing a compromising and compensation mechanism through negotiation and bargaining.

In South Fujian Province, Xiamen and Zhangzhou are of greater potential in social and economic cooperation. Zhangzhou is less developing comparing with that of Xiamen. However, this city is able to provide required labour to Xiamen and absorbs some lower trend industries that are labour intensive. It is also located on the upstream of Jiulongjiang River, the catchment area of which is defined as the water conservation area to protect drinking water for Xiamen. In order to protect water quality, Zhangzhou has to implement a strict protection policy to control the environmental quality of the river catchment area. It is because of the austere controlling policy along the river catchment area that development is forbidden. Economic development, typically the GDP increase of Zhangzhou, has suffered from the strict controlling policy as a consequence. Nevertheless, Xiamen benefits from the stringent controlling policy, otherwise drinking water support to the city would have been in severe pollution risk. There have been negotiations with regard to compensation by Xiamen to Zhangzhou between the two municipal governments. Xiamen did provide compensation to Zhangzhou for its loss in GDP and economic development. However, the local residents and government officials of Zhangzhou are not satisfied by the levels of compensation. During the survey by the author, the residents and government officials to Zhangzhou argued that the total volume of compensation had not considered the market price. Several issues had not been considered in the compensation, including how many years Xiamen should pay for the compensation, the total value of loss in economic terms by strict control of Jiulongjiang River catchment areas for securing water quality, and the components of compensation. It is obvious a significant event in China to establish a rational and appropriate mechanism for compensation, typically when the approach of Development Priority Zones has been implemented. It needs innovative thinking in regional coordination. Until now, most Chinese cities lack experience and knowledge in decisions and the calculation of the costs for compensation, typically the cost of environmental protection in economic terms. The approach and format of compensation calculation is still a critical issue that needs to be decided.

Lessons from the Cities Alliance in South Fujian Province

To deal with global and national competition, the megalopolis region of South Fujian Province established a Cities Alliance among Xiamen, Zhangzhou and Quanzhou. This innovative mechanism is notable for competitiveness promotion, for attraction of inward investment, for addressing the challenges of these three cities and for releasing the development pressures from the rapid development of the Yangtze River Delta and Pearl River Delta areas. It also provides a development framework to coordinate with Taiwan.

The Cities Alliance is a trial-and-error approach for regional coordinating development and an innovative mechanism in regional governance in China, especially in metropolitan regions. From lessons and experiences of the Cities Alliance in the South Fujian Province, it is able to conclude that this innovative mechanism has performed generally well from the perspectives of information exchanges, joint development of regional infrastructures, especially in regional transport facilities, and regional tourist marketing. This mechanism can work much better among the cities with mutual benefits and interdependences in social and economic development or resources. However, there is an unsolved issue and a deficit of the Alliance in the formulation of a reasonable and realistic compensation mechanism for the losers. The further development of the Cities Alliance will be impacted without appropriate solution to compensation.

Cities in a region may have consensus in coordination and development objectives. However, competition among them seems more crucial than coordination at the present Chinese development stage owning to its social political institutions. There are many different interests when it goes into detailed operational issues and policy implementation. The regional collaboration can be better operated by a self-organizing coordination system among the cities with more mutual interests and mutual interdependences instead of the collaboration approach assigned by higher hierarchical government from top down, such as the Cities Alliance of the South Fujian Province. Self-organized mechanisms can provide opportunities for individual cities to decide on the approach of coordination, format of compensation, and consensus of development objectives.

In the procedure of regional coordination, benefits of each partner should be balanced in the market. Interests of each partner should be considered and reflected in the coordination. In a city alliance similar to the South Fujian Province, each city should be able to consider its own cost and benefit from cooperation in the alliance. All participants may achieve their benefits through negotiation and bargaining with other partners.

For effective coordination and implementation, a mechanism for discussion and negotiation between cities in a region should be executed at various levels of a hierarchy. These levels include meetings for information exchanges among mayors of each city in making strategic decisions, in specifying general targets and in deciding the coordinating approach. The consensus seeking in defined projects and programmes should be operated through regular meetings and negotiation among the heads of governmental departments from different cities in a city alliance. The thorough implementation of these programmes and projects should be carried out by low level staff that should take responsibility for the daily operation of all the designated coordinating projects and programme.

In addition to cooperation among municipal governments, a city alliance as a type of regional coordination mechanism should include a partnership between the public and the private. An appropriate and reasonable form of partnership between the public and the private may require considerable time to establish in present-day China. However, it is necessary to form such a partnership between the public and the private, or among various stakeholders in the market economy. The negotiation and communication between the public and the private in China have already existed but without an open, transparent and systematic approach.

In the process of city and regional social economic development, it is important to provide opportunity for private business and local communities to express their views, interests and objectives. It is necessary to produce public policy that is more accessible and transparent. By doing so, all the stakeholders in society are able to participate in the policy-making and delivering process. Public policy is therefore able to be known and understood by different stakeholders. The involvement of private businesses and local communities is not only to have a full and efficient use of resources in the market, but also to reduce the operation in the 'back door' or in 'bellows', which creates opportunities for corruption. The precondition of corruption is obscure decision-making and operation. Nevertheless, it is important to realize that without an appropriate channel to express their interests and objectives, without transparent decision-making, private businesses or some interested groups may attempt to influence policies in order to change public policies and decisions through improper and illegitimate methods. Transparency and the right for facts knowing and participation are the only possibility to avoid corruption.

Integrate Urban System (Megalopolis) Development Plan of Liaoning Coastal Area: A Case Study

Liaoning Province and its Coastal Urban Area

Liaoning Province is located in northeast China where to boundaries with Japan, Russia, North Korea, and Mongolia. The coastal area to the south part of Liaoning is the place accessible to the outside world through seaways in northeast China. This location provides a great opportunity for Liaoning to become a transport hub.

According to 1 per cent sample census in the year of 2005, there were 42.2 million permanent residents, among which 24.77 million were urban residents in Liaoning Province. The total territory of Liaoning is 148,000 square kilometres. The coastline of Liaoning is about 2,920 kilometres.

The average development speed of Liaoning Province is generally higher than that of the whole country. Manufacturing and other secondary industries in Liaoning have played a significant role in its economy. The economic structure among primary, secondary and tertiary industries in the province is 10.3:53.1:36.6, while the national average ratio is 11.3:48.6:40.1 in the year of 2007 (China State Bureau of Statistics 2008; Liaoning Provincial Department of Statistics 2008). In term of employment structure, the labour force in non-agricultural sectors in the province is higher than that of the national average. In the year of 2007, the percentage of the Liaoning labour force in both secondary and tertiary industries was 67.6 per cent, which was 8.4 per cent higher than the national average of 59.2 per cent (Liaoning Provincial Department of Statistics 2008). These figures illustrate a higher urbanization process of the area than that of the national average.

Liaoning has established rail and road networks that link three provinces of northeast China and Inner Mongolia. These networks provide services for 70 per cent of the import and export cargo transport in northeast China. With the establishment of seaports network of Liaoning coastal area and the support of the network of so-called 'three transversal and three perpendicular corridors',[2] the capacity of international trade will be further expanded.

Liaoning coastal area, which consists of six main cities of Dalian, Yingkou, Panjin, Dandong, Jinzhou and Huludao covers an area of 56.5 thousand square kilometres (Figure 4.1). It is about 38 per cent of total provincial territory and 42 per cent of total population of Liaoning with 17.46 million residents living in the coastal area. GDP of these six cities was 398 billion yuan, which was around 45 per cent of total provincial GDP in the year of 2005 (Liaoning Provincial Department

---

2 Three perpendicular transport corridors of rail and road are Harbin to Dalian, Dalian to Jiamusi, and Chaoyang to Qiqihaer. These three vertical corridors link the major and developed cities in northeast China. Three transversal corridors are three urban development belts with each provincial capital as its centre and link the main cities in three different provinces.

of Statistics 2008). These figures illustrate the importance of the coastal area in Liaoning in terms of its economy and population.

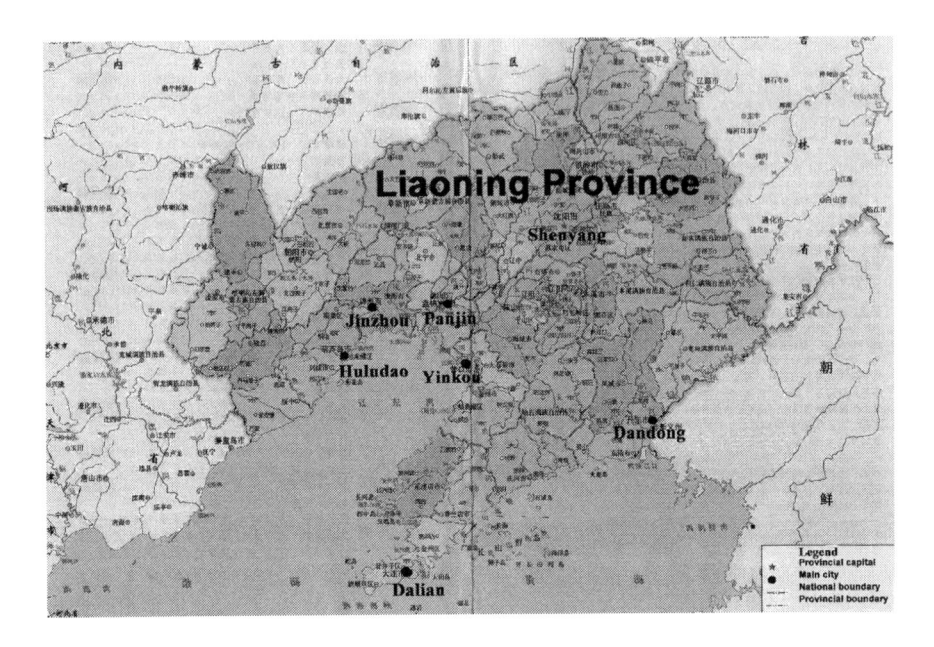

Figure 4.1   Map of major cities along the coast of Liaoning Province
Source: Adapted by the author.

Planning Background

Since the late 1990s and beginning of the twenty-first century, one development priority of the Chinese central government has been to regenerate old industrial bases in northeast China, and to promote development in Bohai Bay Rim. In order to follow this development policy and to play a leading role in the strategy, Liaoning provincial government decided to formulate a regional urban system development plan for the coastal area in Liaoning.

In addition to the importance to the local economy and urbanization, the cities and towns of Liaoning coastal area are significant to the regeneration process of the whole of northeast China because of their location at the main seaports and gateway to northeast China and its links to domestic and international trade. This area is also allocated by various economic technology development zones, export-oriented industrial zones, free trade zones and other development zone (or parks) that enjoy special incentive packages in favourable taxes and land use polices. It is one of the most economically prosperous areas in Liaoning and the whole of northeast China.

The formulation of Integrated Urban System (megalopolis) Development Plan of Liaoning Coastal Area was to provide spatial development policies for delivery of a major development strategy of Liaoning Province, i.e. further opening up of the coastal area while stimulating the regeneration of old industrial bases of northeast China that mainly locate in Liaoning Province. This plan focuses on the spatial development of so-called 'Five Dots and One Belt', which refers to one development belt that links the five major co-urban areas in the coastal area of Liaoning (Figure 4.2).

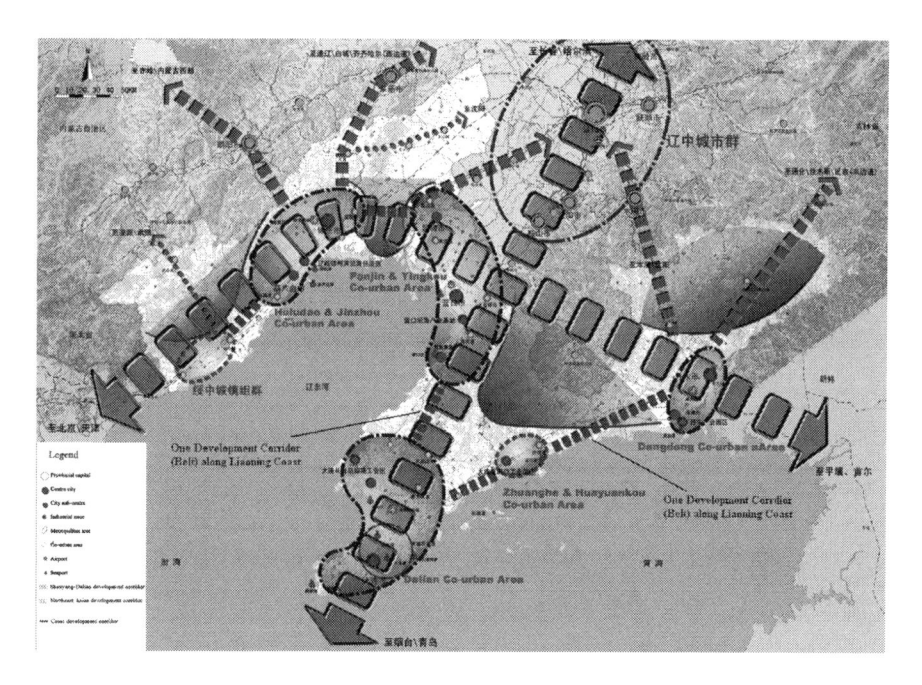

Figure 4.2 **Spatial development of 'Five Dots and One Belt' along Liaoning coastal area**
Source: CAUPD 2006.

Basic Contents of Integrated Urban System (Megalopolis) Development Plan of Liaoning Coastal Area

Guiding concepts for plan formulation While making this plan, although the 'Scientific Outlook on Development' had been regarded as the underpinning concept to guide the formulation of the plan, the promotion of competitiveness of the Liaoning coastal area was clearly the priority. The planning objective was to build Liaoning as a national strong economic development province through regeneration of old industrial bases of northeast China and further opening up of its coastal area. It is suggested in the plan that achieving the objective should fully consider the opportunity of developing Dalian as a navigation centre to northeast Asia. It was argued in the plan that the development of Liaoning had to be driving by industrialization and urbanization (CAUPD 2006).

Main tasks of planning and proposed development period There are four tasks that this planning had to address. First of all, an essential task of this plan was to predict potential population and urban system changes, since China was still under its rapid process of urbanization, the size of population will greatly impact the hierarchic human settlement system. Depending upon the forecast of population and urbanization speed, the plan should secondly identify potential urban system spatial structure, i.e. scale and numbers of the defined metropolises, large cities, medium and small size cities and towns, and their population target and required demands of land use; development proposals for each rural county and co-urban areas should also be specified in the plan. Third, the plan should define the main functions and development proposals for each rural county and co-urban areas should additionally be specified in the plan. Fourth, the plan should define the core functions and development objectives of the coastal area of Liaoning Province and to allocate regional land use for decided social and economic development. Finally, the plan was required to specify policies and guidelines for general land use allocation, transport management and resources protection for major development zones, e.g. economic development zones, high-tech industrial development parks in the coastal area.

The proposed development period of this plan was from 2006 to 2020, which was divided into three phases, i.e. the short term from 2006 to 2010, the mid-term from 2011 to 2020 and the long term after 2020.

Forecast of urbanization and urban system changes The forecast of potential population and urban system changes is the most important task of any regional plan in China. It is the same for the Integrated Urban System (megalopolis) Development Plan of Liaoning Coastal Area.

The approach of population forecasting was based on the existing data and development tendency. The Integrated Urban System (megalopolis) Development Plan of Liaoning Coastal Area predicts that the population of Liaoning coastal area should achieve 42.4 per cent of total provincial population in the year 2010 and 48 per cent of total population by the year 2020 (CAUPD 2006).

For the potential changes of hierarchical urban structure, the plan (CAUPD 2006) clarifies that Dalian, the only metropolis according to the hierarchical definition of the Chinese urban system, should continue to be at the top of the system with the maximum population of 3.4 million by 2010 and possible further increasing to 4.8 million by the year 2020. Regardless of the fact that there was not any city with a population between one and three million by the year of 2010, two such cities are possible to emerge by the year 2020. They are Yingkou with population of 1.5 million and Panjin with population of 1.05 million. There will then be three metropolises in the region.

It was proposed in the plan that there should be seven cities with populations between 0.5 million and one million, the size of which is defined as a large city in the Chinese urban system, in Liaoning coastal area by the year 2010. Even with the promotion of two cities, i.e. Yingkou and Panjin, to the higher hierarchical level in the urban system with more than one million, the total number of cities at this level would remain as seven because of the other two cities' promotion from the lower level in the hierarchy. The cities with populations between 0.2 and 0.5 million, defined as a medium size city in the Chinese urban system, will increase to 10. The number of defined small cities with populations less than 0.2 million will decrease to four in the year 2020 from 10 in the year 2010. The population forecast and changing urban system shows that the urbanization process will continue at a rapid pace in the region, as more population will concentrate in metropolises and large cities. The results of the forecast provide references for government to decide infrastructure development and spatial strategy.

Priority Development Policies Proposed by the Plan

The Integrated Urban System (megalopolis) Development Plan of Liaoning Coastal Area recommends the priority development policies to define the main functions and objectives of the coastal area, to encourage the development of major zones, and to create a balanced pattern in the province. The priority development policies cover major social, economic and environmental aspects (CAUPD 2006).

The plan suggests integrating development between the coastal area and the whole province. This policy suggests that the development of coastal area should coordinate and integrate mainly with central and south of Liaoning Province, and be extended to northeast China and the Ring Bohai Bay Rim to have comprehensive and integrated development between the coastal area of Liaoning and the whole region.

This policy of promoting economic development while protecting ecological environment at the same time tries to promote the coordination between economy development and natural resources and environment protection. It proposes that the existing method has to be changed to follow a sustainable development approach. Environment and natural resources should be protected in the process of economic development.

The interaction between industrialization and urbanization is a key regional development policy. The plan (CAUPD 2006) proposes to establish an urban development belt along Liaoning coastal area. The economic development objectives and targets of the urban development belt are identified and then distributed in terms of space. The policy is seeking to establish a mutual support and benefit relationship between industrial development and port construction, and between urban and rural areas. The underpinning concept for this policy is that industrialization should be treated as a driving force for urbanization whilst urbanization should promote the upgrading of industrialization and its sustainable development. This policy is also seeking a harmonious development between economy and society, and between city and countryside.

Building a harmonious society and encouraging rural economic development through integrated urban and rural development has been a foremost policy of the Chinese government. A regional plan should attempt to reduce development and income polarization and to create a balanced development pattern within its territory. The main polarization in China is the gap between urban and rural areas. It is recommended in the plan (CAUPD 2006) to extend urban infrastructure and services to rural areas by increasing public expenses in the countryside. Promotion of rural economic growth is suggested to follow the approach of sustainable rural development. County towns and other small towns should be developed as hubs to service their rural hinterlands. The planning policy also raises the crucial requirement of providing job skill training for rural labour. Planning policy suggests providing incentive policies package from the governments to employ more local rural labour force.

Regional planning should consider both long-term development objectives and short-term construction targets. To coordinate between short-term development targets and long-term development objectives is recommended in the Integrated Urban System (megalopolis) Development Plan of Liaoning Coastal Area (CAUPD 2006).

The plan proposes to decide the phases of development according to spatial layout of 'Five Dots and One Belt', e.g. five co-urban areas and one development belt. It is recommended in the plan that the first phase should begin with the five 'dots' (co-urban areas) and then to form one 'belt' which consists of both hierarchic urban settlements and varies economic development zones. The development along the 'belt' should be further extended to its hinterland area. Short-term development targets are predominantly associated with defined projects by the Eleventh Five-Year Economic and Social Development Plan (2005 to 2010); whilst long-term objectives should be considered the demands of sustainability of Liaoning coastal area, typically the natural resources protection.

In addition to the long-term development objectives and short-term construction targets, an overall vision of the area should be raised when making decisions in priority development projects. The policy requires a balance between developed areas and less developed areas to provide equal opportunities for all.

China is under its process of transition to the market economy. It is no longer possible for the government to decide and to provide all development resources. The government has to cooperate with the private market, and to depend on its mechanism to generate vitalities and activities of society and economy. The planning policy supports the establishment of partnership with the market force. However, it is also suggested in the plan that the government should influence development direction by planning, and to apply the government's resources to avoid malignant competition. It is mentioned in the plan that forming a coordination relationship between the public and the private is important in the market economy.

Roles of Co-urban Areas in Development along Liaoning Coast

Co-urban areas or urban regions are significant players in regional social and economic development. The Integrated Urban System (megalopolis) Development Plan of Liaoning Coastal Area (CAUPD 2006) defines the main functions of the five co-urban areas in the coastal area of Liaoning Province.

The planning policies (CAUPD 2006) suggest that a co-urban area should function as a leading space to promote regional development and old industrial regeneration of northeast China. The Liaoning coastal area should play a key role in international cooperation and attraction of foreign direct investment to restructure local economy and upgrading existing industries, especially in petrochemicals, automobiles, ship building, electronics, machinery and manufacturing equipment production. The local economic restructure should establish industrial clusters to increase its international competitiveness in manufacturing.

It is further suggested in the plan (CAUPD 2006) that economic development of co-urban areas on Liaoning's coast will benefit to help the regeneration process of old industrial bases that are mainly located in Liaoning Province, and to encourage the economic transition of the whole of northeast China. In order to optimize regional layout of productivity for competitiveness promotion and integrate regional economic development, the plan proposes that the Liaoning coastal co-urban areas should become an essential partner to Binhai New Development Zone in Tianjin, which is also located in Bohai Sea Rim Area.

Under the guidelines of the national strategic policy of redeveloping the old industrial bases in northeast China, Liaoning coastal co-urban area, with Dalian as its leading seaport city, has been regarded as the crucial development pole in northeast China. In order to achieve this objective, Dalian co-urban area and other seaport city regions should fully consider the advantages of their locations. The seaport cities should be targeting at becoming the Northeast Asia navigation centre and actively involving in economic cooperation and international market competition in East Asia.

Provision of enough job opportunities has been a critical problem in Liaoning Province for many years. It is especially since the application of the market economy in China that many old industrial and resource-dependant cities suffer from high unemployment and underemployment in the process of their economic transition. In the resource-drained cities, the unemployment rate has increased to over 15 per cent. One approach to address this problem is proposed by the planning policy (CAUPD 2006) to transfer labour forces from these resource-drained cities to co-urban areas along the coast line of Liaoning. By delivering this policy, it is also able to avoid the potential social problem because of increasing unemployment. This policy is proposed based on considering two preconditions. The first condition is that the resource-drained cities are unable to provide enough job opportunities due to the deficiency between the large population in them and the very high unemployment rate locally. The second condition is the increasing demands of labour force, especially the shortage of skilled labour forces in the coastal co-urban areas due to its rapid industrialization and urbanization. However, it is recommended in the plan (ibid.) that the process of transformation should be operated within the market mechanism instead of government's order. Government's function in the process is to provide favourable policies to support the transformation.

Spatial Development Strategy

The spatial development location has been decided in the Integrated Urban System (megalopolis) Development Plan of Liaoning Coastal Area (CAUPD 2006). The plan suggests that Dalian should be a leading city within the spatial structure of Liaoning coastal area. The areas that include the two predicted metropolises of Yingkou and Panjin along the motorway from Shenyan (the provincial capital) to Dalian, which is also regarded as the most active developing and major transport corridor in Liaoning Province, should be the major development sub-region of the province. The west part of the Shenyan to Dalian motorway corridor will be a development axis between Jinzhou and Huludao. The east part of the motorway is Dandong. The west to east axis between Jinzhou and Huludao and Dandong is regarded in the plan as two wings in the spatial layout. Within this spatial structure, other small cities and towns will work as hubs along the comprehensive transport networks (see Figure 4.3). This economic and spatial pattern is formed to link with other two economic spaces in Liaoning Province, i.e. Liaoning west area and Liaoning central area; and to coordinate with Beijing, Tianjin and two additional provinces in northeast China.

Several development strategies are proposed according to this spatial structure (CAUPD 2006):

- The social and economic functions and activities along Shenyan to Dalian development and transport corridor that forms one part of Harbin to Dalian transport and economic corridor should be further strengthened. The main industries, population and major cities in northeast China have been concentrated along this corridor. The development of this corridor will play a significant role in regional development of the entire northeast China.
- Dalian as the leading city in the region should be emphasized and extended to be a new business and industrial centre in northeast China.
- Development of Yingkou and Panjin co-urban area should be encouraged due to their advanced location. Yingkou is located on the Shenyan to Dalian corridor and convenient for its links with Liaoning central and east areas. The development of Yingkou and Panjin co-urban area will provide opportunity for coordination between Liaoning coastal area and Liaoning central area.

- Development of Jinzhou and Huludao co-urban area and Dandong urban area, as two wings of this spatial development pattern, should be promoted. Jinzhou and Huludao co-urban area, which is located to the west of Liaoning Province, is able to establish coordination links with Beijing and Tianjin. Dandong to the east part of Liaoning Province is at an appropriate location to link with Russia, Japan and Korea. These two wings should also function as hubs to serve economic activities in their hinterlands to have a regional integrated development.
- Activities of small towns should be encouraged by building radial transport corridors to link rural areas to have an integrated urban and rural development.
- The whole spatial development should follow the principle of building an ecological environmentally friendly urban system.
- Existing industrial parks (zones) should be restructured for the purpose of industrial clusters establishment in order to have mutual support and interactive development between urbanization and industrialization. Capability of enterprises in innovation should be the priority in the restructure process in order to become a national modern industrial base.

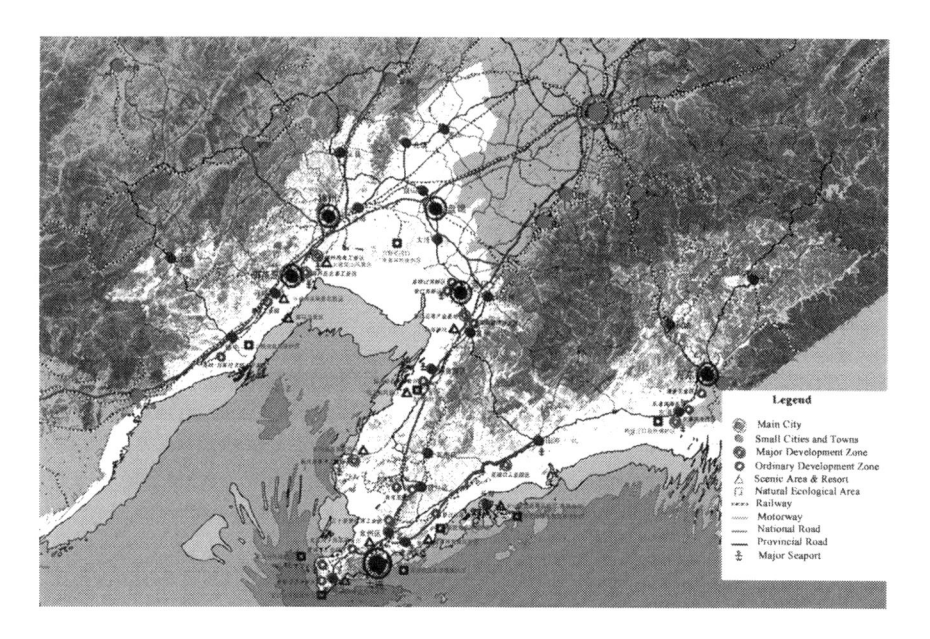

Figure 4.3 Integrated urban system development plan of Liaoning coastal area

Source: CAUPD 2006.

Phases of Spatial Development

Three phases are defined in the plan for spatial development. The short term, which was decided as same period with the Eleventh Five-Year Plan (2005 to 2010), should have five 'dots' of co-urban areas as development priority. The target was to have an interaction and mutual support between urbanization and industrialization in these five co-urban areas through restructuring existing industrial parks (zones). It was recommended in the plan (CAUPD 2006) that it was imperative to promote the comprehensive capacity of the urban labour force, and to concentrate population and industries to the five 'dots' of the co-urban areas in this development period.

The middle term, which is defined as the period between the years of 2011 and 2020, should emphasize and promote the further development and coordination of co-urban areas. The target is to form a highly integrated urban system along the coastal route. Establishing the transport network, the restructuring of seaports and industrial bases, and rapid development of major cities and towns are the principle conditions to achieve this target.

The long term, defined as the period after the year 2020, should extend the development to hinterland areas through the support by urban hierarchic system in coastal area. The target is to strengthen the development links with hinterlands of northeast China for an all-round opening up market and to establish a rational human settlement network.

Urbanization Strategy

Establishment of an appropriate and active urban settlement system   The plan (CAUPD 2006) suggests that it is necessary to think different but reasonable roles and functions of various hierarchical cities and towns in urban settlement system in the process of urbanization. The promotion of external competitiveness of major cities, typically metropolis of Dalian and other large cities should be regarded as strategic priority to play a leading and comprehensive role in the region. All these urban settlements in Liaoning coastal area should be functioning as spaces for focused industrialization and urbanization. Economic activities of small towns should be promoted to form a network to link its rural hinterland. Industrial development in small towns is encouraged to provide job opportunities for rural labourers.

Encouragement of labour intensive industry in urbanization to increase employment opportunities   Promotion of labour intensive industry development should be the priority of economic development strategy to address the severe unemployment problem. This policy will be implemented along with the process of urbanization in Liaoning Province. The urbanization process in Liaoning Province has suffered from its transition to the market economic from former higher centrally planned economy, particularly the high unemployment rate as a consequence of competition in the market and bankruptcy of former state-owned enterprises owing to their inefficiency. Promoting a high quality of urbanization can be associated with economic development. In order to address the high unemployment rate, the development of real estate is proposed since the real estate development is able to generate many labour intensive industries, e.g. industries in building, steel and iron, cement and glass, as well as real estate management, community service. In addition to providing job opportunities, the real estate development can also provide opportunities for recreation and holiday tourism at seaside and beaches with development of resorts along the coastal area.

Policy-led concentration of population to coastal area There are numerous potential capabilities of Liaoning coastal area for further development, especially in job provision. In order to reduce pressure of unemployment from resource-drained cities and to increase potential development capability to the coastal area, the policy encourages economic development and urbanization of the coastal area to absorb labour forces from resource-drained cities. The re-allocation of population will help to address the contradictions and mismatch between labour forces and job opportunities.

Industrial Development Strategy

Implementation of all-round opening up strategy and innovation mechanism The Integrated Urban System (megalopolis) Development Plan of Liaoning Coastal Area (CAUPD 2006) is suggesting implementing an all-round opening up development strategy by starting from the so-called 'Five Dots' of co-urban areas. The all-round opening up development strategy will be extended to hinterland areas and cover the entire territory of the province. The attraction of FDI is still treated as strategic policy but to focus on the investment in the R&D, modern agriculture, machinery and manufactory equipment industry, chemistry, high-tech industry, civic and transport infrastructures, logistics, services and finance.

Industrial development policies of the plan (CAUPD 2006) encourage cooperation in development and exploration of energy, raw materials, mineral materials among the Northeast Asian countries. The cooperation in trade, tourism and export/import processing industries is also a fundamental development policy in the plan (ibid.).

Transition of economic development pattern for a sustainable development approach This policy is targeting for promotion of recycle economy, i.e. recycle energy, recycle ecological agricultural development, and cleaning industrial production approaches. Ecological re-circle industrial clusters in metallurgy, petrochemicals, coal mining, and electricity are encouraged by planning policy to form an integrated clean production activity.

The application of energy efficiency, reduction of carbon emissions and reducing waste should be the key strategy in industrial development. Application of clean and renewable energy by wind, solar and tide is promoted in the plan (CAUPD 2006).

Conclusion

The role of region has been playing an important function in social and economic development in global economy. Regional planning in China has attracted interests and attention by different ministries in the central government. There are two types of regional plans which are the different responsibilities of two different ministries. The Ministry of Housing and Urban-Rural Development is responsible for formulation of the statutory plan, Integrated Urban System Plan, which has been specified by City and Countryside Planning Act 2008. Despite the fact that the plan for Development Priority Zone has not been identified as a statutory plan by any act, it has been operated by the State Council in an administrative system. Plan for Development Priority Zones defines critical regional development policies by the National Development and Reform Commission (NDRC) that is responsible for national and local economic and social development.

The planning approach of the Development Priority Zones has impacted traditional Chinese regional planning such as in the approach of spatial structure proposal. More critical, the contradictions and overlapping responsibilities between different government organizations in regional planning may create problems in regional development policies.

The importance of a region in the global and national competition has created a new regional governance mechanism, the Cities Alliance. Regardless of the fact that the Cities Alliance in the South Fujian Province was initiated by the provincial government from top down, it provides a framework for local cities to share resources in the coordination. After the development the Cities Alliance in South Fujian Province, this innovative regional governance mechanism has been operated in several provinces in China. However, this mechanism is still at its beginning stage. It needs to be improved incrementally in its operation, typically the establishment of a compensation mechanism in the collaboration.

The case of the Integrated Urban System (megalopolis) Development Plan of Liaoning Coastal Area provides details of the contents of a regional plan in China. It is clearly illustrated in the case of Liaoning that there are several fundamental issues that a regional plan should attempt to deal with, such as the balanced development among different sub-regions, or between urban and rural areas to build a harmonious society. A regional plan should promote integration between industrialization and urbanization, and distribute rationally the different resources through spatial planning. More important, sustainable development policies, including energy efficiency and uses of renewable energy, have to be recommended as a key policy in the development.

Chapter 5
Statutory City Comprehensive Plan

Introduction

A City Comprehensive Plan, which is produced by the local municipal governments, is seen by Chinese planners as the most important plan in the system. This chapter reviews the content of the plan, policies and proposals for urban development put forward in it, e.g. a description of the nature of the city, development objectives, the land use structure, zoning, comprehensive arrangement for different types of development. Some cases and examples are used to illustrate that although the structure of City Comprehensive Plans is the same, the policies and approaches vary depending on circumstances and the value system of the planners producing those plans.

Statutory Role of a City Comprehensive Plan and its Contents

A City Comprehensive Plan is defined as a statutory plan by both the first City Planning Act 1989 and the City and Countryside Planning Act of 2008.

A City Comprehensive Plan expresses the policies and proposals for urban development. Generally speaking, a City Comprehensive Plan should include a description of the nature of a city, development objectives and its scale, general development norms and index, land use structure, zoning and comprehensive arrangement of different development. As a comprehensive plan, it consists of a set of different special plans, e.g. civil engineering plans, short-term construction plan, transport plan, landscaping and greening space plan and others.

A City Comprehensive Plan should consist of two parts, i.e. urban system plan of the city region and plan for 'urban planning areas'. The planning target period for a Comprehensive Plan is 20 years. It is a requirement that the plan should be reviewed and revised once every five years. It is required in the 'Regulations for Urban Plans' Formulation' (MOC 2005) that a Comprehensive Plan Brief, which defines the key principles for Comprehensive Plan formulation and the bases for the plan's policy-making, should be established at the outset to help decide:

1. Brief of Integrated Systematic Plan of City Region, which should consist of integrated urban and rural development strategy; comprehensive targets and demands of protection of eco-environment, land and water resources, historical and culture interests; spatial management principles; forecast of

total population within region; definition of population, function spatial layout of cities and towns, and city regional transport strategy.
2. Definition of urban planning areas.
3. Development objectives, natures and functions of a city.
4. Proposal of boundary of forbidden, control and suitable development areas.
5. Forecast of urban population.
6. Research on spatial increasing boundary of the central city and proposals of land use and construction scales.
7. Principles of transport development strategy and outward (inter-city) transport facilities.
8. Development objectives of major infrastructures and public facilities.
9. Principles and construction policies of establishing a comprehensive system to disaster prevention.

After the approval of the City Comprehensive Plan brief by the municipal government and then approved by a higher hierarchical government, the City Comprehensive Plan can then start the process of preparation. The contents of a City Comprehensive Plan are comprehensive. It should include, but not be limited to, the following aspects:

1. Analysis and definition of city natures, functions and development objectives.
2. Forecast of urban population.
3. Definition of forbidden development areas, control development areas, suitable development areas and built up areas, as well as policies of spatial control.
4. Principles and policies of rural village development and its control. Decisions of selecting villages and townships for further development, forbidden development and demolition. Standards of villages and townships construction.
5. Allocation of construction land use, agricultural land use, ecological land use and other uses.
6. Research and decision on spatial increasing boundary of central city, land volume for construction, and boundary of construction land.
7. Definition of spatial layout of construction land uses, and proposed control codes of land use for controlling zones and relevant control index, e.g. building density, height, plot ratio, population density.
8. Specification of city centres and district centres and their location and scale, as well as layout of major public facilities.
9. Transport development policies, urban public transport general layout, policies of bus priority; and location decisions for major outward (inter-city) roads and transport facilities.
10. Policies of greening system and its layout; definition of various functional green lands and their protection boundaries (so-called 'Green Line');

decision of water, i.e. lake conservation boundaries (so-called 'Blue Line'); and principle of coastline uses.
11. Conservation policies of historical and cultural interests; boundary defining for historical and cultural conservation areas and protection area of historical buildings (so-called 'Purple Line'); defined boundaries for different standards of historical heritages.
12. Research of housing demands to make housing polices, construction standards, layout of housing land use; it is important for the plan to decide standards and policies of affordable housing and low-cost housing for low-income people.
13. Development objectives and general layout of telecommunication, water supply, drainage, gas, heating and sanitation.
14. Objectives of protection of eco-environment, and policies and approaches of pollution control and treatment.
15. Establishment of a comprehensive public security and disaster prevention system, and policies to protect against flood, fire, earthquake and other natural disasters, and re-construction principles.
16. Definition of old town or district within a city, principles and approaches of regeneration; standards and demands for improving production, living and environment within the old district and towns.
17. Principles and policies of development and uses of underground space.
18. Definition of development phases, and approaches for plan implementation.

A City Comprehensive Plan should consist of a written document and the relevant appendix, and explanation of planning proposals and collected database and information. The plan maps should include the Existing Layout of Cities and Towns under the municipal government's jurisdiction, Existing Urban Land Use Map, Land Value Evaluation Map, Urban System Planning Map of the city region, City Comprehensive Planning Map within the Municipality, Road Network Planning Map, Specific Professional Planning Maps of the infrastructure and environment, and the Proposed Short-Term Construction Map. The ratio of the maps will differ depending upon the size of the city. The map ratio for large and medium sized cities should be between 1/10,000 and 1/25,000. The map ratio of small cities should be between 1/5,000 and 1/10,000. The ratio of the designated town should be 1/5,000.

Municipal Integrated Urban System Plan

In order to have a harmonious and equitable city and town development within a municipality region, the Integrated Urban System Plan of a city region was introduced in the 1990s (MOC 1994). This plan forms one part of the City Comprehensive Plan. The elements of the Integrated Urban System Plan of a city region should include the following aspects (MOC 2005):

- Proposal of integrated urban and rural development strategy, typically spatial development pattern, major infrastructures, public facilities, eco-environmental protection between the central city and its nearby region.
- Development targets and demands of protection for the environment, land and water resources, energy, cultural and historical interests.
- Forecast of total population of the city region and urbanization level, definition of population scale, functions, spatial layout and construction standards of each city and town within the region.
- Development direction, land use scale and control areas for development of major cities and towns within the region.
- Transport development strategy of city region, principally decisions on regional transport, communication, energy, water supply drainage, flood protection, solid waste treatment, major social services, location selection of storage and production of hazardous articles.
- Definition of urban planning areas according to the demands of urban development and management of resources.
- Proposals of implementation of plan.

Relationship with Other Plans

Five-Year Social and Economic Development Plan

The City and Countryside Planning Act 2008 specifies that 'Preparation of a city comprehensive plan and town comprehensive plan, rural plan and township plan should be in according with national economic and social development plan and linking up with land general plan' (Article 5, NPC 2008).

The national economic and social development plan is the most important plan in the Chinese planning system. The plan covers all the aspects of social economic development of the entire country, or a province of a provincial economic and social development plan, or a city for municipal economic and social development plan. An economic and social development plan consists of development policies of agriculture, industry, tertiary, urban and rural development and social development.

In an economic and social development plan, the guiding principles and development goals of a period of five years are clarified. Policies of addressing rural economic development, agriculture and living quality of villagers have been the priority of government during the last two decades. Development of modern agriculture, increasing rural villagers' income, improvement of the countryside, investment in agriculture and rural areas and rural reform is the main contents.

In the chapter of industrial development of the Eleventh Five-Year National Economic and Social Development Planning period (2005–10), it was titled as 'Optimizing and Upgrading Industrial Structure'. It was suggested that high-tech industries and energy resources industries are the development priority. Industrial structure and the layout of raw resources should be considered to be adjusted. Internet communication technology was regarded as an aspect for active promotion.

In regarding to the tertiary industry, it was divided into two types of productive service and consumption service that should be expanded and enriched separately. Accelerating the service industry was specified as development policy.

Reducing the unbalanced development in regions was the other important policy in the Eleventh Five-Year National Economic and Social Development Plan. In order to promote balanced development among regions, three tasks were defined. First was to deliver regional development strategy; the second was to establish main regional functioning zones; and the final one was to promote urbanization in a healthy approach.

Development of recycle economy, conservation and restoration of natural ecology, enforcement of resources management and rational use of maritime and climate resources were the polices to have a 'resource-conserving and environment-friendly society'.

The achievement of Chinese economic development is definitely a consequence of reforms and open door to the outside world. Continuing reform in various aspects has always been the main policy of the Chinese government. This was clearly illustrated in the five-year economic and social plan. The Eleventh Five-Year National Economic and Social Development Plan, which was ended by the year of 2010, clarified the reforms in the economic mechanism, fiscal and tax system, revenues system and a modern market system. The further and deepening institutional reform was explained in a special chapter.

Open door to outside world for mutual benefit and the 'win-win' possibility was the main content of the chapter of 'Carrying out an Opening up Strategy Featuring Mutual Benefits and Win-Win'. However, it had been realized by the Chinese government that the existing approach of international trade might not be sustainable forever. It was necessary to make some changes. The encouragements of changing international growth approach and emphasizing quality of foreign direct investment then became the policies in international economic cooperation.

Social development in terms of population, health, safety and public security, and social management system were the focus points of the Eleventh Five-Year National Economic and Social Development Plan for the establishment of a harmonious society. With social and economic development, democracy is irrevocably becoming a tendency of society. In the Eleventh Five-Year National Economic and Social Development Plan, Chapter 11 was about 'Promoting Socialist Democratic Politics'. Similarly in the promotion of cultural development, the development policies are defined in Chapter 12 of 'Promoting Socialist Cultural Development'. The next two chapters were about 'National Defence and the Army' and 'Implementation Mechanism of the Plan'.

A City Comprehensive Plan is to define the nature, scale and development tendency of a city, and rational use of land, allocation of urban space according to the objectives of the economic and social development plan.

General Land Use Plan

Regarding the relationship between a City Comprehensive Plan and a General Land Use Plan, Shi (2008) has raised a view that a General Land Use Plan defines the land use functions within an administrative boundary, i.e. the whole country, or a province, or a city; a city plan rationally specifies spatial functions of various urban development activities on the defined urban development lands by a General Land Use Plan. The overlapped areas of these two types of plan are on the defined urban development land.

China is a country with a shortage of cultivated land. China has about 21 per cent of the world's population, but only 6.4 per cent of its land resources (Yu 2009). The strict control of cultivated land has then been the priority policy similar to control of population – 'family planning'. A General Land Use Plan is trying to provide allocation and arrangement for development, use, reclaim and protection of land resources according to the National Economic and Social Development Plan by targeting a rational and effective use, and supporting social and economic sustainable development from the perspectives of national long-term strategy and protection of cultivated land.

It is specified in the Land Administration Acts of 1986 and 1999 that a City Comprehensive Plan, a village, or a township plan should connect to a General Land Use Plan, and the land use allocation by city planning, village and township planning should not break the limitation of land available for urban construction and development as defined by a General Land Use Plan (Article 22, NPC 1999).

In order to regulate the formulation and approval of a General Land Use Plan, the Ministry of Land and Resources has published a Circular (MOLR 2009) to define its contents. Of a national General Land Use Plan, the plan should consist of following items (Article 16, MOLR 2009):

1. Evaluation of delivery of existing plan.
2. Assessment of planning background, and land demand and supply of land.
3. Strategy for land use.
4. Definition of planning objectives, e.g. protecting volume of cultivated land, total area of agricultural land, volume of construction land, and arrangement of land reclaim.
5. Optimizing schemes of land use structure, allocation and land intensive use.
6. Differential policies of land use.
7. Responsibilities and methods for plan delivery.

In a provincial General Land Use Plan, it should cover main items of (Article 17, MOLR 2009):

1. Implementation assessment of national General Land Use Plans relating to the province.
2. Schemes for key land use issues.
3. Main tendency of land use for each region within the province.
4. Adjustment policies of land use for each municipal government.
5. Land allocation for major projects.
6. Implementation mechanism innovation of General Land Use Plan.

Article 18 (MOLR 2009) specifies the contents of a General Land Use Plan for a municipality. It should focus on items of:

1. Implementation assessment of a provincial General Land Use Plan.
2. Land use scale, structure and allocation.
3. Land use zones and zoning ordinances.
4. Land use control policies of city centre.
5. Adjustment policies for land use of each county within jurisdiction.
6. Land allocation for major projects.
7. Responsibilities of plan delivery.

For the purpose of effectively control land use, the government's Circular has also specified the contents of land use of a county, which is the rural governmental administration:

1. Implementation assessment of a municipal General Land Use Plan.
2. Land use scale, structure and allocation.
3. Land use zones and zoning ordinances.
4. Definition of boundary of each town and village.
5. Policies of major development areas and land reclamation.

It is clear from the contents of the General Land Use Plans at different hierarchy that it is a top-down planning approach by allocating availability and indicators of land use for urban and industrial development and construction. The main principle of planning is to control and supply and guide the demand for land use. Although the relationship between a City Comprehensive Plan and a General Land Use Plan has been clearly specified in the City Planning Act 1989, the City and Countryside Planning Act 2008 and Land Administration Act of 1986 and 1989 that they should connect to each other, the connection between these two types of plan is not easy in practice because of the differences in planning policies as the consequences of different approval and administration process of the two types of plans, and the differences in planning targets and objects. Critically, with the implementation of strict protection of cultivated land and mechanism of land use quota, the land available for urban development defined by city plans and general land use plans may be different.

Plan Making Approaches

It is defined in the 'Regulations for Urban Plans' Formulation' that plan formulation should be 'organised by government, led by experts, coordinated among the departments, participated by the public and policy making in term of scientific' (MOC 2005). Before producing or revising a City Comprehensive Plan by a municipal government, implementation evaluation of existing City Comprehensive Plan, and capacity and conditions of present infrastructures is required. It is also required to undertake relevant pre-studies of a city before applying for approval by the higher hierarchical government. Strategic studies include land use, water supply, objectives of city further development, functions and spatial allocation of a city.

The formulation of a City Comprehensive Plan should only be started after formal approval from the higher hierarchical government. The cities under direct jurisdiction, capital of a province, and other specified cities by the State Council, which are cities with populations of more than 500,000, should acquire permission from the State Council for their City Comprehensive Plan production. Other cities should apply for permission from the provincial or autonomous regional government.

The first stage of production of a City Comprehensive Plan is the brief preparation, which should be submitted for approval by the Ministry of Housing and Rural-Urban Development in the central government, or Provincial Department of Housing and Rural-Urban Development according to the designation of city standards. A City Comprehensive Plan will then be produced according to recommendations suggested by the Ministry of Housing and Rural-Urban Development in the central government, or Provincial Department of Housing and Rural-Urban Development as a condition for approval of the City Comprehensive Plan brief.

Municipal governments often assign, or entrust, urban planning institutes[1] to organize a planning team to prepare the relevant plans. The invitation to tender for a contract to produce the plan has been a popular practice. Municipal governments also set up a steering office, a standing body in some cities, e.g. the Municipal Urban Planning Commission, consisting of members from relevant departments within the government to guide the formulation of a City Comprehensive Plan. Under this steering office, an organization responsible for plan preparation and administration is set up to work closely with the planning team.

---

1 Urban planning institutes carry out urban development research and produce urban plans. They are quasi-government organizations.

Approval Procedure of a City Comprehensive Plan

The City and Countryside Planning Act 2008 (NPC 2008) specifies that the local Municipal People's Congress should approve the City Comprehensive Plan of a city first, before submitting it for examination and approval by the higher hierarchical government (Article 16). The City Comprehensive Plans of the cities under direct jurisdiction of central government, capital city of a province, and other designated cities by the State Council (Article 14) should be submitted to the State Council for approval. The procedure for the preparation and approval of the City Comprehensive Plan is very complicated since it deals with many issues of urban development.

After the completion of the City Comprehensive Plan, the local municipal bureau of planning administration, associated with other responsible bureaus and experts, should initially undertake an evaluation of the plan contents. Generally speaking, there will be several workshops and meetings among those responsible bureaus and experts in order to have a detailed discussion. Depending on the comments and suggestions from these workshops and meetings, the plan will be revised and then a report submitted to the municipal government. With the approval of the municipal government, the plan will be submitted to the People's Congress of a city for examination and approval. After approval by the People's Congress of a city, the municipal government will then submit the plan to the higher hierarchical government (the central government, or the provincial government) for examination and approval. As soon as the City Comprehensive Plan has been approved, the municipal government will make an announcement to the public.

During the last few years, with increasing consciousness of urban planning and concerns about living quality by ordinary residents, and a graduated democratic process in China, public participation has been a requirement for many people. Article 26 of City and Countryside Planning Act 2008 specifies the requirement of public participation in city and countryside plans production. It defines that the plan producing administrative organization should publish draft plans by public hearing inquiry and other methods of consulting local residents and experts. The public participation should not be less than 30 days. It is also required to submit the summary of the opinions and their adoption in the plan making when submitting the plan for approval by higher hierarchical governments.

Definition of 'Urban Planning Area' and its Impact on Planning

According to the City and Countryside Planning Act 2008 (NPC 2008), any development or construction activity within the defined 'urban planning area' requires permission from the planning administrative authority and obeys the Act (Article 2). According to the Act, the City Comprehensive Plan should only address the development within the so-called 'urban planning area'. The areas outside the boundary of a defined 'urban planning area' should not be allowed to have an urban development.

A defined 'urban planning area' should be located within the boundary of a municipality. 'Urban planning area' refers to the areas which should be controlled by planning according to the demands of potential urban and rural development and construction, and built up areas of a city, town and village. A detailed boundary of the area should be defined by a city or town comprehensive plan, or village plan according to the requirements of urban and rural social and economic development.

The initial reason to define 'urban planning area' was to separate urban and rural areas and to protect agricultural land and countryside. The dual-system of land ownership, i.e. the state ownership in the urban area and collective ownership in rural areas, is the underpinning of this concept. Whenever there is a demand for land for urban development and construction in rural areas, it is required to transfer land ownership to state-owned from collective-owned through land purchase process by local governments. However, urbanization and city expansion in Chinese cities is so rapid that many rural villages have been surrounded by the defined 'urban planning areas' and urban development has been undertaken without transferring land ownership. The phenomenon of the so-called 'village in city' is then appearing. A 'village in city' refers to areas of the land use that is still owned by rural collectives, and the governance and management system of which is also under rural administration. 'Villages in city' are not included within the 'urban planning area'. However, the 'village in city' has already formed one part of the city from physical spatial perspective.

Living conditions and construction standards in a 'village in city' may be lower than a normal level owning to inappropriate planning and planning control. It is outside the specification of 'urban planning area' and without the transfer of land ownership.

Case Study of Xiamen

Basic Information on the City

Xiamen, a port, tourist city and a Special Economic Zone in China has been awarded by the Chinese central government the titles of the 'Cleanest City in China', the 'National Garden City', the 'Best Tourist City', and the 'Model City of Environmental Protection'.

Xiamen is located in Fujian Province (see Figure 5.1), at the estuary of the Jiulongjiang River close to the Taiwan Strait to the southeast of China. Xiamen consists of Xiamen Island (see Figure 5.2) (including Gulangyu Isle) and the north part of Jiulongjiang River along the coast. The total land area of Xiamen is 1,565.09 km^2. The sea area within its boundary is about 300 km^2. The main part of the city is Xiamen Island with an area of 131 km^2. The topography of Xiamen Island is a gradual descent from the south to the north, with the northwest part being relatively flat and the southern part mountainous and hilly. As an island city, there are 234 km of coastline. Several scattered islands outside the port function as a natural

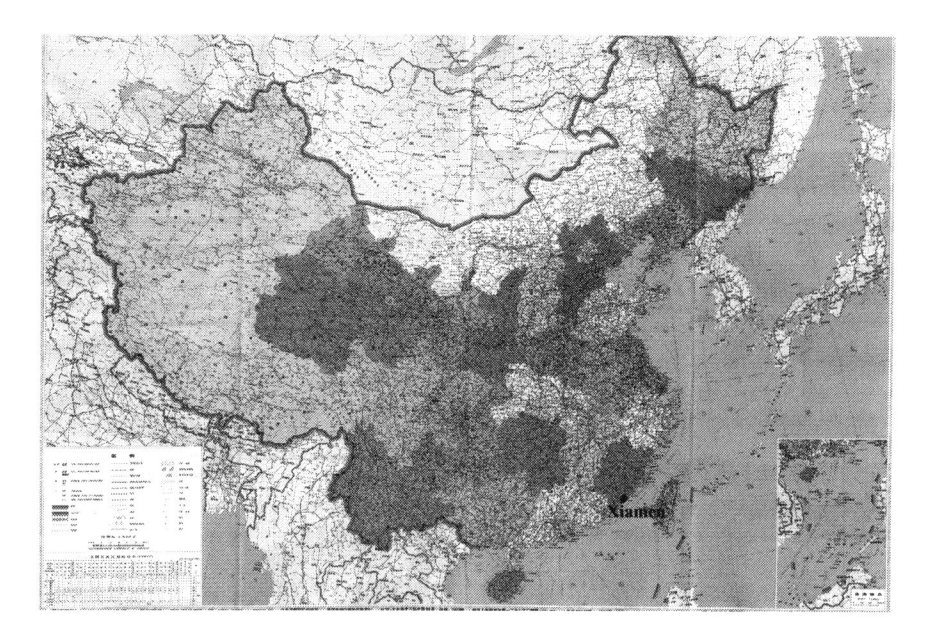

Figure 5.1   Location of Xiamen in China
Source: Adapted by the author.

breakwater. The navigation lane within the port is more than 12 metres deep. Xiamen has a subtropical climate. The average temperature is about 21°C. The average rainfall is about 1,200 mm. There are typhoons from July to September.

Economic Development

As a national defence coastal city for nearly 30 years after the founding of the People's Republic of China in 1949, in order to prepare for war with Taiwan, development and construction in Xiamen had been restrained. At the beginning of the reform in the 1980s, Xiamen was very poor in both its economic capacity and industrial base. In 1980, the total GDP of Xiamen was only 640 million yuan. Revenue was 183 million yuan. Industrial output volume was only 943 million yuan. The investment in fixed assets was 122 million yuan.

Since the establishment of the Special Economic Zone, economic development has been regarded as the priority of the local government. There has been great progress in economic and urban development in Xiamen. In 2000, GDP was 50.187 billion yuan, an increase 28.9 times over that of 1980, with an annual average increase rate of 18.3 per cent. The average GDP per capita was 37,447 yuan (about 4,600 US dollars). The proportions of primary, secondary and tertiary industries in GDP were 4.2 per cent, 52.8 per cent and 43.0 per cent respectively. The total industrial output value was 77.64 billion yuan, which was 76.2 times

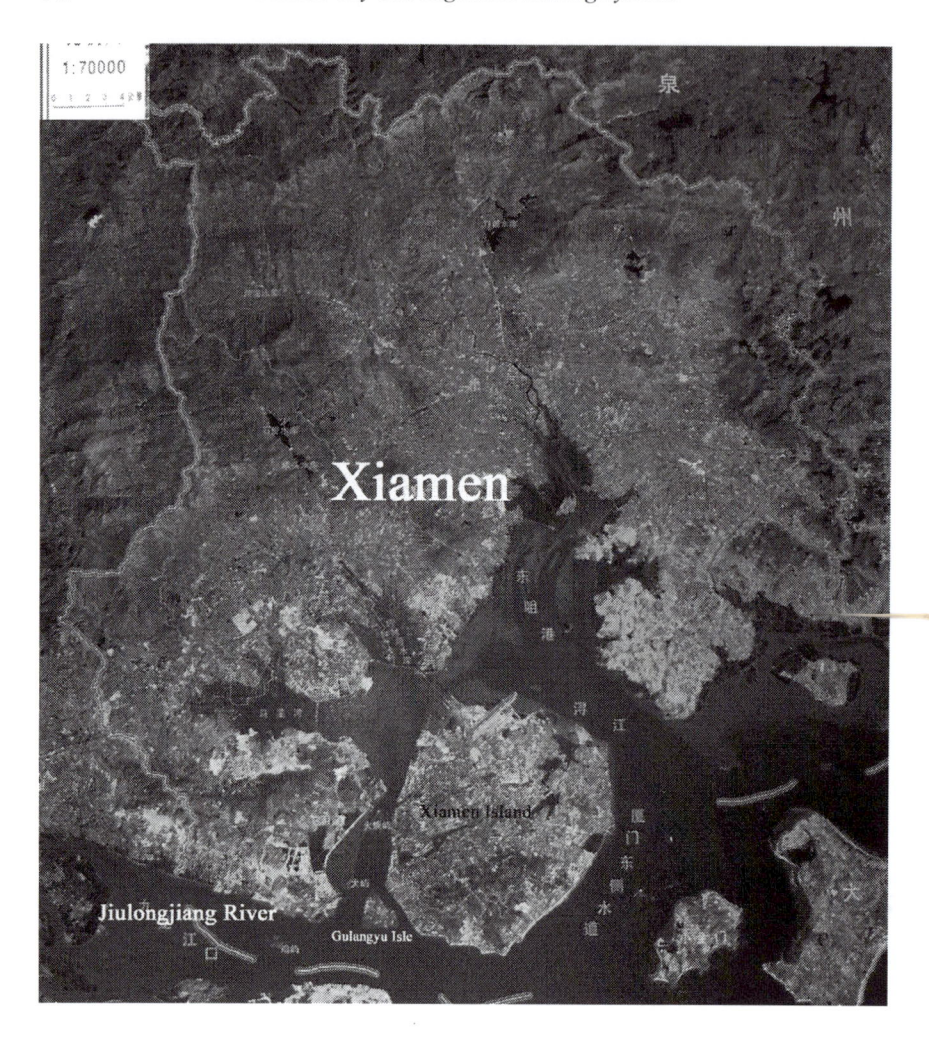

Figure 5.2 Map of Xiamen
Source: Photo taken by the author.

that of 1980. Total revenue in that year amounted to 9.15 billion yuan, which was about 49 times more compared with that of 1980. The objective of economic development proposed by Xiamen municipal government was to reach 10,000 US dollars per capita by 2010. According to Xiamen Economic and Social Development Statistics Annual Bulletin 2009, the GDP per capita of Xiamen in the year of 2009 was 64,413 yuan (about US$9,613).[2] From 1981 to 2001, the total investment in fixed assets was 148.7 billion yuan, among which, 33.8 billion,

---

2 At the exchange rate of US$1 = 6.7 yuan.

or about 22.7 per cent, was the input in infrastructure construction (Xiamen Municipal Development and Reform Commission, ,2001). In the year of 2009, the investment in fixed assets was 88.212 billion yuan, which was near equivalent to the total infrastructure investment of Xiamen for two decades from 1981 to 2001 (Xiamen Statistics Bureau 2010).

Since the open door and reform policy in 1980, economic development in Xiamen has mainly been based on secondary and tertiary industries contributing less pollution because the city is defined and controlled as a scenic city. In 1980, secondary industry contributed about 58 per cent of the total GDP. The main industries were food processing, textiles and machinery. Nevertheless, during the industrialization process in the last two decades from the 1980s, the main industries have changed to the three dominant ones of electronics, machinery and petrochemicals. Tertiary industry has also increased rapidly.

Globalization

One of the characteristics of Xiamen's economy is its high integration with the international economy, evidenced both in international trade (see Figure 5.3) and in investment contribution by FDI. The largest and most important enterprises in Xiamen are FDIs.

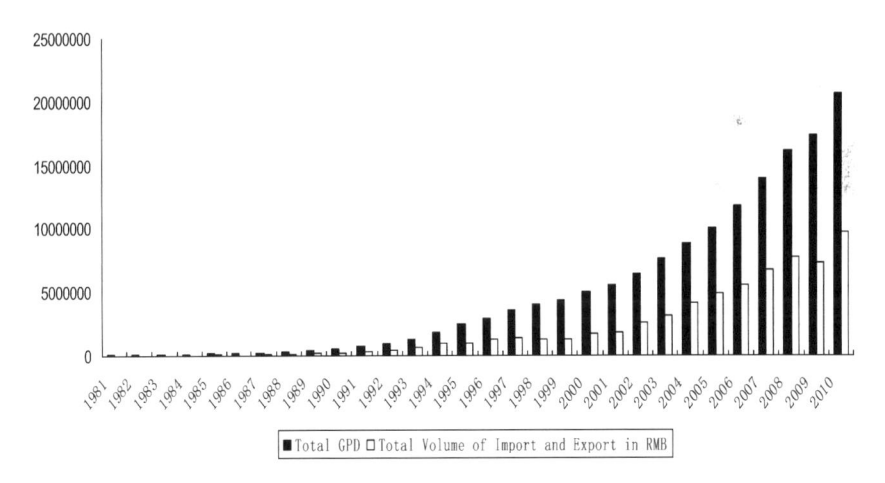

Figure 5.3   Relationship between GDP and export and import volume
Source: Xiamen SEZ Statistic Year Book 2011.

Xiamen was starting its economic development as a typical export-oriented and FDI economy. For its initial development stage, international cooperation and investment have made a great contribution to the economy. Many well-known multinational companies, including ABB from Sweden, Linde from Germany, Matsushita from Japan, Taikoo from Hong Kong, and Kodak and Dell from the United States, have established either main production enterprises, or branches, in Xiamen. According to the Xiamen SEZ Statistics Year Book 2001, by the end of 2000, there had been 11.452 billion US dollars of FDI, generating industrial output value of 59.355 billion yuan in 2000. In the same year, revenue from FDI reached 3.38 billion yuan, accounting for 37.21 per cent of the total revenue. In 2000, there were 11 enterprises whose output had reached 10 billion yuan annually, eight of which were FDI. Their output was about 84.83 per cent of the major enterprises. FDI has been deployed in the fields of electronics, the chemical industry, textiles, foodstuffs, electric power, real estate, finance and others.

According to Xiamen National Second Economic Census Bulletin 2010, by the end of 2008, 23.9 per cent of total enterprises in Xiamen were from overseas; the total capital of these FDI enterprises have reached 52.122 billion US dollars; they have provided 44,650 jobs in the secondary industry which is about 62.7 per cent of total employment in this industry. In the tertiary industry, there were 403 FDI enterprises providing 17,720 jobs, which is around 11.5 per cent of employment in the sector.

In 2000, the total volume of exports and imports in Xiamen reached 10.049 billion US dollars, comprising an export value of 5.879 billion US dollars and import value of 4.169 billion US dollars, respectively. FDI enterprises, including joint ventures, cooperative business and exclusively foreign-owned enterprises, contributed 6.614 billion US dollars to international trade, with exports of 3.626 and imports 2.988 billion US dollars, respectively.

Domestic enterprises contributed 3.435 billion US dollars to international trade. Among these domestic investment enterprises, non-state-owned, i.e. collective, privately owned enterprises grew fast in their international trade with an export value of 340 million US dollars. This was the result of marketization in Xiamen.

However, the situation has changed since the start of the twenty-first century, although FDI still contributes an important role in local economic development, the function of the domestic private enterprises has been crucial. In 2008, according to the Xiamen Municipal Government Annual Report 2009, there were 2.04 billion US dollars of FDI; the total volume of international trade accounted for 45.39 billion US dollars, among which export occupied 29.39 billion US dollars and import contributed 15.99 billion US dollars. Among the total international export and import trades, domestic private, non-state-owned enterprises contributed 13.157 billion US dollars in international trade; while FDI enterprises contributed 13.741 billion US dollars (Xiamen Statistics Bureau 2010). Although both FDI and private enterprises have all increased and significantly contributed to the local economic development, it is obvious that the percentage of domestic private enterprises has increased remarkably.

Sea and Air Ports

Xiamen port is one of the important seaports in China and also one of the oldest international trade ports in Chinese history. This port began its first trade with the world in 1680. By 1830, the port had already established international trade relationships with more than 30 countries in the world. Xiamen port was well known for tea exports to Western countries. The port has continued to develop since the beginning of the 1980s. It is the sixth largest port in China. In 2001, the capacity of the port was 21.95 million tonnes of cargo, and 1.29 million standard containers. There are seven docks with 77 berths within Xiamen port. By the end of 2009, there were 97 berths, with import and export capacity of 110.96 million tonnes of cargo and 4.68 million standard containers (Xiamen Statistics Bureau 2010). It has navigation links to the United States, Europe, the Mediterranean, Japan, Korea, Singapore and Hong Kong. The port has greatly contributed to the development of export-oriented industry and integration with the world economy.

According to a survey by Xiamen Airport Ltd in 2002, there were 33 domestic and international airways offering services on 53 domestic routes and eight international routes through Xiamen International Airport. The number of passengers was 3.38 million with a capacity of 98,200 tonnes of cargo, luggage and parcels, amounting to 56.7 per cent of total air flight passengers and 64.14 per cent of total air freight in Fujian Province.

Tourism

Since the first City Comprehensive Plan, Xiamen has always been defined as a tourist or scenic attraction city because of its beautiful landscapes, i.e. mountains, beaches, gardens and local features. Historically, there are 24 interesting scenic areas, divided into three categories: 'Eight Major Landscapes', 'Eight Minor Landscapes' and 'Landscapes Beyond Landscapes'. The most important ones are Gulanyu Isle (see Figure 5.4), and Wangshiyan Rock and Botanic Garden. They are listed as National Scenic Attractions for conservation and tourism.

Tourism is an important industry in Xiamen. Numbers of tourists increase incrementally year by year. In 2001, there were 573,000 overseas visitors and 8.5 million domestic tourists. Income from tourism was 311 million US dollars. In 2009, there were 25.28 million visitors to Xiamen, among which 1.36 million were from other countries and 23.88 million were domestic tourists. The income from tourism was 48.57 billion US dollars (Xiamen Statistics Bureau 2010).

Figure 5.4   Gulandyu Isle
Source: Photo taken by the author.

Planning as a Learning Process to Cope with Uncertainties in the Market Economy

Xiamen, as one of the Special Economic Zones for the purpose of introducing overseas experiences of science and technology, was the first city that adopted the market economy in development. Urban planning in Xiamen initiated from the centrally planned economy had to cope with the changes and new ideas.

The first City Comprehensive Plan of Xiamen was prepared in 1956 within the Second Five-Year Plan period in the 'Great Leap Forward'. The nature of the city was defined as a port city, a national scenic and resort city, as well as a national defence city (due to its proximity to Taiwan). The proposed population was 500,000 with a total area of 31.86 km². It was intended to extend the urban area incrementally to the east and north of Yuandan port, which was a commercial seaport. Similar to the situation in the whole country, implementation of the plan was stopped until the open door and reform policy at the beginning of the 1980s.

Since the transition to the market economy in China, the urban planning of Xiamen has attempted to cope with the uncertainties in development. However, uncertainty as the main feature of the market economy had not been readily accepted by all. A number of uncertainties at the national, regional and local levels would then have to be addressed in Xiamen Comprehensive Plans, if the plans for the future of the municipality were to be robust. Compared with the former planning period when China was greatly influenced only by the government defining physical, economic and social realities, the uncertainties in the transition to the market economy derive mainly from the adoption of a market economy, and from influence of government, as well as from the investors and developers who use their resources as power.

In 1981, 'Xiamen City Comprehensive Plan 1980 to 2000' was formulated. The reason for the plan's formulation was that Huli, with the area of 2.5 km², had been defined as one of the four Special Economic Zones by the National People Congress of China in 1980. This plan was approved by the Fujian provincial government in October 1983. This City Comprehensive Plan tried to extend and implement the policies and proposals of the Municipal Economic and Social Development Plan (Five-Year Plan). It tried to allocate land resources for development in the Special Economic Zone (SEZ), and other parts of Xiamen. Xiamen, as the window of China opening to the world, began to experiment with the market economy.

In March 1984, the National People's Congress of China decided to expand the Xiamen SEZ to the whole of Xiamen Island, including Gulangyu Isle, making a total area of 131 km² compared to the original 2.5 km² in Huli. Administrative uncertainty affected and drove urban planning. It was due to this uncertainty that the amendment to the Master Plan of 1981 was proposed. A development study of the whole of Xiamen Island was undertaken by an international consulting firm (from Singapore) in 1984. The study suggested that Xiamen should use its crossing to the West area – Haicang – for future development. This was the first time that a bay pattern city of Xiamen was proposed to the municipal government.

Xiamen Municipal Bureau of Urban Planning Administration and Xiamen Municipal Institute of Urban Planning and Design subsequently amended the City Comprehensive Plan of 1980 to 2000 because of administrative changes. The amendment scheme of the plan was completed in 1988. It was approved by the provincial government in June 1990. In this amendment plan, the nature of the city continued to be defined as a seaport, a scenic attraction city and Special Economic Zone. The population of the whole city was to be limited to 1.1 million (510,000 in SEZ) by 1990, 1.3 million (640,000 in SEZ) by 2000, and 1.6 million (700,000 in SEZ) by 2020.

A potential opportunity for major development arose in Xiamen in 1990. With the relaxation of the relationship between Taiwan and Mainland China, investment from Taiwanese business persons in the Mainland increased rapidly. An investment of 10 billion US dollars to establish a petrochemical industry in Xiamen was proposed by a Taiwanese businessman. It was considered that the population of Xiamen would increase by one million because of the labour force demands of this project. The project would greatly impact on the economy, industry, nature, structure and scale of Xiamen. After discussion with the central government, the investor made an announcement in January 1990. This project was then called the '901' project. Local politicians and urban planners assumed this project would happen and be located in Xiamen because under the experience of the centrally planned economy, when a project was proposed for a certain place, it would normally happen. Under such a situation, the urban plan would have to be adjusted and amended to meet the expected changes. This was the beginning of the production of the 'City Comprehensive Plan 1995'. The plan formulation process took more than 10 years because of changes during the development process of the market economy. In discussion with the planners of the project team and local planning officers, it was generally agreed that this planning had been a learning process to understand the characteristics of planning in the market economy.

Historically, Xiamen was a city whose features were its scenic attraction, seaport and trade. There had been little industrial development. The proposed '901' project with its huge petrochemical industry would seriously affect the existing features. The one million or more population could be a nightmare without a plan-led process. It was decided that an urban plan was required to deal with the changes.

In order to produce the plan to meet the changes, China Academy of Urban Planning and Design was entrusted to take on the work with the Xiamen Municipal Institute of Urban Planning and Design. The work started in March 1991, immediately after the existing plan had been approved by the provincial government in June 1990. Considering the recent approval of the City Comprehensive Plan, the Municipal Bureau of Urban Planning Administration proposed formulating a new master plan for Haicang instead of amending the City Comprehensive Plan for the whole of Xiamen. Nevertheless, it could not produce a master plan only for such a huge project without amending the Comprehensive Plan. The planners of the project team from the China Academy of Urban Planning and Design suggested that the City Comprehensive Plan should be amended at the same time. However, this proposal had not been accepted for the reason that: 'The formulation and approval of the City Comprehensive Plan is a serious issue. The City Comprehensive Plan is a statutory document to guide urban development. It is wrong to change the plan frequently. The royalty of the plan should be maintained' (the opinion of Director of the Municipal Bureau of Urban Planning Administration interviewed on 10 July 2003).

The argument was that the City Planning Act 1989 had been only formally implemented for less than one year. The City Comprehensive Plan was regarded as a statutory plan with legal authority. From this understanding, frequent changes in the urban plan were thought by the planning officer to lose its 'authority'. Nevertheless, the Municipal Development and Planning Commission supported the idea of amending the City Comprehensive Plan. They provided the necessary funds for the amendment process. The amendment of the plan was then formulated in a very unusual and informal way since the work had not been formally approved by the local municipal urban planning authority.

It was not until 1994, when a new mayor with a new term of government came into force, that revision of the City Comprehensive Plan was formally undertaken under the process of approval by the Municipal Bureau of Urban Planning Administration because the new mayor supported the amendment of the existing City Comprehensive Plan.

Various Planning Methods in the Process of City Comprehensive Plan Formulation

In the market economy, although it is said that planning is dealing with uncertainties, variables of uncertainties are changeable. The uncertainties of development will be different in different contexts and at different times. It is not only that planning should be flexible, but the programme and actions adopted should take into account possible changeable uncertainties as a consequence. Christensen's learning prototype of 'unknown technology but agreed goal', or bargaining prototype of 'known technology, no agreed goal' conditions of uncertainty (Christensen 1985) could be applied to the Xiamen case.

Examining the Xiamen case, it was found there were agreed objectives by local government. At the beginning of the 'City Comprehensive Plan of Xiamen 1995', the objective was to provide available land and infrastructure to attract investment, particularly the '901' project. The 'means' of achieving this were far from certain. It was not certain if there would be inward investment for the proposed development. There was a need to solve major uncertainties relating to (1) general economic development changes and international political affairs, mutual government relations, particularly relations between Taiwan and Mainland China, (2) devolution and marketization, in the transition. These uncertainties therefore necessitated the adoption of an appropriately contingent approach when producing urban macro plans for the future in a situation where:

- The objectives (ends) are not agreed or may change: as might be the case if the relationship between Taiwan and Mainland China improved or worsened.
- The direction for devolution and marketization is unclear since the transition to the market oriented economy.
- Some of the knowledge and information exists, but not all, to achieve a variety of objectives.

The Xiamen municipal government and urban planners met and learnt the lessons from the uncertainties in the market economy very quickly, and even before the completion of the plan formulation, the situation had changed.

In 1994, China's President Jiang Zeming visited Xiamen. He made several comments about the development of Xiamen. He also clarified that the proposed large investment project in petrochemicals by the Taiwanese businessman, the '901' project, had been cancelled because the Taiwanese authority had forbidden it for political reasons.

Although the initial reason for the City Comprehensive Plan's amendment no longer existed, politicians and planners were still concerned that Xiamen would be subject to the most rapid development era in history because of the continuing open door policy, urbanization, and the relaxation of the relationship between Taiwan and Mainland China. It was realized by local planners that the characteristics of development would be very different from those under the centrally planned economy. The loss of the '901' project provided an opportunity for planners to experience uncertainty in the market, and the uncertainties from politics and the global economy. Characteristics of uncertainties of development in the market economy had begun to be realized and the planners had to try to cope with them. This could be treated as a learning process for urban planning to adapt to the market economy. Under such a situation, the formal process of the usual planning approach was inappropriate. New approaches had to be introduced.

It was clear that planning notions of the centrally planned economy were inappropriate. At this time, systematic and rational planning approaches had been introduced to China. The planning approach was a compromise between following the concepts of the centrally planned economy, and using objective rational methods to deal with development in a market economy. In a pluralistic stakeholders' market economy, other stakeholders may not always accept or support planning rationality. Some real estate developers argued that it was reasonable for developers to break the regulations and policies of urban macro plans during the development process. It was their view that developers would know the 'Hot Spot' of the development and investment depending on their sensitivity to the demands of the market, and their assessment and understanding of the market. The 'Hot Spot' here refers to potentially profitable investment and development areas and projects. This view expresses the conflicts existing in the development. It can be argued that consensus seeking among decision-makers should be required in the planning process. Urban planning should function as a means of communication. The 'pragmatic' planning approach will be applied in an uncertain market and development mainly depends on private investors. Even so, planners still have to undertake careful data and information collection for definition of present issues. These issues and problems need detailed analysis by means of disaggregation into clusters of sub-problems and objectives.

With clarification of the problems, alternative strategies or schemes can be generated. Within this systematic and rational approach to planning, the planners proposed the following to cope with development:

- According to the special geographic pattern of Xiamen, the plan would define the reasonable urban land volume with the aim of controlling land use within the urban planning area, in the light of thresholds of water resources, environmental capacity and energy supply.
- To meet the uncertainties in the market economy (dividing the possible developmental levels into normal development, super rapid development, and a level of development between the two), planning research would concentrate on the analysis of development potential and capacity.
- The technical principles of plan formulation would decide spatial priorities and phases according to different development speeds. The tasks of planning were no longer only to detail and extend the social economic

development plan, but also to cope with diversity and multiple development in the market economy.
- In order to meet the changes in the market economy and shift the focus to economic efficiency instead of total GDP volume, planning would concentrate on development quality instead of the development scale.

Systematic and rational approaches to planning mixed with the concepts of urban planning in the centrally planned economy It was clear that the philosophy of urban planners is mainly affected by concepts of urban planning in the centrally planned economy. It was an idea of Chinese planners that urban planning should concretize the development policies and proposals of the municipal economic and social development plan. There was clear evidence in the City Comprehensive Plan of Xiamen to support this argument. It was clearly explained in the plan that the task of the plan was to interpret the objectives of the municipal economic and social development plan and the regional strategy into physical planning policies, and to ensure these mutual linkages and collaborative development (CAUPD 1995). The analysis of planning proposals and development schemes of the City Comprehensive Plan shows that urban planners have been trying to allocate physical space for the municipal economic and social development plan. There was no analysis and forecasts of social and economic development by urban planners while producing the City Comprehensive Plan. Rather they used data and information from the Five-Year Municipal Economic and Social Development Plan. This planning approach was very typical of urban planning practice under the centrally planned economy. The definition of the nature of the city and the definition of the city scale of the plan are other examples of planning notions from the centrally planned economy. It is impossible for urban planners to control urban development. The possibilities of inward investment and industry will change the nature and scale of the city as defined by the planners, particularly when development of the city has to depend upon inward investment, and 'development is the absolute principle' is the main policy, since these will affect the values of government. After the former Chinese urban resident registration system had been relaxed, migration from rural to urban areas was very difficult to control and to predict. Many contingencies would affect the city. For example, if the Xiamen '901' project had been implemented, with the large petrochemical industry, the proposed nature of Xiamen as a tourist and resort city would no longer exist. In addition, the proposed population of the last City Comprehensive Plan (1988) of 1.6 million by 2020 would have been achieved in a short time with the project.

The approach to spatial allocation for designated development and schemes decided in the municipal economic and social development plan was also implemented by the centrally planned economy. The urban planners of the city comprehensive planning team tried to follow the Five-Year Plan, and allocate the available land use to the proposed development.

'The Ninth Municipal Economic and Social Development Plan of Xiamen' and 'Social and Economic Development Plan to the Year 2010' (1995) formulated by Xiamen Municipal Development and Planning Commission (The Economic Development Planning Authority), provided basic development policies for the City Comprehensive Plan of Xiamen. The City Comprehensive Plan of Xiamen used the data and prediction of these social and economic plans to forecast the GDP annual increase as 20 per cent, the population increase as 5.6 per cent annually between 1995 and 2000, and 15 per cent GDP annual increase and 6.1 per cent annual population increase between 2000 and 2010. The prediction of total population is explained in Table 5.1.

Table 5.1   Prediction of population and urbanization in Xiamen (in thousands)

	Existing population	Plan predicted population	
	1994	2000	2010
Temporary population and migration	322	415	590
Urban registered population	458.6	610	900
Actual urban population*	780.6	1,025	1,490
Urbanization (%)	70.6	81.3	90.1

Note: * The actual urban population refers to those residents who actually live in the urban area. It comes from both the 'temporary population and migration' and 'urban registered population'.
Source: 'Social and Economic Development Plan to the Year 2010' (1995).

The two plans provided general social and economic development policies for Xiamen municipality. The two plans dealt with the following aspects, for which the City Comprehensive Plan of Xiamen should allocate available land for development:

- Agricultural development should continue to service the urban residents and produce export oriented products with the target of a 'High Quality, High Production and High Efficiency' modern export oriented agriculture system.
- For the secondary industry, the economic planners suggested that the existing industry needed adjustment and optimization to improve technological standards, and to establish a technological and capital intensive industrial system, which would be export oriented. The future development tendency should depend on the mechanical, electronic, petrochemical and electricity industries, the four dominant industries, as well as the construction industry.
- In the tertiary industry, eight systems were proposed. They included a market system open to home and abroad; a financial system with multiple channels of resources and diversity; a transport system with sea, air and road networks; a telecommunication system with the latest technologies and good services; a tourist system with local characteristics, and large

scale and high standard services; a scientific research and education system, one of the best in China; an information consulting service system with convenient, complete and global linkages; and a resident service and social security system to assist the local residents to enjoy a better life.

In order to allocate land resources and to deliver the proposed economic development designated in the two social and economic plans, the urban planners divided the whole municipality into four economic zones according to geographic, social and economic characteristics. While allocating land for the spatial layout of development, the main issues of concern would be the existing capability of infrastructures, the existing development and land use characters of the proposed areas and their available land resources. In addition, the proposed development programme and short-term construction schemes decided by the various authorities were important matters for consideration while allocating land resources. Economic activities were specified in each zone with integration between social economic development and spatial development in the Comprehensive Plan (see Table 5.2).

Table 5.2   Four zones and their economic activities

Zones	Functions	Area (km^2)	Population	Density (person/km^2)
South Economic Zone – Xiamen Island	Industry, trade, finance, tourism	131	414,160	3,188
West Economic Zone – Urban Expansion	Industry, trade, finance, tourism, transport	423.77	226,786	535
North Economic Zone	Agriculture, industry, trade	702.76	275,417	329
East Economic Zone	Fishing, industry, trade, tourism	375.79	258,571	688

Source: CAUPD (1995).

As regards the targets of the plan formulation, during interviews and focus group discussions planners indicated that the urban macro plan formulation should achieve the social and economic development objectives of local government. This was also clearly defined as the targets of the City Comprehensive Plan (CAUPD 1995). The rational planning approach would then be the most appropriate methodology for the planning process within the political, social and economic context.

During the formulation of the City Comprehensive Plan of Xiamen, through discussion with local politicians and analysis of Xiamen's social, economic and environmental context, the planners of the planning team made the conclusion that there were two major tasks in formulating this plan. One was to cope with the interaction between different development strategies and scenarios. It was considered that the plan should provide for the availability of infrastructure during the rapid economic growth while trying to reduce the risks of investment. The second task was to select, understand and decide the relevant development issues, i.e. spatial development phasing, structure, direction, to create a physical spatial plan to satisfy different development possibilities between the slowest and the fastest economic development speeds. To achieve the tasks, the rational planning approach was applied.

Using the rational planning approach, it was believed that planning could function more appropriately and powerfully and achieve comprehensive aims as a consequence. While producing the City Comprehensive Plan of Xiamen, the planners hoped that they could control the events of the development in an orderly, systematic and comprehensive process. The planners tried to seek the most effective course of action to achieve the objectives. Figure 5.5 shows a typical City Comprehensive Plan formulation's process. The planners first established and defined the goals and objectives. They then selected and evaluated alternative schemes, which they thought were reasonable to reach the proposed goals and objectives. There were many possible ways. Through analysis, e.g. cost and benefit, and others, the appropriate scheme could be decided. The selected course of action would then be proposed to be carried out.

In the City Comprehensive Plan of Xiamen 1995 (CAUPD 1995), it was suggested that the nature of the city should be defined as a 'seaport, scenic city, and one of the central cities along the Southeast coast of China'. It further commented that while maintaining the 'seaport' nature, it was necessary both to expand the scale of the seaport, and to develop the airport and cyber port, due to rapid recent growth.

Communication within authorities – a coordinated planning approach A City Comprehensive Plan relied on the mediation and coordination skills of the urban planners. Achievement of mediation and coordination would require a compromise solution. It was clearly illustrated in the field research that communication and negotiation with government bureaus had been the main tasks in the planning process. A City Comprehensive Plan sought to try to coordinate all the programmes and resources of different bureaus to ensure the possibility of plan implementation, since planners lacked the power and resources to implement the plan. The coordinated planning approach was executed as a stage in a City Comprehensive Plan.

In the City Comprehensive Plan of Xiamen production period, the headquarters of the local government usually asked all governmental bureaus to assist and to provide all necessary data and information to the planning team. However, the bureaus within the local government normally had their own development plans and programme, leading to inevitable conflicts among various sector plans. There were usually discussions and negotiations between the urban planning bureau and

Statutory City Comprehensive Plan

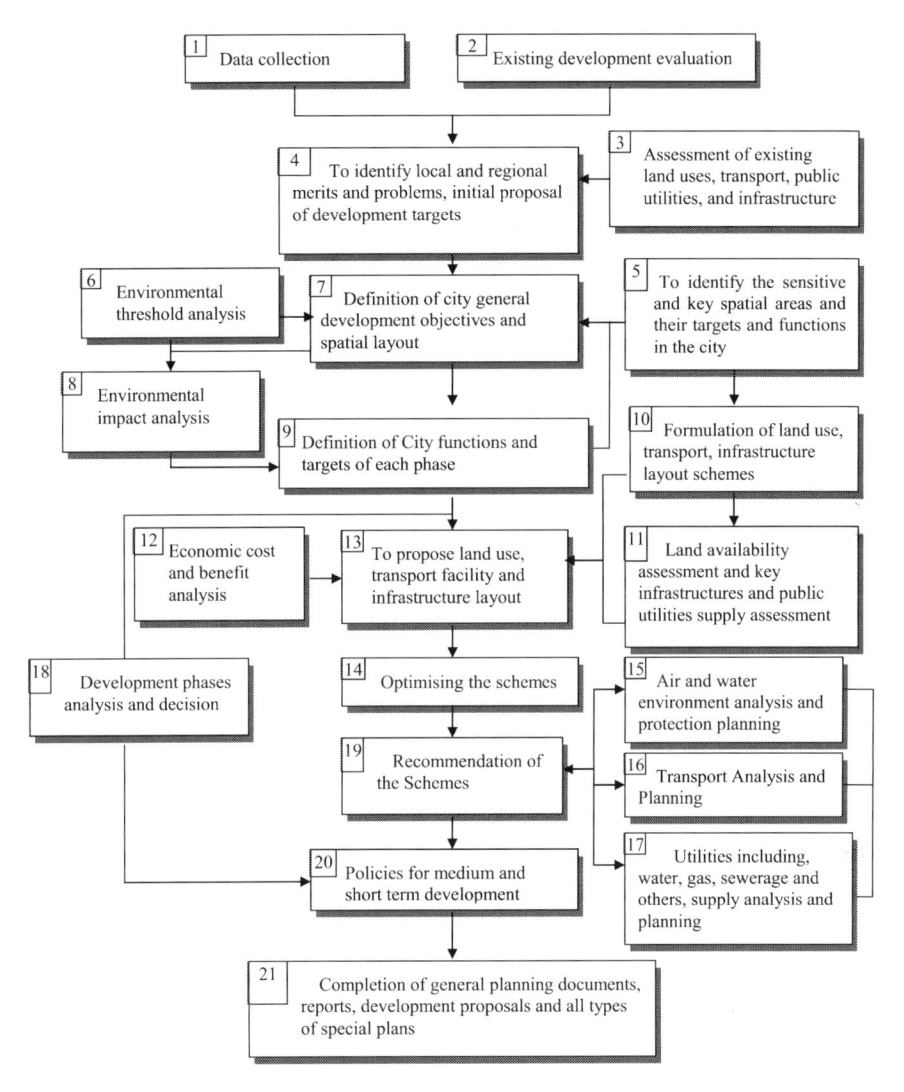

Figure 5.5 City Comprehensive Plan formulation process
Source: Summarized and designed by author from the survey.

other bureaus if there were any conflicts. Both sides tried to persuade the other to accept its scheme. Generally speaking, urban planners found it very difficult to persuade other bureaus of the merits of these schemes, but had to compromise, especially when the funding of certain projects had been decided. This situation often happened when the investment was made by the government. The conflicts were created by the sectoral administrative system. The critical problem is that urban planning has not the authority to control investment projects. Urban planners have to discuss and negotiate to ask these authorities to coordinate with land use proposals in a City Comprehensive Plan. It is sometimes very difficult since these authorities view the construction and development of their projects from either their professional aspects, or from their interests as stakeholders in development. The location selection of the electric power station in Songyu was such a case in the formulation of the Xiamen City Comprehensive Plan. The electricity authority suggested the power station should be located at Songyu, which is near to the seaside of Haicang. After construction of a special dock, coal could be conveniently discharged for the power station. However, urban planners strongly objected to this proposal because Songyu is opposite to Gulanyu, a national tourist attraction, and development of a power station would greatly affect the beauty of Gulanyu as a tourist attraction. Urban planners suggested another place for the proposed power station. The electricity authority refused to change its development scheme. The discussion and debate of site selection of power station took a long time. Debates and arguments between the electricity supply field and the urban planning profession were serious. Xiamen had never been in a city without a shortage of electricity supply. Economic development needs made this power station necessary. Urban planners failed to persuade mayors of Xiamen and electricity authority to relocate the power station to another place.

Short-term infrastructure development projects from 1995 to 2000 with different investment channels from various authorities were then included in the City Comprehensive Plan of Xiamen as its short-term plan element. They consisted of the following projects including a wide range of aspects:

- 143 million yuan investments in a sewage treatment plant in Caikou, 160 million yuan in a sewage treatment plant in Songyu, and 189 million yuan investment in a solid waste comprehensive treatment plant by the Municipal Commission of Urban Construction.
- 4.45 billion yuan investment in Haicang, Dongdu, Songyu and Xianyu docks for 20 new berths constructed by the Xiamen Bureau of Seaport Administration and foreign direct investment.
- The electricity power station in Songyu divided into 2.3 billion for the first phase, to be completed by 1996; four billion for the second phase to be completed by 2000; and 21.3 billion for the third phase; and the construction of sub-power station projects by the Xiamen Electricity Corp.
- Water supply projects consisting of the second phase of Xiamen Island of 2.1 billion yuan investments, and 78 million yuan investments in Jimei by the Xiamen Water Supply Company.
- 405 million yuan investments in Xiamen Airport's expansion for a capacity of 8.95 million passengers by Xiamen International Airport Corp.
- 2.4 billion yuan for telecommunication expansion by the Xiamen Bureau of Telecommunication.
- 4.1 billion yuan investments in a gas supply project by the Xiamen Bureau of Public Utilities.

- 160 million yuan investment in the Malan Dye improvement project.
- Several railway development projects, e.g. Xiamen railway station expansion, Xinlin cargo railway station expansion, and the North cargo railway station expansion by Xiamen Railway Bureau, direct investments by central government.
- Several road and bridge construction projects, e.g. 2.7 billion yuan for Haicang Bridge comprising a six-lane dual carriageway of 5.5 km; 500 million yuan investment in Xiamen section of Xiamen–Zhangzhou Motorway; 130 million yuan investment in the access to Xiamen–Quanzhou Motorway, and around 2.22 billion yuan investment in the construction on city roads in Xiamen Island, and 700 million yuan to establish the road network in Haicang by the Xiamen Commission of Transport, and Xiamen Road and Bridge Construction Corp. (Note: the exchange rate in 1995 was approx. 1 pound sterling = 11.5 yuan.)

These investment projects would no doubt greatly affect the decisions of planners in finalizing their proposals for the spatial layout of development in the City Comprehensive Plan.

Public participation in plan formulation is only at its beginning. There was no defined stage of public participation in urban planning. Communication in plan formulation mainly occurred among government authorities. There was no communication with other stakeholders, e.g. developers, investors and local communities. However, in the production phase of the City Comprehensive Plan of Xiamen, after withdrawal of the '901' project by the Taiwanese businessman, other international investors began to negotiate with the local municipality, local planners participated in the process of negotiation with investors regarding investment. A pragmatic approach was applied. Communication with international investors emerged.

Pragmatic planning approach in operation It is generally agreed that the poor performance of planning results from the length of the process and the degree of detail of the plan formulation methodology. Nevertheless, there are some exceptions. There exist some possibilities of alternatives to avoid poor performance in practice under certain situations in China. The possibility of applying different planning approaches comes from the loose wording in the City Planning Act 1989 and government ordinances of plan formulation. There is no strict definition of the planning approach to be used. Moreover, the City Planning Act 1989 encourages the introduction of new technologies in the urban planning process. The City Planning Act 1989 (NPC 1989a) only regulated that the plan should be in line with the specific conditions of China (Article 5). Planning approaches for plan formulation will then depend upon the understanding of planning concepts by planners. It provides discretion for planners to use different methods to deal with development issues, e.g. uncertainties in the transition to the market economy. The City Comprehensive Plan of Xiamen illustrates such an example. A pragmatic planning approach was utilized in the City Comprehensive Plan of Xiamen. Nonetheless, it is not a commonly accepted approach in the Chinese context yet, though there may be other places with similar experiences. Thus, this case study pointed to a gap and contradiction between the planning system and the actual planning practice in China. With pressure from local politicians and developers, planners have had to find a compromise method, otherwise, the existing planning would not function well or even be useless in the market economy.

It was agreed by local politicians and planners that the majority (at least 80 per cent) of the development followed the general principles, policies and land use structure proposed in the City Comprehensive Plan of Xiamen. The plan has played an important role in controlling the major land use. The road network has been constructed and completed according to the proposals in the plan. The city spatial layout proposed in the plan has been generally followed in the development. Two large sized land use functions have been changed to the northeast of the city. One area, Fanghu, has changed its land use function to housing from an initial proposal for industry in the plan, whilst Xianghou has changed from housing to industry. There are also some changes in various land plots, which depart from the initial land-use proposal in the plan. These changes are typically from proposed green land to profitable commercial land uses of retail, housing and office blocks. Nevertheless, these changes were agreed and decided by government during the process of land leasing to private developers. They were not changes without planning permission, but were approved by the municipal bureau of urban planning administration. These changes are predominantly within the areas near to the Convention Centre. According to the City Comprehensive Plan, this area is defined as public greening and tourist land use. After the construction of the Convention Centre, particularly as the area has perfect landscaping of a mountain at the back and the sea in front, the area has attracted many real estate developers' interests in luxury real estate development. An increase in tourism in Xiamen, particularly since the implementation of tourism economy and encouragement of 'long holidays' by the central government, the development of tourist facilities, and a high standard of housing have been boosted in the area. Private developers undoubtedly follow this up.

From the Xiamen case, it was surprising to find that the achievement of the majority of policies and proposals of the City Comprehensive Plan of Xiamen came from the pragmatic project oriented planning approach. Although the cancellation of the proposed investment by the Taiwanese businessman in the petrochemical industry in Haicang had impacted on the preliminary development objectives, the project proposal had always provided a marketing opportunity for Xiamen. Additional investors from Southeast Asia had come to negotiate with the local municipal government. Urban planners, particularly the manager of the project team, had often been asked by the municipal government to participate in discussions and negotiations with major investors and developers. He provided advice to mayors and other senior politicians of the municipal government when there was any proposal for an inward investment project. During the negotiation and bargaining process, planners explained the available land and supporting infrastructures associated with the land for proposed development projects. If there were any changes or proposals raised by investors regarding land use, or location, the plan was amended immediately. The available land for proposed projects would then be allocated in the urban plan.

After discussion and negotiation, the plan would be revised according to the agreement between investors and the municipal government. This is a very typical pragmatic project oriented planning process. The principle underpinning this process is that urban planners make planning policies and development schemes according to their beliefs. Of course, what they do and what they believe should not contradict laws or regulations. Urban planning practice has been greatly influenced by planners' understanding of planning and their discretion. Moreover, it was the devolution of investment and development power to the municipal governments in the transition to the market economy, and competition among cities that forced the municipalities to apply the notion of entrepreneurialism. This requires planning to be flexible, project oriented, and to consider the demands of the market.

*Proposed Development Corridors for Different Functions
to Address Uncertainties*

Regardless of the fact that the pragmatic planning approach could be used for rapid economic development and urbanization with many social and economic changes, it is important to have planning rationality as the basic concept of urban planners for development plan formulation. Faludi (1986) suggested that the significance of flexibility is that it functions as a bridge for the gap between uncertainty during plan making and the actual situations in which ensuing decisions will be taken. A plan should be of a loose framework to be filled in later. However, it is not enough to be merely a loose framework, detailed research should be undertaken by urban planners to find all the potential capabilities of the city. It will then be the task of urban planners to negotiate, to bargain through a positive role to promote the development with realistic, affordable and sustainable methods. With full research on the potential capabilities, especially the city's natural resources and ecological environment during the planning process, planners could act as 'gatekeepers' to avoid the probable development schemes of others damaging sustainability. This is particularly important in the competing developing world when development is regarded as a priority and absolute principle to override other issues, including equity and sustainability.

Since there are so many uncertainties in the market economy, and impacts from central government, as well as competition from other cities, especially the eastern coastal cities in China, it is very difficult to forecast what kinds of investment and industries may come to Xiamen. It is exceptionally difficult to predict changes during rapid economic development in a developing country like China. It is the situation of Christensen's (1985) prototype of 'unknown technology, no agreed goal'. As suggested by Christensen (1985), it is a situation in chaos. Planning in this case is trying to establish order to clarify the agreed and clear goals, or workable technology.

As explained by the project manager of the City Comprehensive Plan of Xiamen 1995:

> the initial reason for formulation of Xiamen plan was changed and adjusted before the completion of the plan because of cancellation of the "901 project". We did not know who would come to invest and what type of investment would come. What we could do was to provide various kinds of schemes depending upon the local characteristics to meet the demands of inward investment. (Interviewed on 3 July 2003)

Under such a situation, the planners of the project team suggested nine development corridors with different land use functions for diversity of future economic development initiated by the different kinds of inward investments. No matter what type of inward investment came to Xiamen, there would always be one possible corridor ready to allocate potential investment. In their view of urban planning, a location is ready for any investment; the task of urban planners is to distribute land use and space. This illustrates a planning concept in China. A senior planning officer from the Ministry of Housing and Urban-Rural Development held a similar view. It is his view that:

> a City Comprehensive Plan is the general means to assign and comprehensively coordinate resources, typically spatial resources within urban areas. This basic function of urban planning cannot be carried out by any other kind of profession, or other non-urban planning authorities. (Interviewed on 15 July 2003)

The features of the development corridor were designated according to the existing land use, infrastructure provision and potential future development. This is a compromised planning approach of the centrally planned economy and rational planning. Urban planners did not consider whether investors would accept and follow the proposed corridors. The nine corridors are summarized below:

1. Ring Island Road with a Scenic and Tourist Development Corridor; this corridor links different scenic attractions and interesting tourist areas within Xiamen Island.
2. Bingnan Retail and Business Development Corridor; this corridor would set up retail, business and other functions in the city centre.
3. Bingbei Public Buildings Development Corridor; this corridor connects the municipal government administrative centre, some national banks' buildings, and recreation and sport facilities.
4. Huli Hi-tech Industrial Development Corridor; this starts from the High-tech Industrial Park to form a belt of technology intensive, high added

value, small scale land use, low transport demand, energy efficient, and non-pollution industry; it is the only industrial area within Xiamen Island.
5. Tongji Tourist and Industrial Development Corridor; the corridor starts from Jimei to Tong'an town; tourist interests are mainly concentrated in Jimei; water efficient and low polluting industrial development with labour intensive but high skilled labour demands would be allocated in the corridor.
6. Jiguan Industrial Development Corridor; this corridor, which is not too far away from the centre, would include industry of the labour intensive kind, but which also needs close links with a railway and motorway for transport purposes (utilization of North Cargo Railway Station, and convenient access to the motorway).
7. Maluan Industrial Development Corridor; this would be concentrated in the existing Xinyang Industrial Zone available for those large enterprises needing large areas of land.
8. Haicang Industrial Development Corridor; this is the largest existing industrial development zone in Xiamen; the features of the industrial zone would be able to meet the demands of large scale industries, requiring large demands for ship transport, large energy consumption.
9. Songyu Infrastructure Development Corridor; the corridor starts from Songyu, and regionally significant large infrastructures, i.e. electricity power station, sewage and drainage treatment plants, seaports and railway station, would locate along the corridor.

The proposed nine development corridors provided a planning framework for Xiamen municipal government to undertake infrastructure construction according to the features of each corridor. Through infrastructure provision, development of the city and the spatial layout might be led by the planning policies and schemes. In this case, the rationality of urban planning can make a contribution to development, such as seafronts along the 'Ring Island Road with Scenic and Tourist Development Corridor' and 'Haicang Industrial Development Corridor'. There was neither proposed industrial development, nor necessary infrastructure provision by government to encourage industrial investment in the corridor of 'Ring Island Road with Scenic and Tourist Development Corridor'. Rather the City Comprehensive Plan proposed the location of industrial development, especially heavy industrial development, along the 'Haicang Industrial Development Corridor', where government had invested in providing necessary transport facilities and other infrastructures. This demonstrates that the planning polices of government guide the development process of infrastructures.

The nine corridors radiate inward from the west coast of Xiamen towards hinterland areas. The concepts underlying the proposal of these nine corridors are to cover as many different types of development in Xiamen as possible depending upon the local characteristics and potential development direction. Xiamen could then construct along the different corridors depending upon different time periods, different milieus, diverse economic development stages, for different functions and natures of the city. The planning idea seeks to avoid conflicts among the separate developments and to provide as many development opportunities as possible.

Conclusion

Statutory City Comprehensive Plan has been explored and analysed in this chapter. It is found that planning in Xiamen has been a learning process in the transition from the centrally planned economy to the market economy, from clear decisions and investment in development projects by the government to uncertainties in the market.

It is evident from the case study that the traditional city comprehensive planning process has been influenced by the notions of centrally planned economy and rational planning approach. These influences are illustrated in the people's concepts and understanding of urban planning, the planning procedure, and the usage of norms and standards in the plan's formulation. There was evidence that the urban plan had to follow the policies of the Municipal Social and Economic Development Plan. This is a typical concept of urban planning in the centrally planned economy. Nevertheless, the market had taught the local municipal government and urban planners the lesson that development was difficult to predict and uncertain. It was difficult for planners to control. There were many uncertainties and changes for the future caused by different aspects. Under pressure of competition, urban planners, driven by the local politicians, had to produce the plan to meet the demands of investors and the market by a pragmatic planning approach with the feature of being project oriented. The local municipal government required inward investment. They needed planning to support this objective.

The case of Xiamen Comprehensive Plan demonstrates that the designated planning process and its contents are complicated, since social and economic changes are so rapid and uncertain. If there had not been any innovation in planning, the plan would not have been able to guide the urban development. However, the proposed policies and schemes for specific development in the urban plan may be out of date immediately after, or even before, the completion of the plan. The existing procedure of plan formulation or amendment is time consuming. A different type of urban plan is needed with greater flexibility and formulated in a short time to cope with rapid changes. This is particularly so when new concepts and ideas have been introduced but are rapidly changed as a result of science and technology development. If plan formulation were to last for more than two to three years, the latest ideas and concepts applied while formulating the plan would be out of date. The expectation that planners should be able to predict and forecast changes in the future is impossible and unreasonable.

Although the format, structure and process of plan approval have been regulated, the planning approach and methodologies have not been specified in either the City Planning Act 1989 or the City and Countryside Planning Act

2008. The application of the planning approach will be dependent on planners' discretion. The majority of urban planners are applying planning approaches of the physical master plan, or rational planning, or a mixture of both approaches, depending on their education, understanding of planning concepts and knowledge. Additional methods and technologies can be used in planning practice. Generally speaking, there are also some innovations in Chinese planning approaches. The case study of Xiamen illustrates that Chinese planning has gone further to meet market demand with a pragmatic planning approach owing to pressure of competition in the market. The application of the pragmatic planning approach in Xiamen seems to work properly in the situation of rapid change and competition. However, both local government and planners may simply consider interests of developers and investors. It is critical to consider equal opportunity for all and protection of the local community's interests, particularly those of disadvantaged groups, when applying this approach.

Chapter 6
Statutory Regulatory Detailed Plan and Planning Norms

Introduction

A regulatory detailed planning approach was initiated by local governments to deal with the demands of controlling development since the delivery of land use reform in the process of China's transition to the market economy in urban areas. It is the view of some Chinese academics that a Regulatory Detailed Plan is a main tool in planning management by local planning authorities to implement the City Comprehensive Plan, to protect public interests and to coordinate different interest of stakeholders (Li 1999; Li et al. 2009). It is an important means of public policy. This planning approach has been used to guide and facilitate the achievement of an ideal society by controlling development (Shao and Duan 2010) and distributing resources to ensure the social and economic intention, and public welfare of a society (Yu et al. 2010). It has been used to secure and coordinate the interests of stakeholders through a clarified management and technical process (ibid.). Meanwhile, a Regulatory Detailed Plan has been and will continue to be an important tool for local entrepreneurism. It is regarded as the main reference to decide the price of land use rights lease (Yuan and Hu 2010).

Zoning ordinances of the United States were used as a reference of land use control for producing a Regulatory Detailed Plan at its beginning experimental stage in the 1980s and 1990s (Dong 1992; Wang and Zhou 2000; Yang 1993; Zhao and Le 2009). After more than two decades in practice, the Regulatory Detailed Plan is now defined as one of the statutory city and countryside plans in addition to the National Integrated Urban System Plan, Provincial Integrated Urban System Plan and City Comprehensive Plan according to the City and Countryside Planning Act 2008. This chapter explores and analyses its initial approach of plan making and its relations with the City Comprehensive Plan and planning management (development control). Several national and professional planning norms and standards, which are the main materials that should be used when defining relevant controlling indexes, codes and ordinances of land use in a Regulatory Detailed Plan, are discussed. These norms and standards are important in the Chinese city and regional planning system. They should not be ignored.

Initiation of Regulatory Detailed Plan

In the later 1980s, the market economy had already introduced and impacted Chinese economic development due to China's open door policy, which was mainly the countries of the market economy for the attraction of inward investment. The influences of the market economy stimulated the former free land use system that had been in operation for decades in the centrally planned economy for amendment and reform. The transition to a chargeable land use system by leasing land use rights started its operation in the year of 1989 (Xie et al. 2002; Zhu 2004). The reform of land use system and application of the market economy created different stakeholders in the development process. These changes required an innovative approach in urban planning to establish a mechanism that can regulate different responsibilities, benefits and accountabilities of various stakeholders in development process while protecting the public interests. The former detailed planning approach, which was formed in the centrally planned economy when all investment, capital and resources were controlled and allocated by the central government, was inevitably unable to cope with the new phenomenon of the market economy. The Regulatory Detailed Plan was initiated as the consequence of these changes and an innovative mechanism to cope with changing land use system and delivery of the market economy in China.

Experimental Practices of Regulatory Detailed Plans in Shanghai, Guangzhou and Wenzhou

Soon after delivery of open door and reform in China since the late 1970s, the cities in the east and the southeast areas of China, which were first opened to outside world, have been challenged by the pressures of demands for land use leasing rights to private developers who were from other countries. Chinese local governments also confronted new phenomena creating by the delivery of the market economy, e.g. the appearance of different interested groups, profit seeking by private developers and approaches of protecting public interests in the market economy. These were all new issues that Chinese local governments had never met but had to address since the founding of People's Republic of China. The local governments in the east and southeast of China had to find an appropriate solution to cope with these changes. Innovative approach in urban planning to deal with the new challenges was sought by local municipal governments instead of the usual top-down system from the central government.

In 1982, Shanghai Municipal Institute of Urban Planning and Design prepared a plan to regulate land use in the Hongqiao Development Zone, an area that contained several consulates of different countries. The land had been leased instead of allocated to the consulates. The principles of zoning ordinances from the United States were adopted in the plan in order to provide guidelines for international developers to undertake construction in the zone. However, it was the first plan that had included controlling codes and ordinances for land use in

China. The relevant controlling code and ordinances defined in the plan were decided according to a sample survey of different types of land use and capability of construction volume on a piece of land at different periods in the history of Shanghai. According to the survey and data analysis, the plan defined regulatory indexes, e.g. the land use function allowed the building area, plot ratio, building density, buildings set-back line, height of buildings, access road for vehicles and pedestrians, and parking spaces. This plan attracted great interest from Chinese professional planners when it was presented at the National Symposium of Planning and Design Experiences in June 1986.

In 1987, in order to deliver the City Comprehensive Plan, Guangzhou municipal government entrusted the local planning institute to prepare a so-called 'Neighbourhood Block Plan', which covered an area of 70 square kilometres in the city centre. The objective of the plan was to function as a planning tool to guide and control urban rehabilitation and development in the old city centre. The contents of the plan were very similar to that of zoning ordinance. Through surveying and analysing the existing development situation and information, the plan identified six types of planning control indicators for each neighbourhood block within the designated 70 square kilometres. These six types of indicator included: (1) indicators of land use function for each plot of land in a neighbourhood block; (2) indicators of road network; (3) indicator for public facilities and service facilities that should be provided according to the definitions of City Comprehensive Plan; (4) indicators for population, greening areas and provision of public facilities; (5) controlling indicators of plot ratio, density and distance between buildings; (6) other indicators for control guidance and ordinances, e.g. protected historical buildings, or any items of historical and cultural interest.

The significance of Guangzhou's experience is that two by-laws of 'Approaches of City Planning Management for Guangzhou' and 'Detailed Regulations for the Plan Delivery' were issued by the Guangzhou municipal government after completing the formulation of the 'Neighbourhood Block Plan'. However, the most significant event in the evolution of the Regulatory Detailed Plan was the preparation of the 'Regulatory Plan for Old Town Renewal' by Wenzhou municipal bureau of planning administration in 1989. This plan was regarded as a milestone of the Regulatory Detailed Plan in the Chinese planning system.

Wenzhou has a reputation for its private business and the market economy oriented development. The market economy has actively played an important role in local business even in the era of the centrally planned economy. This phenomenon was very unusual in China. Although activities of the market economy were forbidden and controlled strictly by the Chinese government in the whole of China, they had never totally stopped in Wenzhou. The so-called 'black market', which means the activities of the market economy operating in an illegal and hidden approach, was popular in Wenzhou even during the Cultural Revolution. Since the implementation of the reform and open door policy in China in the late 1970s, the activities of the market economy in Wenzhou had been resumed and developed much more quickly, and flourished, compared to other places in China, especially at the beginning of Chinese reform. Privately owned houses in Wenzhou had survived the Cultural Revolution. Some housing repairs were needed after they had been ignored for several decades. Some Wenzhou citizens were intending to operate private real estate development, or private business at their private houses. The rapid development of private businesses especially in real estate in Wenzhou created conflicts of different interests among local residents. This situation forced the local government to consider the application of appropriate approaches to tackle these conflicts. At the same time, the requirements of housing repairs by local residents in Wenzhou old town also impelled the local municipal government to adopt an innovative mechanism.

In order to deal with the conflicts and to regulate the process of old town rehabilitation, a Regulatory Detailed Plan for urban neighbourhoods in the old town was proposed by the local municipal government at the beginning of 1988. After completion of the plan formulation, it was formally approved by Wenzhou municipal government by the end of 1988. The contents of the plan consisted of a local government 'Circular for Rehabilitation Plan of Wenzhou City Centre'; a written document of 'Regulatory Plan of Neighbourhood Rehabilitation in City Centre', and varies special plans, ordinances and controlling indexes, e.g. codes of land use plan, density of buildings, plot ratio, allocation of roads, infrastructure of water supply, sewage, electricity, telecommunication. Two by-laws of 'Trial Regulation of Planning Management (Control) of Neighbourhood Rehabilitation in City Centre' and 'Technical Regulations of Land Use and Building Management of Neighbourhood Rehabilitation in City Centre' were also promulgated by Wenzhou municipal government (Yang 1989).

Before the approval of the plan by the local municipal government, there was a public consultation with the local citizens. The plan proposed different approaches to control various functions of land use by definition of categories. For example, buildings of historical and cultural interest, parks and green areas, seaport and blocks that required unitary repairing and improving were defined as the Category One. Category Two included the land uses along the city distributor roads or main roads, public buildings, and public utilities. Other neighbourhoods and land use functions were listed as Category Three. All land plots within the old town were designated by three types:

- available for development;
- not suitable for development; and
- available for development with certain conditions.

Comprehensive planning control indexes were established for planning management.

This innovative planning approach of the Regulatory Detailed Plan in Wenzhou and other cities for trial experiments was introduced to cope with changes in the process of transition to the market economy from the centrally planned economy. The significance of the Wenzhou experience was that it was the first time representatives of local citizens were invited to be involved in the plan producing process as a new mechanism of public participation in planning which had never happened before.

The whole city centre of Wenzhou was divided into three different levels of density district for land subdivision. The regulations of land use and development were in very detailed ordinances and control indexes set. This set covered nearly all elements of land use and its development, e.g. different types of land use categories, density, plot ratio, the distance between buildings, minimized land for housing construction, and facade of buildings. The regulations of private housing development and private shops development were also specified according to the local feature of Wenzhou for conservation purposes.

Based on experiences from different cities, typically Wenzhou, the Ministry of Construction promulgated the 'Regulations of Plans Formulation' (Circular No. 12, MOC 1991) and the 'Planning Management for Leasing of State Owned Urban Land' (Circular No. 22, MOC 1992). These circulars specified the requirements of Regulatory Detailed Plan formulation and clarified the contents of a Regulatory Detailed Plan.

Contents of a Regulatory Detailed Plan

A Regulatory Detailed Plan, as a tool for planning management and development control, should follow the policies decided by a City Comprehensive Plan. It should regulate the development of a specific piece of land by interpreting policies of a City Comprehensive Plan through detailed indexes and codes, ordinances, and plans for land use and buildings.

It is regulated by the City and Countryside Planning Act 2008 (NPC 2008) that a Regulatory Detailed Plan should function as the material for consideration for issuing planning permits, and the state-owned land right lease. From this perspective, the functions of a Regulatory Detailed Plan should consist of the contents that are available for technical regulations, planning management, land management and project management (Zhou 2012).

According to 'Regulations of City Plan Formulation' (Circular No. 149 of Ministry of Construction, 2005), the contents of a regulatory detailed plan should cover the following items:

1. Boundary of different land uses within a development plot, specification of suitable or inappropriate building types.
2. Specification of codes and indexes, and ordinances for buildings and development, e.g. height of buildings, density, plot ratio (total building floor space/total floor space of land plot), greening ratio (total greening space/total space of land plot), description of required public facilities, distance between buildings, building setback and required accompanying infrastructure.

3. Urban design guidance, e.g. volume of construction, architecture style and colours, and others.
4. Definition of access of road and traffic arrangement, parking spaces, land use and location of bus stop, detailed section of road, types of junction and its linear, pedestrian and other traffic modes according to traffic demand analysis.
5. Definition of location, diameter of pipe and boundary of engineering work, as well as demands of use of underground space if necessary according to the volume of development and construction.
6. Specification of relevant codes for land use.
7. Regulations for application of incentive plot ratio and compensation.
8. List of indexes of land uses control.

All the indexes can be divided into two categories. One category is the indexes with prescriptive or regulatory function. These include land use, construction density, height of the building, plot ratio, greening ratio, location of transport access, parking space and required accompanying public facilities. The other category is for discretionary indexes. They can be population density (person/hectare), styles, patterns and colours of buildings, and other environmental related standards.

Outcomes of a Regulatory Detailed Plan should have a written document, plans (and maps), a set of tables of indexes and the appendixes that consist of the explanation of planning policies, basic data and information, and the reports of pre-studies.

The set of indexes normally includes a written explanation, tables of indexes and maps. Indications of roads, defined land use functions, controlling codes of land plot, boundary of each plot, setback of buildings, location of traffic access points, and pedestrian areas should be clearly illustrated on the plan (see Figure 6.1). The contents of tables (see Table 6.1) should describe regulatory indexes of plot ratio, greening ratio, height of buildings, total area of each plot, floor space of buildings, population, distance between buildings and parking spaces. The written documents should define the total volume of land use and development policies of the planning area.

The important role of a Regulatory Detailed Plan in the process of planning permits and leasing the state-owned land, typically as an important planning document that associates with the leasing contract of the state-owned land, the Regulatory Detailed Plan is a powerful legally binding document, and a type of development 'contract' between local government and the project developer (or investor) (Yuan and Hu 2010). The contents and functions of the Regulatory Detailed Plan have been discussed and debated widely in China.

The common criticisms of the Regulatory Detailed Plan in Chinese literature is that it is too strict, complicated and detailed without the necessary resiliency to cope with the uncertainties in the process of the plan delivery (Lu 2009; Xu et al. 2011; Yao 2007; Yu et al. 2010). As a consequence, the Regulatory Detailed Plan

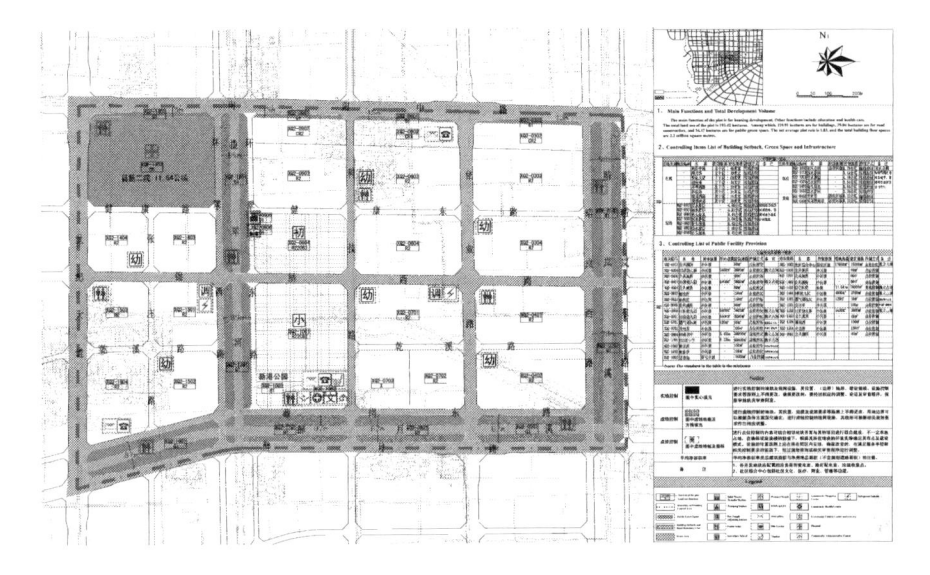

Figure 6.1 Regulatory Detailed Plan
Source: Hefei Huaxiang Planning and Architecture Design Ltd, 2011.

may have to be revised regularly (Yao 2007). Some researchers (Lu 2009; Yao 2007) comment that the contradictions between the rigid control of development and uncertainties in the market economy will make planning management and plan delivery more difficult.

Nevertheless, it is argued by Yuan and Hu (2010) that as a public policy, a Regulatory Detailed Plan should adopt different controlling policies for different development areas or different development stages of a city. It should be resilient in the newly developing area to consider the principle of effectiveness, but more strict in other sensitive urban built-up areas.

It is the view of Yu and his colleagues (2010) that the outcomes of a Regulatory Detailed Plan are unable to fit the dynamic feature in urban development, typically since Regulatory Detailed Plans are required to cover the whole planning areas of a city (or a town). They should also target a development period of 20 years. The complicated and detailed contents of a Regulatory Detailed Plan inevitably create contradictions to the uncertainties and different interests of stakeholders in the market economy.

Formulation Approaches of a Regulatory Detailed Plan

When preparing a Regulatory Detailed Plan, it is necessary to review the planning policies and defined land use function of a City Comprehensive Plan on a specific plot of land or an area. It is also necessary to review its neighbouring plots of

Table 6.1 Land use planning control index in a Regulatory Detailed Plan

Land plot	Land use		Building	Housing within the area	Building			Green land	Plot ratio	Housing		Parking space
		Area	Floor space	Floor space	Density	Height control		Ratio	Ratio	Household	Person	
No.	Function	Ha.	M²	M²	(%)	Metres		(%)	(%)	No.	No.	No.
01A-1	G12	0.84						90				
01B-1	U11	24.11						40				
01B-2	U11	1.00				9		15				
02A-1	G22	1.03						90				
02B-1	R21	3.68	55,200	55,200	25	24-		30	1.5	552	1,932	276
02B-2	R24	1.55	500					90				(646)
02B-3	R21	4.94	74,100	74,100	25	24		30	1.5	741	2,594	370
05A-1	R21	2.56	33,280	33,280	28	18		30	1.3	277	970	138
05A-2	R22	0.35	3,000		45	9		15	0.85			10
05A-3	R14	1.10	500					90				(322)
05A-4	R21	3.39	44,070	44,070	28	18		15	1.3	368	1,285	185
05B	G22	1.22						90				
06A-1	R12	1.02	4,000		15	18		35	0.4			3
06A-2	R11	4.44	46,620	46,620	30	15		30	1.05	311	1,090	311
06A-3	R14	1.81	500					90				
06A-4	R12	0.35	2,500		30			30	0.75			3
06A-5	R11	4.84	50,820	50,820	30	15		30	1.05	339	1,190	339
06A-6	R12	0.33	2,800		45	9		15	0.85			10
06B	G12	0.58						90				
07A-1	R11	3.51	36,855	36,855	30	15		30	1.05	246	861	246

07A-2	R14	0.87	500				90		
07A-3	R12	0.31	2,635		30	9	30	0.85		3
07A-4	R11	2.71	28,455	28,455	30	15	30	1.05	190	667	190
07B	G12	0.49					90			
08A-1	R21/C21	1.01	15,150	12,650	35	24	25	1.5	362	724	181
08A-2	R21/C21	1.23	17,500	14,500	35	24	25	1.5	415	830	208
08A-3	R12	1.71	9,000		15	24	40	0.55			4
08B-1	C21	0.94	7,520		20	24	30	0.8			25
08B-2	C65	2.55	5,000		15	18	40	0.3			17
08C	G12	0.47					90			
09-1	R12	0.29	2,300		45	9	15	0.85			8
09-2	R11	4.08	42,840	42,840	30	15	30	1.05	286	1,003	286
09-3	R14	2.11	500				90			
09-4	R12	0.3	2,500		35		30				3
09-5	R11	3.62	38,010	38,010	30	15	30	1.05	253	889	253
09-6	R12	0.28	2,200		45	9	15	0.85			7

Source: Created by the author based on information provided by Hefei Huaxiang Planning and Architecture Design Ltd, 2011.

land which have already been controlled by Regulatory Detailed Plans or plans of which are under production. The required data and information for a Regulatory Detailed Plan formulation should include present land use and its defined categories specified by the National Planning Norms of Land Uses; present and forecast population and their existing and proposed locations; existing buildings and their functions; property right of the buildings; floor space, architecture quality and any significance for conservation of these buildings; existing public facilities and their location; existing infrastructure and pipelines; land economic information including price of land, different land use prices, development approaches of land; and any historical and culture protection and conservation interests.

According to the 'Regulations for Plan Formulation' (MOC 2005), the process of formulation of a Regulatory Detailed Plan should follow several stages.

Pre-studies

At this stage, the main work is to understand the higher hierarchical plans, typically the City Comprehensive Plan, as main policies to be interpreted in the process of making a Regulatory Detailed Plan. It is also required to analyse existing data, information and conditions from survey and investigation to generate the planning principles and ideas of a Regulatory Detailed Plan for a specific piece of land. Special research should be done before the plan is formally made. The research should consist of urban design policy, land economy, traffic impact analysis, conservation of historical and cultural interests, ecological environment protection, and provision of the required public facilities. The research provides basic concepts and background for the formulation of a Regulatory Detailed Plan.

Land Use Plan and Engineering Plan

Based on the results of the pre-studies, planning schemes are then able to be decided through a comprehensive way for the allocation of various land use functions, public facilities, road network, greening and landscaping, and civil engineering design. The proposed schemes should follow the policies defined by a City Comprehensive Plan. However, it is possible to make necessary adjustments according to local actual conditions.

Land Use Subdivision

After completion of planning schemes, the next stage is to subdivide the whole land plot to form various basic units for development and construction under the control of a Regulatory Detailed Plan. When making decisions on land subdivision, it is necessary to consider existing land use functions and other special plans that are attached to a City Comprehensive Plan, such as environmental plan, historical and cultural conservation plan, greening and gardening plan, and disaster prevention plan. Land price, characteristics and physical boundary of the land plot should also be investigated in the process of land subdivision. Land plots through subdivision are the controlling objects of the regulatory detailed planning.

Definition and Decision of Controlling Indexes

Establishment of controlling indexes to each land plot and its land subdivision is the core work of a Regulatory Detailed Plan. Definition of land uses should follow National Standards of 'City Land Use Categories and Building Land Use Categories' (GB137-90) that was published by the Ministry of Construction.

In the production process of a Regulatory Detailed Plan, planners should undertake a compatibility study for different land use functions and availability of public facilities provision. The definition of compatible land use functions and public facilities should be clarified in a Regulatory Detailed Plan according to local conditions and relevant studies. The definition of other required indexes is established by simulation from modelling, summary and induction of empirical work and from approaches of investigation, analysis and comparison by planners.

Modelling simulation is an approach similar to urban design. Based on the application of outcomes of pre-studies, modelling simulation provides a rough layout. Indexes and specific control ordinances are then calculated and specified by this approach.

Summary and induction from empirical work is an inductive approach from existing completed and executing Regulatory Detailed Plans and experiences of planners from their former work. The technical and economic indexes of existing plans can be used as materials to be analysed and then introduced to the plan that is being produced.

The methods of investigation, analysis and comparison are adopted closely associating with summary and induction from empirical work. The application of experiences and indexes of other Regulatory Detailed Plans has been a main approach in decision of indexes. However, the adaptation of indexes to local conditions should be analyzed and considered. It is then critical to have a detailed investigation and survey of the planning area, and to compare the data and information about the investigation of the characters of the referencing areas that the proposed controlling indexes are introduced from.

Approval and Amendment of a Regulatory Detailed Plan

It is the responsibility of the municipal (town) bureau of planning administration to organize formulation of Regulatory Detailed Plans. However, the bureau of planning administration does not undertake plan making itself but invites planning institutes or planning consultants to produce the plans. The task of the planning bureau is to manage the making of the plan.

A Regulatory Detailed Plan should be approved by the municipal (or town) government and submitted for record by higher hierarchical government and people's congress of the municipality (or town). Before submission for approval, it is regulated by 'Regulations of City Plan Formulation' (Article 26, MOC 2005) that the draft regulatory detailed plan should be noticed for public consultation. Public hearing, demonstration of plans, and other methods of public participation should be organized by the municipal (or town) bureau of planning administration. The process of public participation should not be less than 30 days. The opinions, suggestions and comments from public participation should be considered and assessed by the plan-making organization. The amendments to the plan and the reasons for the amendments, as well as rejecting opinions according to public participation, should be enclosed with the plans when submitted for approval.

When an approved Regulatory Detailed Plan needs revision because of social and economic changes or any other reasons, the municipal (town) bureau of planning administration should organize evaluating workshops among stakeholders to decide if the revision of the plan is really necessary. A report of application for plan revision should be submitted to the former plan approval government. With the approval of the application, the revision of the present Regulatory Detailed Plan is able to take place. However, if revised items of a Regulatory Detailed Plan have been specially defined as the compulsory control items (or policies) of a City (town) Comprehensive Plan, it is then necessary to revise the present City (town) Comprehensive Plan before the revision of the Regulatory Detailed Plan (Article 48, MOC 2005).

The Relationship with a City Comprehensive Plan

In city and regional planning, the most important resources that can be allocated are the rights to use or the land development rights, and different kinds of urban space relations that are established on urban land use rights (Sun 2003). The establishment of different controlling indexes decided by the Regulatory Detailed Plan is actually the intervention of local governments to control the land use rights or to grant the land users (developers) the way of development rights (Tian 2007).

In the Chinese planning system, a City Comprehensive Plan and a Provincial Integrated Urban System Plan, as the regional plan, are unable to be applied directly to control and manage urban development and construction projects. It is required to be operated through Regulatory Detailed Plans.

It is regulated by the City and Countryside Planning Act 2008 (NPC 2008) that a Regulatory Detailed Plan should follow and intensify the policies of a City Comprehensive Plan, and to interpret the policies and guidance into various detailed regulations, ordinances and codes of land use in development process. A City Comprehensive Plan consists of a city's macro development policies and guidance, which refers to population, job provision, economic and industrial development, land use, transport facilities, landscaping and environment, and infrastructures. A City Comprehensive Plan targets the whole city by defining general development tendency and layout of various systems of a city, i.e. system of hierarchic human settlements, system of river and water resource and others. A Regulatory Detailed Plan should follow the general development and land use policies, and interpret and embody these policies into detailed control indexes and ordinances to control the development of a plot of land. A Regulatory Detailed Plan includes various controlling codes, ordinances and policies of land subdivision, land use function, development intensity, provision of public facilities and infrastructures, and guidance for city design for a specific plot of land. It is used as a controlling tool for land use and its development by specifying on small scale land and defining relevant controlling indexes. For example, as regarding to population, a City Comprehensive Plan should forecast population change and its allocation within a city. A Regulatory Detailed Plan should then define development volume, environment capacity, facilities provision through development controlling codes and ordinances for a specified plot of land according to population prediction by a City Comprehensive Plan, and then breaking down to the specific area of designated land plot.

While deciding land use pattern and its function, road network, development volume by a Regulatory Detailed Plan, the development objectives defined by a City Comprehensive Plan should be interpreted. For example, in the process of urban rehabilitation or development in a city centre, where there may be plenty of historical features and high density of population in China, a City Comprehensive Plan may specify policies and objectives of rehabilitation or development as historical conservation and improvement of living quality for local residents, a Regulatory Detailed Plan should then define the controlling codes and ordinances through careful study and survey of local architecture features, their styles and quality; and specifying different categories of conservation areas and buildings as well as historical landscapes in order to operate different treatment approaches for buildings, e.g. demolishing, repairing and protecting. Regulatory Detailed Plans may decide to decentralize local residents to a certain level to provide appropriate public open space. They will also decide the provision and location of necessary public facilities, bus stops and greening areas.

However, some policies of the City Comprehensive Plan may not be able to be followed by a Regulatory Detailed Plan, such as integrated urban and rural development and urbanization. Physical oriented planning policies are generally able to provide material for consideration and being interpreted in a Regulatory Detailed Plan (see Figure 6.2).

Although a Regulatory Detailed Plan should follow the policies of a City Comprehensive Plan, there are some mismatches between these two types of plans that generate problems in operation. The mismatches appear, first of all, in the planning period. A City Comprehensive Plan usually aims at 20 years and requires review and revision every five years. It is also able to amend or revise the plan if there are any significant changes in economic development, political context or administrative boundary. A Regulatory Detailed Plan usually targets short-term development since the relevant ordinances are specific and detailed. It is difficult and unnecessary for a Regulatory Detailed Plan to target long-term development, e.g. 20 years, especially as development is uncertain. However, whenever a City

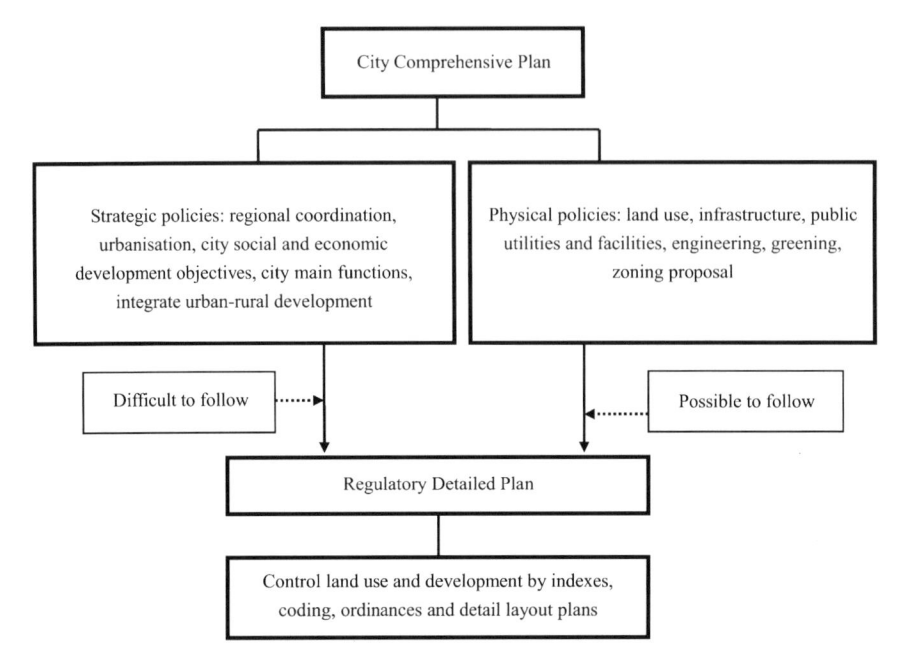

```
                    ┌─────────────────────────┐
                    │  City Comprehensive Plan │
                    └─────────────────────────┘
                         │             │
    ┌────────────────────┴──┐   ┌──────┴─────────────────────┐
    │ Strategic policies:    │   │ Physical policies: land use,│
    │ regional coordination, │   │ infrastructure, public      │
    │ urbanisation, city     │   │ utilities and facilities,   │
    │ social and economic    │   │ engineering, greening,      │
    │ development objectives,│   │ zoning proposal             │
    │ city main functions,   │   │                             │
    │ integrate urban-rural  │   │                             │
    │ development            │   │                             │
    └────────────────────────┘   └─────────────────────────────┘
              │                                │
    ┌──────────────────┐              ┌──────────────────┐
    │ Difficult to follow │ ┄┄┄▶   ◀┄┄┄ │ Possible to follow │
    └──────────────────┘              └──────────────────┘
                         ▼            ▼
                    ┌─────────────────────────┐
                    │ Regulatory Detailed Plan │
                    └─────────────────────────┘
                                │
                    ┌───────────────────────────────────┐
                    │ Control land use and development   │
                    │ by indexes, coding, ordinances     │
                    │ and detail layout plans            │
                    └───────────────────────────────────┘
```

Figure 6.2   Relation between City Comprehensive Plan and Regulatory Detailed Plan

Source: Created by the author.

Comprehensive Plan is required to be revised, the Regulatory Detailed Plan should also be amended as a consequence.

The mismatches are also illustrated in the different interpretation and understanding of planning policies of a City Comprehensive Plan. The process of intensifying and interpreting the policies of a City Comprehensive Plan by a Regulatory Detailed Plan is complicated. This can be shown in several aspects. In terms of interpreting from macro policies to micro policies, the two types of plans not only differ in the scales and technologies, but also in planners' value and applicable approaches. For example, the definition of public facilities in a City Comprehensive Plan should consider interests of all citizens of a city. It is evident that all the citizens are able to benefit from the public goods from the provision of public facilities. The macro definition and provision of public facilities might be easier to achieve consensus among citizens. However, within a specific area which is controlled by a Regulatory Detailed Plan, the definition of a public goods, and consensus in provision of a public facility is difficult to reach but often in contradiction among different stakeholders, i.e. local government, local residents, and other persons who may be impacted by the development as they may have very different understandings and requirements for public goods and public facilities.

A critical issue in the relationship between a City Comprehensive Plan and a Regulatory Detailed Plan is balance between local interests and general interests of a city as a whole. It is usual to focus on local interest and development targets by ignoring general targets of a city during the planning process, especially when discussing and consulting with local residents and developers. Before the City and Countryside Planning Act 2008, a Regulatory Detailed Plan is often led by project-oriented approach in China. It is common to formulate a Regulatory Detailed Plan only when the development project and investment for a specific plot of land have been confirmed. The specified ordinances and controlling codes of a Regulatory Detailed Plan often ignore necessary coordination and association with neighbouring plots of land. It sometimes happens that controlling codes and ordinances, even development policies between two neighbouring plots of land, are in contradiction and conflict.

The other critical mismatch between a City Comprehensive Plan and a Regulatory Detailed Plan is generated by the time-consuming nature of plan formulation and approval under the rapid social and economic development scenario. This problem is very significant since China is under such rapid development that there are great changes in a short time. When a Regulatory Detailed Plan is considered to produce after the completion of a City Comprehensive Plan, the process of formulation and approval of which may take a couple of years, the social and economic background of a city may have already changed. A City Comprehensive Plan is unable to provide a more certain context that a Regulatory Detailed Plan would require. Vice versa, in the process of producing a Regulatory Detailed Plan, planners may need to change some policies and guidance defined by a City Comprehensive Plan that may be out of date to negatively impact the development after plan making and approval for several years.

In principle, the relationship between a City Comprehensive Plan and a Regulatory Detailed Plan is rational, logical and appropriate. Since a Regulatory Detailed Plan is the only planning tool that directly controls development, a Regulatory Detailed Plan can significantly influence the quality of the built environment and overall strategic policies of a city. A Regulatory Detailed Plan must be formulated and operated within the framework of a City Comprehensive Plan by following the principles and policies of the City Comprehensive Plan. However, there is often a departure from the original design of the system in actual practice. The regulation role of a City Comprehensive Plan is comparatively weak. It is usual for Regulatory Detailed Plans to drive amendments and adjustments of policies of a City Comprehensive Plan instead of following and interpreting policies of a City Comprehensive Plan. Since the implementation of the City and Countryside Planning Act of 2008, in order to avoid the impacts of Regulatory Detailed Plans to policies of a City Comprehensive Plan, and to avoid the contradiction between neighbouring plots of land, as well as to avoid potential corruption in planning and land leasing process, it is required by the Planning Act and government circulars that the Regulatory Detailed Plans should be produced to cover the whole 'planning areas' before the actual development and investment to a specific piece of land has taken place.

Relationship between Regulatory Detailed Plan and Planning Management (Development Control)

It is because the policies defined by a City Comprehensive Plan are usually of a general and macro nature for development, the policies and proposals of a City Comprehensive Plan have to be followed and embodied in detail in a Regulatory Detailed Plan to guide city and countryside development and construction through the development planning management process, within which Regulatory Detailed Plans are the main material for consideration. This statutory status of a Regulatory Detailed Plan has been further enforced through the implementation of the City and Countryside Planning Act 2008. Regulatory Detailed Plans have become the most important material for consideration in the process of implementing city and countryside plans, and the main documents for the decision of planning permit applications or leasing the state-owned urban land use (ELRD 2008). The promotion of their status is unpredicted. However, this creates a risk that the task of project planning management (development control) has then been put on Regulatory Detailed Plans (Yin 2008). The high claim for the so-called scientificalness and rationality of city planning, especially regulatory detailed planning, has been the topic that has attracted considerable attention recently in Chinese planning literatures (Du 2008; Duan 2008; Yan 2008).

Planning management as a tool for implementation of a City Comprehensive Plan is operated in China as a process of application for development project planning permission from local planning authorities through a certificates system of so-called 'two permits and one permission notice', i.e. 'Land Use Planning Permit', 'Building Construction Engineering Permit' and 'Permission Note for Location'. Under this mechanism, all the development and construction projects should apply for planning permission from the local planning authority. The local planning authority should evaluate the development or construction proposals according to detailed controlling ordinances, indexes and codes which are the contents of a Regulatory Detailed Plan for a piece of land development site, and then issue the certificate of 'Land Use Planning Permit'. However, it is because of the dual systems of land use distribution, i.e. administrative assignment approach and commercialized approach of land use right leasing tendering in Chinese cities, the process of land allocation for development projects are different. As the consequence, the process of application for planning permission for a development project is also different. Although the Regulatory Detailed Plans have similar functions within the two land allocation approaches, it is operated at different stages because of the different processes of application for planning permission.

The approach of land administrative assignment is still operated for land uses of public facilities which provide public services, e.g. land uses of government administration, transport facilities, education, health facilities, sports and infrastructures. The approach of commercialized land leasing is adopted mainly for commercial land use, e.g. development of real estate, retail and business, and industries (MOLR 2002).

There are different features of these two approaches of land allocation. For approach of administrative assignment for land use, it is usual to have projects first and available land later. This means that the development project is confirmed first of all and then application of land use assignment from the local bureau of land administration later. In the process of planning management for this type of land allocation, it is necessary to decide the site location of the project according to its characteristics and policies of a City Comprehensive Plan. After location decision for development project, a 'Permission Note for Location' is issued by the local bureau of planning administration. The project developer should then apply for a 'Land Use Planning Permit'. At this stage, the main controlling indexes and ordinances, e.g. plot ratio, development density height of building, that are all defined by a Regulatory Detailed Plan for the specific development site should be used as main material for consideration to assess the application documents and proposed development (construction) schemes raised by the project developer before formally issuing the 'Land Use Planning Permit' by the local bureau of planning administration.

The feature of the approach of land use rights leasing tendering is to have the available land and potential development project ready at the same time. Since China executed its land use reform in urban areas in the late 1980s, in principle, the land uses for commercial development have gone through the process of tendering for the land use rights leasing. Local municipal government should normally prepare a piece of land ready for proposed development. It is required by the central government's circular (Circular 22, MOC 1992) that when a piece of land is ready for tendering, it is necessary to have a Regulatory Detailed Plan which defines detailed planning indexes and ordinances for development and construction of the specific piece of land for tendering. The process of tendering for land use rights leasing and then the contracting documents for the land as well as the actual development on the ground after the contract of land use rights leasing is signed should be fully considered and within the context of controlling indexes and ordinance of the Regulatory Detailed Plan that defines land use boundaries, development functions, plot ratio, greening ratio, height of buildings, density of development, parking space, building setback line, and access of roads for the special piece of land. The development on the leasing land should have to follow these controlling indexes and ordinances that should be formed as one important part of the contract for the land use rights leasing. After successfully obtaining land use rights from tendering, the developer should employ planning consultants to produce a construction detailed plan and architecture design according to these controlling indexes and ordinances. The development or construction project is allowed to start only after construction detailed plan is submitted and approved by the local bureau of planning administration.

The local bureau of planning administration is able to refuse planning applications if the construction detailed plan produced by the developer has departed from the defined planning indexes, development controlling codes and ordinances decided by the Regulatory Detailed Plan associated with the land use rights leasing contact. Under this situation, the developer can either amend the development and construction details proposed in the construction detailed plan to fit the development standards of the Regulatory Detailed Plan, and then re-submit construction detailed plan for approval by the local bureau of planning authority. This is the usual method for the most development projects. It is also possible for the developer to appeal for amending relevant controlling indexes, codes and ordinances of the Regulatory Detailed Plan that impact the typical development project on certain piece of land if developer can raise appropriate rational reasons to argue that the controlling indexes, codes and ordinances of a Regulatory Detailed Plan are unrealistic and irrational. However, the process of amendment or revision of a Regulatory Detailed Plan is complicated. It needs a comprehensive review of the report and then submit for approval by the government that approved the existing plan. This complicated approach is time consuming. It is not a common practice for developers to appeal.

It is possible for the developer or the winner of the tendering for land leasing to transfer the development right of a piece of land after the tendering or auction process. However, controlling indexes, codes and ordinances of the Regulatory Detailed Plan binding on that piece (plot) of land should be transferred together with the land use rights.

National and Professional Urban Planning Norms

National and professional norms are important material for consideration during plan formulation, especially in the process of producing a Regulatory Detailed Plan for which development ordinances and controlling indexes that guide real estate have to follow either national or professional planning norms. These norms are also critical in decision-making in the process of planning management when assessing development project planning permit application. It is not possible to understand the methodologies of regulatory detailed planning production without understanding the national and professional planning norms. Although there is a series of norms promulgated by the Chinese central government or professional organizations to cover different planning aspects, it is not possible to show all of them, but some important norms are illustrated as examples.

The main purpose of application of these planning norms is to guide the provision of the necessary infrastructures and public facilities as well as to establish a comfortable and sustainable built-up environment during the urban and industrial development. The underpinning principles of establishing these norms are based on the concerns to establish a rational and better-off living quality for residents.

Urban Land Use Categories and Planning Land Use Norms (GBJ137–90)

This national norm defines different urban land use into three hierarchical categories. The top category is 'General Land Use' in urban areas. It defines 10 different land use functions (see Table 6.1), i.e. residential land use (R), public facilities land use (C), industrial land use (M), warehouse land use (W), land use for external transport facility (T), road and square land use (S), public utilities land use (U), green land (G), special land use (D), and water resources and other land use (E).

The general urban land use category is further divided into 46 different types of the so-called 'Medium Category of Land Use' and 73 different types of land uses defined as 'Small Category of Land Use'. The lower the category in the hierarchy, the more detailed the land use type will be.

The urban development land uses should consist of all types of land use in the general category except 'water and other related land use (E)'. However, there is a problem while calculating planning land use standards so far, in that they only consider the urban population, but rural to urban migrants have been ignored. This creates two problems. One is the unfair treatment of migrants from rural areas; the second problem is the unreliable and unrealistic calculation since the rural to urban migrants have been a common phenomenon in most Chinese cities. In some cities, e.g. Dongguan, the population of rural to urban migrants are higher than the local citizens. The unit of calculation of land use is based on square metres per person for different land use type.

The norm defines urban development land use per person which varies from 60.1 m^2/person to 120 m^2/person according to four different types of city in a hierarchic city system depending upon their sizes but regarding present development land use per capita as material for consideration. It is possible to increase land use per person within the allowed limitation of the land use index. It is similar to the process of producing a Regulatory Detailed Plan, the land use per capita is depended upon cities at different hierarchic levels and its size. For example, it is regulated that housing land use should be 18.0–28.0 m^2/person; road and square land use is arranged as 7.0–15.0 m^2/person; greening land use should be minimized at 9.0 m^2/person (among which, the public greening land should be minimized to more than 7.0 m^2/person). The larger the city is, the smaller space per capita will be.

The Planning and Design Norms for Urban Housing Areas (GB 50180–93)

This norm should be applied in the process of the residential development plan. It is defined in the norm that urban housing area is categorized as three hierarchical standards of residential areas, small residential areas and community area. The size of each level is defined as 10,000–15,000 households with 30,000–50,000 residents for a residential area; 2,000–4,000 households with 7,000–15,000 residents for a small residential area; and 300–700 households with 1,000–3,000 residents for a community area.

The norm regulates land use development controlling indexes for housing areas depending upon different types of city and different hierarchies of residential areas, as well as different stories of residential buildings.

It is specified in the norm that the design of residential buildings should consider comprehensively local land use conditions, types of buildings, orientation of windows, distance of each building, greening and open spaces, stories and density of buildings, layout pattern, housing groups composition, and spatial context.

The distance between buildings is defined according to hours of sunshine within rooms of the house (flats) according to the seven types of climate areas in China, and the size of cities. Within the old urban built-up area, considering the existing density of population, it is able to reduce certain standards of planning and design for residential development. However, the period of sunshine in rooms of flats (houses) should not be less than one hour in the winter. Flank side between residential buildings should not be less than 6 metres for multiple stories buildings, and 13 metres for high-rise buildings.

It is defined in the norm that residential buildings with more than six stories should provide lifts. It is also regulated that maximum controlling indexes for residential development net density and net density of total housing floor space should be controlled at the level of 32 per cent or 22 per cent for multiple stories buildings and high-rise buildings separately. The maximum net density of total housing floor space should be controlled within $1.90m^2$/hectare, or $3.50m^2$/hectare for multiple stories buildings and high-rise buildings separately.

The norm also specifies the indexes and parameters for the provision of public facilities within the residential areas according to different hierarchies which are designated by the number of residents. Public facilities are defined as eight types, i.e. education, health (hospital or clinic), culture and education, commercial (retail), banks and post offices, public utilities, and public administration (community facilities).

With regard to greening areas, it is decided in the norm that the ratio of greening space should not be less than 30 per cent in the new developing residential areas, or not less than 25 per cent within the old city rehabilitation areas.

The roads within residential areas have been regulated. The width of roads between the setback lines of buildings should not be less than 20 metres in a large residential area; 5 to 8 metres in small residential areas, and 3–5 metres in community areas; and not less than 2.5 metres in width for buildings access. It is also defined that a main road in a small residential area should have at least two entrances. The main roads in a large residential area should have at least two access roads linking to outside main roads.

The Norms of Urban Roads and Transport Planning and Design (GB 50220–95)

An urban transport plan consists of two parts, one of which refers to 'Urban Transport Development Strategic Plan', the other is 'Urban Roads and Traffic Comprehensive Network Plan'. An Urban Transport Development Strategic Plan should propose urban transport development objectives, transport patterns and structure, urban road and traffic comprehensive network, inter-city transports, location decisions for passenger and cargo transport facilities and their land use size, technological and economic policies for implementation of plans, and finally policies of urban development and transport demand management. A Urban Roads and Traffic Comprehensive Network Plan should decide urban public transport system, the width of different urban hierarchical roads, the section of different hierarchical roads and streets, different types of junctions, exchanging and linking system of various traffic patterns, location and land use of large public transport terminals, and finally location and land use of squares, bus stations, bridges and ferry stations.

Regarding public transport, it is regulated in the norm that there should be a standard bus for every 800 to 1,000 persons in the large city. This rate in the medium and small sized city is a standard bus per 1,200 to 1,500 persons. It is regulated that the metropolises with an urban population more than one million should reserve the land for the rapid rail transit development.

It is also defined by the norm that public transport routes density should be 3–4 kilometres in length per square kilometre of various land uses in the city centre, and 2–2.5 kilometres in length per square kilometre of land uses in the urban periphery. The distance of an urban bus route should be controlled within 8–13 kilometres. The distance of an urban rail line as rapid rail transit system is proposed to run within 40 minutes from beginning to the end of the route.

The norm defines the urban road hierarchy that consists of urban express roads, main arterial roads, arterial roads and access roads. The total land use for urban roads of a city should be controlled at a rate between 8 and 15 per cent of total urban built-up land. This ratio can be increased to 15–20 per cent in a metropolis with a population more than two million.

It is specified in the norm that total land provision for a coach station, a railway station, or a passenger dock yard should be provided at a ratio of 0.07 to 0.10 square metre per person based on the total population of a city.

The Norms of Urban Civil Engineering Pipeline General Plan (GB 50289–98)

The main contents of the norm include construction orders for the instalment of civil engineering pipelines underground and its minimum net distance, and the net vertical distance between pipelines and the minimal distances between pipelines underground and the ground floor; plane location of the poles and elected lines, as well as a minimal distance between buildings, roads and others.

The Norms of Urban Flooding Prevention Engineering Planning and Design (CJJ 50–92)

It is regulated in the norm that cities should be divided into four categories depending upon their possibility of flooding impacts and standards of flood prevention. It is defined in the norm that floods should be divided into four types, i.e. river flood, the sea tide flood, torrents, and mud-rock flow. In an urban flood prevention area, different measures and standards for flood preventions are specified. The norm also defines standards of flood prevention buildings.

Measures of calculation for flood flows, tide and detailed regulations for the design of dams, bank revetment, small reservoir, drainage ditch, and other relevant engineering projects are all specified in the norm.

Conclusion

The planning approach of the Regulatory Detailed Plan was initiated because of the introduction of the market economy and the consequence of urban land use reform in China since the late 1980s.

After more than 20 years of development, the Regulatory Detailed Plans have been adapted to cope with the changes of Chinese social and economic development. This planning approach has been operated to meet the needs of planning management (development control) in the process of rapid urban development. In other words, the Regulatory Detailed Plan has met the demands of local government in the simplification of the planning process, efficiency in decision-making, typically in decisions for the period of land leasing and development project (Wang 2000) in order to increase the effectiveness of urban and real estate development.

Regulatory Detailed Plan have been treated as either the necessary documents as one of preconditioning in the process of leasing the state-owned land, the designated urban development land, or direct material for planning decision-making in project planning permission application (LACNPC 2008).

A Regulatory Detailed Plan, providing various ordinances and controlling indexes and codes to guide and control development in the process of planning management, creates certainty in development. This certainty is also the demand of the market. It is only able to create a profit in a more certain situation by developers. However, as China is still under its rapid economic development and urbanization in the process of transition to the market economy, the economic and physical environmental changes are happening frequently. Uncertainties as the consequences of rapid economic development rate, types of investment and projects, and the difficulty of precise forecasts of population because of large amounts of rural migrants have also been the main phenomenon in Chinese development. This phenomenon generates many problems in the process of Regulatory Detailed Plan making, typically when deciding the most important elements of a Regulatory Detailed Plan, the controlling indexes. It is even more critical when the methodology of Chinese Regulatory Detailed Plan production is also dependent upon the summary and induction from empirical work and experience of planners from their former work. This makes the regulatory detailed plans unreliable and unaccountable. The other problem in the production of a Regulatory Detailed Plan is that planners usually ignore the interests of stakeholders and lack appropriate economic analysis for potential development of a piece of land when making a plan. It is very common that planners only focus on the impacts to the physical spatial layout of development projects. Some proposed controlling indexes and codes in a Regulatory Detailed Plan are inevitably broken through in the developing process without appropriate economic analysis.

It is the view of Lu (2009) that the delivery of a Regulatory Detailed Plan in planning management (development control) is far from satisfactory because of the following (Lu 2009):

- The time required to produce plans is often very limited. Planners may not have enough time to carry out detailed site visits and investigation.
- Shortage or poor economic forecast and analysis of the planning area.
- The main indexes of plot ratio, building density and height that are specified by a Regulatory Detailed Plan, which is significant to the economic cost and benefit of a development project, are mainly decided by experience of planners through summary and induction from empirical work.
- Chinese development has been so rapid that there are many uncertainties which may force the proposed technical controlling indexes and ordinances out of date in a short period.

In order to address the problems of the Regulatory Detailed Plan, several amendments and innovation in plan making have been experimented with. Shenzhen, as a Special Economic Development Zone after the Chinese open door and reform policy since the late 1970s, introduced a range of planning administrative and legislative reforms by establishing a so-called 'three levels five stages' planning system (Bruton et al. 2005). Within this system a Regulatory Detailed Plan and the embodied ordinances and controlling codes should be approved by the local people's congress to be a by-law. This experiment has been confirmed by the City and Countryside Planning Act 2008. It is required by the Act that all the planning areas of a city should be fully covered by Regulatory Detailed Plans which should be approved by the local people's congress. Any change of a Regulatory Detailed Plan should report to and go through the approval process by the local people's congress. However, what has happened since the Planning Act of 2008 has been that all local governments have made Regulatory Detailed Plans to cover all its planning areas, but without going through the approval process by local people's congress in order to avoid the changes as a consequence of development uncertainties.

Nevertheless, with the acceleration of China's urbanization and the entrepreneurialism of local governments, the operation of the Regulatory Detailed Plan has become more difficult, especially when the planning is mainly controlled and decided by local governments. The entrepreneurialism of local governments forces local governments to strive for maximization of local economic benefits in a short term. They can fully use the local resources, typically some monopolized resources controlled by the power. The 'public goods' that should be protected by the city and regional planning may sometimes be ignored. The development controlled by planning management through the regulatory detailed planning process has then been regarded as an impacted 'burden' to development. It is common that the controlling index decided by the Regulatory Detailed Plan has been revised (Liang 2006). The failure or ineffectiveness of Regulatory Detailed Plans in the process of plan delivery by planning management has then become a critical problem in the Chinese planning system (Tang 2006).

Some researchers (Yuan and Hu 2010) argue that it is because of the implementation of chargeable land use rights leasing system in China, the income from land lease has become the most important resource for city governments. Without any major changes in fiscal and tax system in China, Chinese cities will definitely develop towards the possibility of maximizing income of local revenues from land lease instead of long-term planning vision, aesthetics or ecological sustainability.

National and professional planning norms are used as a special mechanism in the Chinese city and regional planning system. National and professional planning norms are important material for consideration during plan making and in the process of planning management (development control). The establishment of these norms, which help to define controlling indexes, codes and ordinances for development, including the provision of 'public goods', e.g. schools, local shops, hospitals, and other public buildings, was initiated in the centrally planned economy. In the centrally planned economy, all the necessary resources for development of a city were controlled and allocated by the governments. What planners needed to do was to allocate spaces to accommodate development and to provide 'public goods' according to the controlling indexes defined by the national or professional planning norms. However, since the delivery of the market economy with Chinese characteristics, these norms have caused some problems for their appropriateness in planning because of the uncertainties in the market economy and differences among regions typically that China is a large country with many climate zones and different regions at different developing stages. The amendment of the national and professional planning norms is necessary.

Chapter 7
Non-Statutory Plan:
Urban Development Strategic Plan

Introduction

During the transition to the market oriented economy within which devolution, globalization and marketization are three main features, the municipality has played an important role in urban and economic development. The function of the municipality has been to become the promoter of local economic development. Development has been regarded as the absolute principle and the priority of local municipal governments' agenda. Local politicians have become very sensitive to the opportunities and pressures of development.

After the entrance into the WTO by China, municipalities have been confronting increasing pressures of competition between cities in China and even from other countries. Some coastal cities, particular those of Special Economic Zones, are facing the problem of losing their privileges in tax reduction and other favourable incentive packages. These favourable development policies used to be available to them only but are now available to all cities (Zou 2003b). These new changes in the political and economic context encourage municipal governments to reconsider their industrial and economic structures and their city spatial layout with the objectives of promoting and enforcing the capability of the city. Local governments require an analysis to compare the strengths, weaknesses, opportunities and threats of cities (Li 2003) when they are involved in competition. Local municipal governments hope that with support from urban planners through their research and planning, the city can expand rationally to be larger and stronger through a 'scientific' approach (Zhang 2002).

In the late 1990s, in order to encourage the urbanization process, there were adjustments of municipal structures by the government of China. Some rural counties were designated as districts of urban areas. This was a special case resulting from the existing dual urban and rural administrative system in China. Because of the changes in the administrative system, there were requirements to amend existing city macro plans. All these pressures and changes required planners to study development policies, dimensions and schemes. Nevertheless, most cities had only completed their City Comprehensive Plan formulation, or had them approved by the higher hierarchical governments. The complicated and time-consuming nature of the traditional City Comprehensive Plan amendment and approval process has driven local municipal governments to seek innovation in the city macro plan to deal with their pressures and problems. The application of the Urban Development Strategic Plan is the result of this demand.

In this chapter, the arguments about the application of the Urban Development Strategic Plan are analysed. The Urban Development Strategic Plan is evidently necessary for several reasons, i.e. pressures and rapid changes. It is a type of planning innovation from the bottom up (compared with the central planning system). The analysis of the arguments will help to understand and assess the performance of urban macro planning in China.

The case of Xiamen Urban Development Strategic Plan is used and examined to illustrate what kind of function and how it was operated in local development. The opinions and comments of stakeholders regarding planning policies and schemes, and their influences on the local government's development policies, are also analysed.

The final section of the chapter will be an evaluation of the Development Strategic Plan to understand if the innovation of urban planning is able to meet the demands of pluralistic stakeholders in rapid social and economic development. Lessons from the plan formulation process and the performance of plan will be identified and evaluated from the field research.

Arguments for the Urban Development Strategic Plan

Since the first Urban Development Strategic Plan, Guangzhou Urban Development Strategic Plan, was produced in 2000, there have been many arguments about this new type of planning in the Chinese literature. It is the opinion of Li (2003) that the appearance and application of the Urban Development Strategic Plan is because the traditional City Comprehensive Plan cannot cope with rapid changes. Li (ibid.) criticizes the fact that in the production of the City Comprehensive Plan, urban planners are always seeking the perfect planning written documents and plan making process instead of carrying out actual research into strategic issues. Planners fail to undertake prediction analysis. The plan formulation process has emphasized the control of the scale of the urban population and the protection of cultivated land. These are the criteria for plan approval set by the central government.

Li's comments explain some reasons for the application of the Urban Development Strategic Plan. However, it is more complicated, typically in terms of the contents of the City Comprehensive Plan designated by the government's ordinance – 'Regulations for Urban Plans' Formulation'. The formulation of a City Comprehensive Plan is too complicated and time-consuming whilst the contents of the plan may not meet the demands of stakeholders under the pressure of development and competition. Under the pressures of the speedy development process in China in the 1990s and the beginning of 2010s, Chinese local governments and planners would then have to find an alternative planning approach at planners' discretion, such as a pragmatic approach to planning. However, it has not been commonly accepted. It should be admitted that not many planners have been able to learn from the others' mistakes and experiences. The continuing application of the physical planning approach, typically with the concepts of the centrally planned economy, risks producing urban plans that may not be useful or necessary in the market economy. The result is that urban planning policies and schemes suggested by planners will be ignored by the government or other decision-makers (Liu 1999).

Research on the tasks of the Urban Development Strategic Plan by some academics (Qiu 2003a; Zhang 2002; Zou 2003b) shows that the plan is to provide a spatial structure with long-term, general and comprehensive features. This could be achieved by coordinating the city's social and economic development strategies. Social economic growths and proposals to increase competition have been regarded as the main objectives of the strategic plan. Urban Development Strategic Plans would focus on the role and function of a city from a much wider aspect compared with that of a City Comprehensive Plan. The plan would review the city within the context of the whole country, or the region (Dai and Duan 2003; Luo and Zhao 2003). Multiple alternatives of spatial structure with various scenarios would be the target of planners in the formulation of the plans (Wang 2002b; Wu 2000).

Fundamental to their studies, Luo and Zhao (2003) suggested that the main characteristics of the present Urban Development Strategic Plans in China include rapidity with a short period of plan formulation, efficiency with an approach of problem orientation, and flexibility without any regulation and restraints in plan formulation. The new type of plan emphasizes the macro issues in the region. These characteristics could meet the demands of competition and the rapid social and economic changes.

The application of urban development plan in China was driven by seeking economic achievements, competitiveness for inward investment, and rapid changing social, economic and administrative contexts. The Chinese local governments require a new planning approach to cope with challenges.

It is typically after China entered the WTO that the pressures of competitiveness of the local governments have been increased. It has become more serious among local governments to attract more and continuing inward investment. Local governments require urban planners to allocate more land for potential development, to analyse future challenges for local cities, and to find opportunities for local development.

Technologies and Methodology of the Urban Development Strategic Plan

The objectives of Urban Development Strategic Plan decide the planning technologies and methods of the Urban Development Strategic Plan. There has been an analysis of technologies applied by the Urban Development Strategic Plan by comparing the plans for Guangzhou, Nanjing, Ningbo and Hangzhou (Wang 2002b). Technologies used in the plans are generally summarized as follows:

- regional analysis, which evaluates the role and function within the whole nation and the region, even within the world because of globalization, to assess the role of the city under study;
- industrial development study, which provides the basic foundation for the economic development of the city;
- spatial structure analysis and resources allocation for the whole municipality instead of the defined 'Urban Planning Area', to achieve a rational development layout;
- proposals which include the planning of major infrastructure, particularly transport facilities to establish the spatial structure for an infrastructure supporting system;
- ecological limitation threshold analysis, which analyses the maximum ecological capacity of the city for sustainable development.

The planning process of the Urban Development Strategic Plan follows the traditional rational planning approach with systematic analysis of the existing situation and data. In this rational planning approach, the objectives are given. With the given objectives, planners would seek the most efficient and effective means and technologies to achieve them. These means and technologies would be based on logical thought.

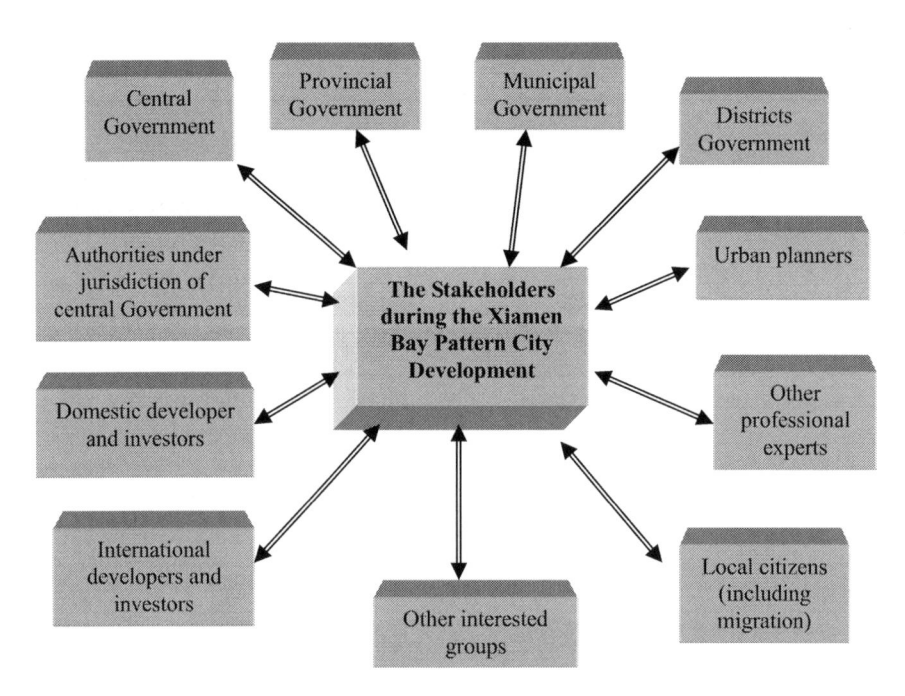

Figure 7.1   Multiple stakeholders in the Xiamen development
Source: Created by the author.

A systematic planning method, which has been used in the marketing process by commercial firms, was significant in the formulation of the Urban Development Strategic Plan. The SWOT analysis, which refers to strengths, weaknesses, opportunities and threats, was used to analyse the competitive capability of the city.

The Urban Development Strategic Plan, as a type of urban macro plan, should coordinate all stakeholders (see Figure 7.1) in the development. Planning policies and proposals cannot be implemented without support and coordination among the stakeholders as planners lack power and resources. Planners have to discuss, negotiate and bargain with powerful stakeholders in the planning process, particularly those authorities with budgets.

Case of Xiamen Urban Development Strategic Plan

The Reasons for Xiamen to Deliver an Urban Development Strategic Plan

After China entered the WTO, the former favourable policies used to attract inward investment to Xiamen as a Special Economic Zone when it was originally established, were applied to the whole country. Xiamen's incentive package thus disappeared. At the same time, economic development along the east coast area of China saw its most prosperous and rapid progress. In the South Fujian region, where Xiamen is located, other cities have been developed so quickly that they have created pressure on Xiamen. It was clearly explained in the Terms of Reference for the Xiamen Urban Development Strategic Plan project that the municipal government would require urban planners to allocate more available land for potential development, to analyse future challenges and to find opportunities for the city. The policies of the municipal government to develop the city as a bay pattern from the existing island type are a means of expanding the city for more urban development land. This city development dimension would create the relevant strategies in spatial layout, land allocation and industries. Under this situation, the local authority decided to prepare a general development strategy, which would be able to fully and efficiently use and control the merits of the city to achieve the objective of an important central city along the Southeast coast.

There was other important change in the Xiamen municipality. In 2001, there was an adjustment of the administrative system in Xiamen. Tong'an County, which used to be the rural area under the jurisdiction of Xiamen municipal government, became an urban district. Due to the adjustment, and the challenges and pressure of the competition with other cities, the approved Comprehensive Plan was seen as out of date and requiring amendment.

With the introduction of an Urban Development Strategic Plan in some other Chinese cities, Xiamen municipal government decided to apply this new type of plan instead of the complicated process of the City Comprehensive Plan amendment. In

April 2002, a project team, which was different from the comprehensive planning team of China Academy of Urban Planning and Design, was organized, entrusted by Xiamen municipal government with the task of preparing the Xiamen Urban Development Strategic Plan. The project was finished in August 2002.

The Plan Formulation Schedule

The project team consisted of 16 members from China and overseas consultants from a variety of disciplines, including urban planning, urban transport planning, economic geography, economics, environmental protection, ecology, urban design and landscaping. There was no local planner in the project team.

Plan formulation lasted four months. It was divided into three stages. The first stage was information and data collection, investigation, and proposal of initial schemes. It lasted two months. The second stage was research into different subjects and report writing. Various seminars, workshops and panel meetings were held at this stage. This took one month. The final stage of writing the general planning document took another month.

During the first stage, the project team visited, discussed and organized workshops with several local government bureaus, including the economic development and planning, urban planning, environmental protection, transport administration, tourist administration, seaport authority, airport authority and the local municipal governments of the adjacent cities of Zhanzhou and Quangzhou. Information and data provided by these government authorities, and their development programmes, were significant to the production of the Urban Development Strategic Plan. They had to be considered by planners before proposing the development policies and schemes.

In the second stage, there were some initial exchanges of opinions and proposals among project team members, and between the team and the local planning authority. The exchange of ideas, particularly a so-called 'brainstorming' session, helped the team achieve a more rational and practical solution.

After plan completion, the project team had to report to the municipal government, including the municipal party chief, the mayor, other senior leaders and directors of the bureaus and the experts from other cities. The project team also presented and reported to the Municipal People's Congress. After the acceptance of the plan, the local municipal government promulgated the 'Implementation Outline of Xiamen Urban Development Strategic Plan' to guide the development of Xiamen.

Competitiveness Analysis of Urban Development Strategic Planning Approach

SWOT analysis and marketing The promotion of city competitive capacity needs city marketing. Urban plans should and could function as an important part of the marketing process. According to Ashworth and Voogd (1990), a city marketing plan is inseparably linked with the urban plan, and can act as a kind of instrument in determining the dimensions of the future development and land use.

An important element of city marketing is the promotion of the city image. According to research by Ashworth and Voogd (1990), the definition of the city-product will heavily depend upon its image. A city is able to attract more inward investment and compete with other cities if it has a perfect image as an environmentally desirable place to live, with open space and green areas, and safe water and drainage systems; a cultured city that is large, structured, modern, lively, clean and safe. All these suggest the city is as a good place to invest in to generate economic development.

Xiamen Urban Development Strategic Plan carried out special research on Xiamen's image. The plan suggested that human beings live in a truly global environment. China's entry into the World Trade Organization (WTO) is an event of significance to the future economic growth of China, in which future economic contingencies compete against contingencies, and cities compete against cities. For Xiamen, international perception – its image, its branding – is of huge importance in the effort to be successful in the competition for:

- trade markets – exports, goods and services;
- investment, which is about attracting inward investment, international capital into Xiamen;
- people, particularly skilled and professional people.

The study, exploration and analysis of SWOT were achieved through discussion among team members after the social and economic analysis of Xiamen. It was a type of brainstorming approach. A planner, responsible for the image study in the team, then summarized the comments of team members and wrote the report for local municipal politicians to consider while making their decisions. The general concepts of Xiamen's SWOT analysis summarized from the plan are as follows:
Strengths

- a rich, strong and comprehensive range of aspects making up the total image;
- the essence of being an island or several islands;
- the wide appeal of its image, domestically and internationally;
- the relative ease of marketing the image, which is easily identifiable and saleable;
- the likelihood of the image enduring over time;
- the success of the city in awards for its environmental programmes.

Weaknesses

- the lack of the sense of the bay as a whole if Xiamen develops from Island City to Bay City;
- the lack of positive marketing of its image;
- possibly only recognized significantly domestically.

Opportunities

- the potential to further explore the wider maritime aspects of Xiamen and the mainland;
- the opportunity to greatly increase the length and vitality of the urban waterfront in its day and night image;
- the potential to define more clearly the role of the historical reputation of Xiamen in the marketing package;
- the opportunity to build on its award-winning success with innovative environmental and sustainable projects, for example, in eco-tourism;
- the opportunity to sensitively exploit the bay as a whole for future activities without compromising the island; and
- the opportunity to market itself as green/dynamic/fresh/modern/cosmopolitan/maritime/cultural city.

Threats

- the over-exploitation of Xiamen for economic development, thus losing its underlying appeal;
- over-industrialization of the bay as a whole bringing risks of pollution and loss of natural resources, such as landscape and biodiversity;
- a speed of growth faster than the infrastructure can support; and
- development exceeding environmental capacity and diminishing natural resources and image.

Regional analysis and study   The SWOT analysis provided the starting point from which to analyse Xiamen's competitive capacity compared with that of potential competitors. Besides the local SWOT analysis, comparative research on the regional context of Xiamen was the other important element in the planning process. After the regional and local context analysis, the planning team would recommend the development policies and schemes as a consequence. Xiamen Urban Development Strategic Plan mentioned that the coastal cities of east China had already established a suitable foundation in secondary and tertiary industries with the capability to absorb new types of industries and growths. On the other hand, there were some common problems along the east coast, in that urban development had contributed to the increase in land price, labour costs and economic growth during the last two decades, and had created problems of land shortage and environmental pollution. It was time for these cities to consider an adjustment to existing industries. In Shanghai, labour intensive industries had already resettled to the nearby provinces of Jiangshu, Zhejiang and Anhui. In Shenzhen, 'the three types of processing and compensation trade'[1] which used to

---

1   Processing industries may process import materials, process products according to customer orders, or assemble product parts supplied by investor or clients. These industries

contribute significantly to economic development at the beginning of Shenzhen as a Special Economic Zone, were no longer able to continue their contracts and resettled to other places according to Shenzhen municipal government regulations. Xiamen was also facing similar challenges and problems for its future development. The initial proposal and forecast by the City Comprehensive Plan to have industrial development as the first phase during the industrialization process and then prepare for the second phase of tertiary industrial development had not happened. Convention and exhibition businesses, real estate and tourism had developed more rapidly than secondary industry, because of the 'Holiday Economy' promoted by the central government.

Capability analysis was one of the key elements of the planning process. An appropriate way to understand the capability of the city is through comparing it with its potential competitors. In the production of the Urban Development Strategic Plan, four other major cities along the east coastline were selected for comparison as potential competitors: Shenzhen, Ningbo, Qingdao and Dalian. These four cities and Xiamen are either Special Economic Zones or Coast Open Cities enjoying similar favourable policies after the reform and open door policy. They are all quasi-provincial-level municipalities in the administrative system. They are seaport cities and have an export oriented economy. The purpose of the comparison was to analyse and find the strengths and weaknesses of Xiamen vis-à-vis its competitors.

From the comparison, it was found that there were less population and lower total GDP in Xiamen (see Table 7.1). Planners agreed that this situation would affect the general demands of local consumption as Xiamen could not establish a large local market. Nevertheless, its GDP per capita was higher. It was only slightly smaller than that of Shenzhen. This was evidence of the wealth of local people and local economic activities. Moreover, the numbers of students in universities were the highest among the five cities. They provided the available high-tech professional labour force in Xiamen, if they could be encouraged to stay through appropriate employment opportunities for these more highly educated personnel. They also contributed to developing the local high-tech and R&D (research and development) industry.

Of these five cities, Xiamen is the oldest seaport engaging in international trade with other countries. Owing to its long history in international trade, Xiamen had already established its reputation in the world especially in Europe and Southeast Asia. The planning team in the discussion thought that this could provide a beneficial opportunity for Xiamen to fully use its international relations for marketing to promote Xiamen above the other four cities.

The five cities are all open cities with significant FDI contribution to economic development (see Table 7.2). Xiamen is the second highest. In one aspect, the FDI would solve the problem of shortage of financial resources, but at the same

---

pay the cost of the loan of machines or material from abroad by the sale of the products they produce.

time, there would be higher risks and great uncertainties in depending upon the international market, particularly in Xiamen because of the political and military uncertainties in the Taiwan Straits. The planners regarded this as a critical issue for future development.

Table 7.1 Comparison of quasi-provincial-level cities along the east coast of China

Cities	Total land area (km²)	Total population (thousands)	GDP (100 million yuan)	GDP per capita (yuan)	Number of university students/ thousand people
Xiamen	1,565	1,312.7	501.89	38,021	18.89
Shenzhen	1,949	4,329.4	1,665.24	39,739	3.26
Qingdao	10,654	7,066.5	1,150.07	16,317	6.52
Ningbo	9,365	5,409.4	1,175.75	21,786	3.75
Dalian	12,574	5,524.7	1,110.77	20,255	15.64

Source: Collection of Statistics of Coastline Cities; China Statistics Year Book 2001.

Table 7.2 Percentage of FDI within total industrial output in 2000

Cities	Percentage (%)
Xiamen	74.95
Shenzhen	76.45
Qingdao	19.79
Ningbo	50.20
Dalian	22.49

Source: Collection of Statistics of Coastline Cities; China Statistics Year Book 2001.

Comparing other regions along the Chinese east coast, it was noticed by the planning team that the South Fujian region, in which Xiamen is located, is situated between the Yangtze River Delta area and the Pearl River Delta area, and opposite Taiwan. The two delta areas are the most important of the three rapidly developing co-urban areas in China. Although the Beijing–Tianjin–Tanshan area, the Yangtze River Delta area, and the Pearl River Delta area only contain 9.92 per cent of the total population in China and 2.04 per cent of the territory, they contribute 31 per cent of total GDP in the whole country. Taiwan, with its rapid export oriented economic development, is regarded as one of the five economic 'tigers' of Southeast Asia. Its total GDP was 326.2 billion US dollars, with GDP per capita of 14,906 US dollars in 2000 (DGBAS 2009). The excellent location provides a base for advantageous opportunities for development.

Taiwan is also looking for business development in Mainland China. Taiwan is short of space and resources, including natural resources and labour resources, and has a limited domestic market. Taiwan's economic development capacity has already reached its maximum on Taiwan Island itself. In the year of 2001, the increasing rate of Taiwan's GDP was -2.25 (DGBAS 2009). Taiwan needs urgently to penetrate the market of Mainland China (Deng 2002). It would be exceedingly dangerous for Taiwan's politicians to restrain investment to Mainland China. Taiwan has to seek opportunities to extend its development. Mainland China would be the best alternative. Since the relaxation in hostilities between the two sides of the Taiwan Straits in 1990, Mainland China has become the largest capital export destination and market for Taiwan. Investment in Mainland China by Taiwan businesses has been profitable. Since 1991, investment from Taiwan has changed from small and medium size businesses to large enterprises; and from labour intensive industry to capital and technology intensive industry and tertiary industry. This change requires good infrastructure, efficient government mechanisms, and highly skilled and professional labour forces. The relationship between Mainland China and Taiwan, and the possible role of Xiamen in it, is significant to the development of Xiamen. It should have been an important element analysed by Xiamen Urban Development Strategic Plan. Nevertheless, according to the investigation and participant observation in the project, because of the shortage of appropriate members in the team with professional knowledge in this aspect, the Urban Development Strategic Plan failed to undertake detailed studies and risk analysis of the coordination between Xiamen and Taiwan, or what kind of role Xiamen should play in the game.

Through comparison, the shared view achieved by the planning team and suggested in the Urban Development Strategic Plan was that the South Fujian region would only be able to function as a sub-centre to the east coast of China. Xiamen should fully play its functions of transfer and value-added industries, banking, seaport and trade development in the Yangtze River Delta area, Pearl River Delta area, and Taiwan. Moreover, as a famous 'Garden on the Sea', with rich tourist resources, Xiamen should emphasize its function as an important tourist attraction. This proposal took into account Xiamen's characteristics of limited land scale, less population, advanced international relations and communication, but poor communication with the hinterland, as well as its sensitive location to Taiwan.

As well as country-wide competition, competition from Xiamen's neighbours was also studied in the planning process. The increasing growth of Fuzhou, the provincial capital and Quanzhou (see Figure 7.2) threaten the central city role of Xiamen. Competition between the cities creates a higher risk and uncertainties to Xiamen.

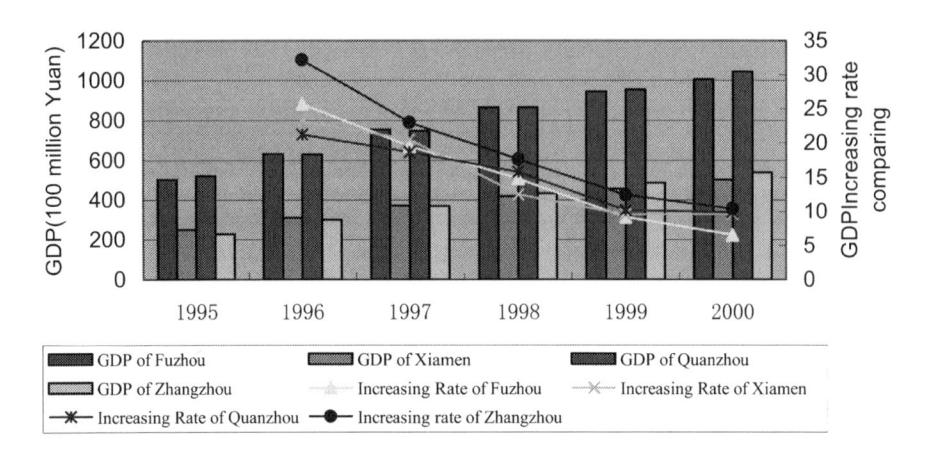

Figure 7.2 Xiamen compared with other main cities in Fujian Province
Source: Xiamen SEZ Statistics Year Book; Fujian Statistics Year Book 2001.

Especially in the absence of an up-to-date and accepted regional plan for the region, and a coordination mechanism as in the Yangtze River Delta area where there had been collaborative meetings among the mayors of Shanghai and governors of provinces within the area, and several workshops and meetings among senior officers and decision-makers from Shanghai and its nearby provinces every year, South Fujian region inevitably faces uncertainty about long-term development and infrastructure. This has influenced long-term strategies for competing cities in the region.

According to the planning team, from the view of the local region in South Fujian, Xiamen was the most important city on account of its reputation, attraction and the high quality social and economic bases and infrastructure. Since 1980, Xiamen has been defined as a Special Economic Zone. It has enjoyed an incentive package for inward investment. Because of its role in the political economy, it had developed earlier than the other two cities. Xiamen had been more mature in inward investment attraction, infrastructure support and industrial development. The problem was that there had not been satisfactory relations for coordination within the areas, but competition. It is necessary to take actions to deal with the problem. Without cooperation in the region, and in the face of competition, it was difficult for the region's future to be bright. This was especially so when comparing it with its two neighbours, the Yangtze River Delta co-urban area and the Pearl River Delta co-urban area. The development of this region was slower than that of the other two neighbours.

In their research, Ashworth and Voogd (1990) concluded that it is possible for an image of a city to be overshadowed by that of its neighbours. Such shadowing could be either positive, if the city reaps the benefits of the attractive image and promotional efforts of its neighbour, or negative, since the city could suffer through spatial association with the poor image of its neighbour. The best approach to deal with the problem is collaboration within the region to promote development to create a better image together. There was no difference of opinion within the project team that it was important to have coordination within the region. In the course of interviews, local politicians and urban planners of the three cities held similar views as to cooperation. The vice-mayor of Zhangzhou said that:

> The South Fujian region should coordinate in economic development. Xiamen should play a leading role. There are great opportunities of mutual benefits and support of each other among the three cities through cooperation. Each city has its own advantages and disadvantages. (Interviewed on 5 July 2003)

It seemed that no one had suggested cooperation before, but competition instead. It was suggested by the planning team formulating the Urban Development Strategic Plan that if urban macro plans of these three cities took account of other cities' development strategies this would avoid conflicts of interests and promote growth opportunities for all and cooperation will mean 'win-win' for all.

Political and Economic Analysis and Economic Development Proposals

The Urban Development Strategic Plan has tried to improve the traditional planning approach in China. It has extended its studies to social and economic development issues. The team included economists undertaking an economic study for Xiamen. They studied the characteristics of the economic growth process since the establishment of the Special Economic Zone to discover how economic growth was promoted and generated. This practice was totally different from that for the City Comprehensive Plan, which followed the social and economic development policies and proposals of the Municipal Social and Economic Development Plan, prepared by economic development authorities. There are two reasons for this change. One is marketization in the transition. The power of economic development and the planning authority has been reduced. In recent government restructuring, the State Development and Planning Commission, which used to be the most powerful authority in the centrally planned economy, has been renamed as the State Development and Reform Commission. Its function has also been changed. The authority will focus on policy issues instead of on concrete social and economic development, which will become dependent on the market. The central government has enforced marketization in the transition. The second reason is that urban planners have realized the weakness of physical planning. Chinese urban planning has to consider and combine social and economic aspects.

Through analysis, the planning team concluded that capital investment, incentives packages and the operation mechanism for inward investment were the most critical issues to promote economic growth. Xiamen had been designated as a Special Economic Zone to execute favourable policies and a Free-Port policy, as well as a delegation by the National People's Congress to the Municipal People's

Congress to promulgate by-laws. The incentive package for inward investment has encouraged foreign direct investment (FDI) to Xiamen for capital and fixed assets. However, since the mid 1990s, the rate of investment and inward capital has slowed down. There may be three reasons for this:

1. the national government's policy shift to focus on the development of the western part of China;
2. competition from the stronger competitors of the Yangtze River Delta and the Pearl River Delta; and
3. main Taiwan investment shifts to the Yangzte River Delta area.

The planning team argued that the decline in inward investment was also due to a market limitation in the region. The market catchment area of Xiamen was limited compared with that of the Yangzte River Delta and Pearl River Delta. The limitation in the market would seriously affect its attractiveness for to inward investment.

In order to increase the catchment area of the market to reduce market limitation, it was proposed in the Development Strategic Plan that the following policies should be seriously considered by local government:

- Short-term goals: through collaboration among governments and commercial organizations, to organize conventions and exhibition activities to strengthen economic cooperation between Guangdong Province, Zhejiang Province, Jiangxi Province, Fujian Province and Taiwan.
- Long-term goals: after establishing the centripetal force to Xiamen in cargo flows, information flows, financial flows and passenger flows, Xiamen should establish its market network with nearby provinces.
- Local region: strengthening the central function of Xiamen in finance, service, trade, science and technology and culture in the local region, and updating the industries in Xiamen.
- Marketing strategy: marketing the South Fujian region internationally by promoting its culture, tourism, ecology, history, and city image as a whole. Becoming actively involved in international cultural and economic exchanges to expand its worldwide reputation.

Regarding policy issues, the planning team suggested that after joining the WTO and further opening up and continuing reform in China, the favourable policies which Xiamen used to have would no longer exist. The hinterland areas would compete with coastal areas in terms of cheaper land and labour costs. Xiamen's problem was its shortage of a higher skilled and professional labour force. With changes in types of inward investment projects, and the proposed high-tech industrial development in Xiamen, the techniques and skills required by these industries would be higher than those of the former labour intensive industries, although, compared with the other four cities at the same level, Xiamen still had a higher percentage rate of university students. The problem is how it can provide a perfect environment and opportunities to keep these graduate students and a highly skilled labour force locally.

Urban planners have recommended several proposals for economic development, typically the industrial dimension in the Xiamen Urban Development Strategic Plan. They are summarized as the 'One Core and Two Centre Strategy'.

A core: New and high-tech industries as dominant industries High-tech manufacturing contributes 27 per cent of total industrial output in Xiamen at present. It was the central government's policies in the Ninth Five-Year Plan (1991–5) to encourage development in electronics, telecommunication and Internet technologies. Based on this background, it was suggested in the Plan that Xiamen should use its authorized power to promulgate certain favourable policies and by-laws to attract a skilled and professional labour force in the high-tech industry, especially those in electronic and computer technologies, to establish an R&D zone in Xiamen. Through collaboration with Taiwanese advanced electronic and computer technologies, there would be the opportunity for Xiamen to have continuing development in high-tech industries.

It was proposed in the Development Strategic Plan that the key industries of Xiamen should be dominated by new and high-tech industries, which emphasized the development in electronic information, software, biotechnology, and aircraft repairing and maintenance. The existing chemical industry should be further strengthened. The machinery could be updated for modern manufacturing. The petrochemical industry should focus on lower stream products instead of upstream ones to avoid pollution.

It was found by the author in participant observation that urban planners had to balance diverse interests. The final proposals were normally compromised ones. The proposal for new and high-tech industries in Xiamen is such a case. There were several workshops to discuss appropriate industries for Xiamen. It seemed that although the petrochemical industry could increase GDP, and there were some investors who had already held discussions with the municipal government, the development schemes of these investors would affect Xiamen as a tourist attraction city and the existing environment from a long-term view. It was difficult to persuade Xiamen's government that development of the tourist industry would be rational depending on the indigenous characteristics. The difficulties came from the demands of local politicians to seek local GDP growth, and more available jobs for local people and potential inward migrants from rural areas. As a developing country, and with traditional policies emphasizing secondary industry development for a long period of time, as analysed in Chapter 2, high-tech industries had always been the priority of economic development. Despite the fact that the contribution of tertiary industry in GDP has increased tremendously, especially tourism in Xiamen, the concept has not been generally accepted. Moreover, tourism is still at the beginning of its growth. It has not matured yet. Some secondary industries have to be recommended for industrial output, GDP and employment. Otherwise, the client, local government, might not accept the plan. The appropriate industry would then be high-tech manufacturing, which would create less pollution but provide more job opportunities and a better contribution to GDP.

Concerning the location of different industries, the planners suggested that the western part of Xiamen should limit the heavy chemical industry and machinery. They would be better located in Tong'an. There were four concerns of planners through observation. One was that industrial land resources within the West were limited; the second was that with the allocation of industries to Tong'an, industrialization and urbanization in the area could be increased to establish the city pattern of a large Xiamen; the third was consideration of the low cost of the labour force in the Tong'an area; the final one was concerns about pollution on the Western coast, particularly the impact of heavy industry on Gulangyu Isle.

Two centres: Tourism centre and logistics centre Through analysis of existing economic development and its components, it was suggested in the plan that Xiamen had already entered the post-industrialization era. Tertiary industry would develop more rapidly from now on. Xiamen would be able to develop as a regional retail and service centre. It was recommended in the plan that the local characteristics to establish a tourist centre should be considered with the targets as:

- a famous international, ecological seaport and tourist city in Pacific-Asia;
- establishing a large tourist enterprise to function as a regional centre and to have a general development of tourist resources within the region;
- marketing Xiamen, increasing its reputation, developing conventional business, and increasing construction of the tourist infrastructure and retail centres.
- Xiamen has a perfect geographic location. It is situated between the strongest economic zones and opposite to Taiwan. Further, Xiamen has merits in its sea and air transport. The seaport with capability for the sixth generation of container-carriers, is the sixth largest in China. The capacity of the airport is the largest in Fujian Province. If the relaxation of hostilities between the two sides of the Taiwan Straits continued, economic relations between Mainland China and Taiwan would further improve. Xiamen would have the opportunity to establish a logistics centre in the southeast Fujian area due to its convenient accessibility between the mainland and Taiwan. To develop as a logistics centre, Xiamen needed to promote its economic capability, and to extend its economic catchment areas to the hinterland, whilst further developing and expanding its seaport. It was proposed in the plan:
- To formulate a unitary plan for the docks development within the whole bay by ignoring the administrative boundary.
- Xiamen port should be developed as an international hub port with containers and miscellaneous cargoes.

Urban Development Schemes Analysis and Spatial Structure Proposals

Even if competitive capacity promotion and social and economic development issues were the foremost parts of the Xiamen Urban Development Strategic Plan, the allocation of land resources for development was still the priority task of the plan. The basic function of urban planning would be to plan the spatial layout, regardless of whether in the City Comprehensive Plan or the Urban Development Strategic Plan.

Rational planning theory regards planners as experts with neutral and apolitical values. The planning proposals would be apolitical (Faludi 1973). It was found in observation that development schemes and proposals were based on planners' knowledge, rationality and sometimes their aesthetic views. Some proposals might be made according to detailed analysis, including cost and benefit. Financial capacity was considered in the assessment of schemes, although this might not always be the case. It was found in the participant observation of the author that during the schemes' assessment, a consultant suggested that a canal be dug to link the lakes to the east and west of Xiamen Island as a tourist attraction. The proposed canal would physically divide the entire city into two parts, while the total cost would be 1.3 billion yuan (about 110 million pounds sterling). There were several serious debates about this scheme among the planning team. It was decided not to recommend this scheme ultimately as a result of cost–benefit analysis. Nonetheless, it illustrates that development policies, proposals and schemes in the plan have often been affected by the values and aesthetic views of planners. They may not always have been accepted by others because of differences in values and aesthetic concepts.

The planning team analysed the land resources for future development as follows:

- Xiamen Island with 133.67 km^2 land, but only 90 km^2 available for urban construction, the rest being mountains, beaches and shoals.
- Jimei, as a cultural, educational, tourism area and one of the four main Taiwan Investment Parks in China, has an area of 213.2 km^2, of which, 18 km^2 of land can be used for urban development.
- Xinlin, as the main industrial zone and access and distribution centre of cargoes in Xiamen, has 234.2 km^2 of land, but only 22 km^2 available for urban development.
- Maluan has a main industrial zone, 'Xinyang Industrial Zone'. There are 30 km^2 for industrial development.
- Haicang, including Haicang new city, industrial zone to the south and port area, has 45 km^2 land for urban construction.
- Liuwudian (the east part of Xaimen Bay), has 180 km^2 of available land.
- Tong'an town and its north area have about 300 km^2 land for development. Nevertheless, water shortages in this area would be the main restraint.

According to existing land availability and potential economic development through analysis by the planning team, it was proposed in the Development Strategic Plan to establish an urban pattern of 'One Municipality with Six Urban Districts'. The geographic pattern of Xiamen, as a seaside city, provides an opportunity for Xiamen to promote as a group pattern city. Separated by the straits between Xiamen Island and the industrial zones located to its other districts outside the island, i.e. Haicang, Xinlin, Maluan and Tong'an, Xiamen Island's environment is protected and the possibility of over-development is avoided. The planners raised numerous scenarios for discussion and analysis.

Through a cost–benefit analysis, the planning team proposed implementing Scenario One in the short term. Moreover, because Xiamen is a scenic and tourist city, they were concerned that the development of Xiamen Island should concentrate on tertiary industrial development. Existing secondary industries and population should be resettled and decentralized at an appropriate time. This proposed policy would not only control the development scale and construction volume in Xiamen Island to protect its environmental capacity, it would also help to stimulate the shift to a bay pattern city from an island pattern.

For the long-term, Liuwudian to the east part of the bay should be promoted as the other tourist, retail and services centre of Xiamen, including administrative, business, finance, information and logistics functions. This area, with its geographic location, would provide seamless linkages and business relationships with Taiwan. This development dimension could be convenient for connection to Quanzhou to form the co-urban South Fujian area.

Concerning the timing and phasing, the urban planners of the team recommended that development should follow the principle of 'Priority developing the West part of Xiamen and reserving the East part for next phase; the South area of the city should be developed first while reserving the North area for the future'. The west refers to Haicang. The main reason for proposing this district first was that in the last decade there had been huge amounts of investment in infrastructure in the area, such as the Haicang Bridge which reduces the distance between Xiamen Island and Haicang. The existing local situation had been ripe for further development. From the rational viewpoint, the development allocated to the area should be in the first phase. Its growth would additionally provide the linkage to Zhangzhou, the nearby city. After the completion of the west, the eastern part of the Bay, Liuwudian, could be improved, unless there was a major contingency, perhaps related to the relationship between Mainland China and Taiwan. The proposal that the north be enhanced later was based a on cost–benefit analysis of infrastructure investment in the Tong'an, a rural area. Regardless of the fact that it had been re-designated an urban district, its main function remains agriculture. There would therefore be great demand for investment in infrastructure to improve the area. Since other parts of the municipality in the south are still available for development, it would be reasonable to maintain Tong'an as a future land reservation, even for subsequent generations. The Development Strategic Plan was written to formulate the policies and scenarios appear justified and reasonable. However, due to many stakeholders' interests in Xiamen bay pattern development (see Figure 7.1), conflicts and argument cannot be avoided because of the different interests and concerns. The local district government, as one of the stakeholders, had a different opinion to that of the planning team.

It was found by the author that the planning team had tried to coordinate the diverse interests and development schemes of government authorities and to take account of their proposed investment schemes. However, the interests of the local communities, developers and district governments had been neglected. There had not been any public participation process in the plan's formulation. Local communities and local residents had not been given the opportunity to familiarize themselves with the planning policies, proposals and schemes. There had therefore not been an opportunity for them to express their opinions. There had not been any mechanism in the planning process for the planning team to communicate with developers and investors. Developers and investors were unable to give their opinions before the plan's approval and publication. The problem then was that the development and construction of Xiamen would need a contribution, particularly investment, from these stakeholders. Conflicts and bargaining would be inevitable in the later implementation stage.

It was also found from the investigation that there had been diverse opinions among the developers regarding the bay pattern city development. This was understandable owing to the different interests. A private real estate developer, from BEACH XIAMEN Real Estate Company, was worried about the increasing price of land in Xiamen Island. It is his opinion that:

> It is acceptable to have more land available for development. The planning of Bay Pattern development suggests strict control over development in Xiamen Island. The problem is that there are still available plots of land, which have been leased but without development. The price of these plots of land will increase. There are numerous opportunities for speculation by those developers who have leased these plots of land. This would be unfair to other developers. As a developer, I would first focus on Xiamen Island. (Interviewed on 25 June 2003)

The other developer, from Xiamen Haicang Real Estate Company, was happy with the bay pattern development. He said that: 'the Bay Pattern City will create more accessible land for Xiamen. Haicang could function as a sub-centre, or even as a second centre of Xiamen with its existing development' (interviewed on 26 June 2003).

The planning team had not even consulted with local district governments (under the jurisdiction of the municipal government). This was because urban planning had been treated as a central government function in the centrally planned economy, and then municipal governments' task after the reform. Planners had not considered consulting district governments since they had never had any power in urban construction and development, though the urban development and construction power had been decentralized to the district government in some super mega-cities, e.g. Shanghai in the transition during recent years. It was because district governments lacked a prior role in urban development and construction in China that their interests had then been ignored. This created another potential conflict for the plan implementation. Collision could emerge in the plan approval process. Local district governments could express their strong opposition to planning proposals if their opinions were not considered when producing the plan.

From interviews with local politicians, planners and citizens, it was evident that the proposal of 'Priority developing the West part of Xiamen and reserving the East part for next phase; the South area of the city should be developed first while reserving the North area for the future' was unpopular with various politicians, especially those from Tong'an. They were unhappy that their development phase was to be delayed. They insisted on equal opportunity for all. Their criticism was understandable because they needed to increase economic capability of their area otherwise their achievements could not be shown and subsequently evaluated. As regards to developing the settlement as a real urban district, local residents would expect to increase their quality of life. The urban development of the area could increase local land values. Economic development, particularly industrial development in the area, would also provide job opportunities.

The planning team emphasized in the report to the Xiamen municipal government that whatever scenario was adopted for the local development, it should be heavily constrained as a consequence of taking into account land availability, shortage of water supply, and the environmental capability of Xiamen because of the nature of the city.

Ecological Environment Threshold Analysis

Although there is no 'neutral role' of planners as suggested by the rational planning theory, because the persons with varying backgrounds and interests will understand problems differently as a result of their different values and different interests, it is still necessary to have rationalization in urban planning. Since there is always a limit to the ecological and environmental capacity of any place on earth, unlimited development is precarious and non-sustainable. It is critical to define the maximum capacity of development and the types of industries through threshold analysis. This is the function of planning as a 'gatekeeper' to avoid the worst situation. In the Xiamen project, the planning team tried to introduce a sustainable development policy by application of Ecological Environment Threshold Analysis.

Analysis of the ecological environmental capacity was significant to the sustainability of Xiamen. It provided information for the decision-makers as to what would be the maximum environmental capacity. To exceed this would create expense in certain costs, including pre-spending of resources of later generations, and the decline in quality of life.

Ecological environment analysis methodology consisted of such elements as:

- Analysing water resources to define the maximum population in the whole area.
- Defining the environmental capability of districts outside Xiamen Island, including the new urban district, Tong'an, with an existing agriculture function but future urban development, by assessing available land resources, their intensive use, and need for environmental protection.
- Determining tourist impact and environmental protection in order to define the maximum capacity of the three urban districts within Xiamen Island and Gulangyu Isle.
- Ascertaining the environmental saturation capacity for Xiamen municipality.

As a result of the ecological environmental investigation and analysis, the planning team pointed out to the local government that there were no large rivers in Xiamen municipality. Xiamen Island had a shortage of fresh water supply. The supply of urban water came from Jiulongjian River and Bantou Reservoir. The average water availability per capita was 963 cubic metres, only 42 per cent of the ratio nationally and 20 per cent of that in Fujian Province. Although the existing water supply could meet the demands of present uses, it would seriously affect and restrain future development, especially peoples' quality of life and industrial development.

It was mentioned in the plan that until now Xiamen had been developed as an island pattern city, and land available for development had been limited. At present, the average land resource was only 266 square metres per capita, only one-third of the national average. The total area of the municipality was 1651.98 square kilometres, of which 919.54 square kilometres were available for development land. This could be divided into 563.29 square kilometres of cultivated land, 254.46 square kilometres of urban built-up area and residential land in rural areas, and 40.62 square kilometres of transport facilities land. There were only 61.23 square kilometres of unused land. The development of the east part in the process of the bay pattern city development would add 300 square kilometres more land for urban and industrial development.

The ecological environment threshold study also focused on water and air quality. After careful and detailed study, it was proposed in the Development Strategic Plan that Xiamen should consider the development and construction of an ecological environmental city as its main principle. Its high quality environment was viewed as the main element that had influenced developers' and investors' decisions on site selection. The plan recommended additional investigation and analysis of the ecological functioning divisions in Xiamen after the Development Strategic Plan.

Through the ecological analysis and investigation, the planning team suggested the maximum population and recommended the optimum population. The ecological environmental capacity for maximum population was 2.9 million. The environmental saturation capacity was about 70–75 per cent of this, at 2.2 million.

The key question after the ecological analysis was what type of development should be promoted. The investigation of the author illustrates that arguments around the planning team arose from conflicting concerns:

1. to protect the high quality environment of Xiamen as a scenic attraction and tourism city;
2. to take economic advantage of further growth; and
3. to attract inward industries, especially since petrochemicals had been listed as one of the dominant industries in the short term and the seaport development was in keeping with the city's existing role.

In this situation, the planning process illustrated Christensen's 'bargaining' scenario. There was no general agreement as to whether the existing high environmental quality and tourist interests ought to be seriously affected by the development of heavy industry, including petrochemicals and the seaport, in the interests of expanding total economic volume and increasing GDP, or should be retained for long-term interests in tertiary industry, education, tourism, finance, retail, even R&D, but at the loss of the GDP increase contributed by heavy industry.

In the wider national and regional context and long-term view, there was a case for retaining Xiamen for environmentally friendly development, e.g. tourism, education and others. However, in the narrower focus of some politicians, the decision was not so clear-cut. A distributional bargaining situation exists with local leaders seeking the short-term objectives of leaving something tangible from their administrative term, whereas sustainable development required the longer term objectives of maintaining the high quality environment area for future generations.

A compromise solution agreed among the team and then proposed to the Xiamen government in the City Development Strategy was to defer the decision about the proposed new industrial development and to impose the requirement that all land currently designated and serviced for industrialization in Haicang should be developed before the east area of Xiamen was commenced. However, as in all instances of distributional bargaining, power eventually decides the issue.

Planning Policies in the Local Government Decision Process

Planning proposals and policies of an urban development strategic plan should help decision-making of local municipal government, the client of the plan, otherwise, the plan may lose its functions in achievement of objectives.

From the investigation, it was found that the Xiamen Urban Development Strategic Plan had influenced the general policies of the local government. Some concepts of the plan were used in the government's decisions, particularly the analysis of Xiamen as a local smaller market, which would affect local consumption demands, and the conclusion that its weakness resulted from the lower total GDP and economic volume. The local municipal government viewed these as the 'opinions of experts' and relied on them to expand the scale of the city. Xiamen's party chief explained that the existing development capacity was inappropriate under the pressures of economic globalization and the aggressive development tendency of other cities and regions in China, because Xiamen is smaller in city size, lower in economic volumes, and has poorer interactive capability in industries. These factors have restrained Xiamen from becoming a central city. The decision of the municipal government was to develop the city to become a super megacity with comprehensive competitive capability through urban expansion and an increase in economic capacity (Xiamen Government, 2002).

After completion of Xiamen Urban Development Strategic Plan, Xiamen's municipal government specified 17 research reports covering most social and economic growth aspects in August 2002. They included infrastructure, industry adjustment, regional development, education, tourism and others. These reports were completed by October and submitted to the headquarters of the municipal government on 30 October 2002. These reports and Xiamen Urban Development Strategic Plan were further developed as a government policy document entitled 'Implementation Outlines of the Xiamen Development Strategy' to guide city development. It was clearly stated in the 'Implementation Outlines of the Xiamen Development Strategy' (XMG 2002) that the objectives of the development are to increase the capacities of the city and to promote new advantages of the city.

The proposed spatial layout in Xiamen Urban Development Strategic Plan was considered and adopted as the government's development policy. It was clarified by Xiamen municipal government that Xiamen Island would be the core of the whole municipality, while expanding its periphery to the east and west to form the pattern of a 'Sea in the City, and a City in the Sea' (XMG 2002). The Outline accepted the idea of 'One Municipality with Six Urban Districts' as the municipal spatial structure. Xiamen Island, as the core, should be developed for high-tech industries and tertiary industries, i.e. tourism, retail, finance and banking. The proposed functions and development dimensions and phases of other districts, e.g. Haichang, Malan, Jimei, by the plan were also generally accepted by the government, except the phasing proposals for Tong'an and the eastern area of Liuwudian.

It was suggested by the planning team that development of Tong'an and the eastern area of Liuwudian should be postponed until the next Five-Year Plan period (2006–10) when other parts of Xiamen will have been nearly fully developed. This development phase proposal of 'Priority developing the West part of Xiamen and reserving the East part for next phase; the South area of the city should be developed first while reserving the North area for the future' was not accepted. Xiamen's government decided in the Implementation Outline that Tong'an should be developed as a new urban district with comprehensive urban functions mainly focusing on ecological tourist development. The existing town will be expanded to the north and the west. The east part of Xiamen, Liuwudian, will be developed as a modern seaside new town.

Although Xiamen Urban Development Strategic Plan turned out to be justifiable and accountable, it was not readily accepted by the pluralistic stakeholders because of their different interests, even though the planning proposals or schemes resulted from 'scientific' and 'objective' research. This is the weakness of the rational planning approach in implementation. It is evident that the bay pattern development of Xiamen is not only the task of the planners and the municipal government but all of the stakeholders in society should make a contribution to it. The lack of communication among stakeholders created the risk that their support might not be obtained. From this perspective, Xiamen Urban Development Strategic Plan failed to become the shared vision of the stakeholders. It was neither accessible to all the stakeholders, nor transparent in its formulation.

From the view of the local municipal government, it was clear from the Outline that the key objective is the city's expansion, for Xiamen to become a super mega-city to increase economic volume. Xiamen Urban Development Strategic Plan had supported the ambitions of the local municipal government via the urban planning profession. The Outline suggested that city expansion should proceed by urbanization and industrialization led by infrastructure construction. The development of industrial zones and new urban settlements should be encouraged. It was the intention of Xiamen's municipal government that, by 2010, the developing belts between Haichang new city and Maluan area, Jimei and Xinlin, and Tong'an and Jimei should be completed. The total urban built-up area will reach 200 km^2 with urbanization of 75 per cent (XMG 2002).

In order to stimulate the development of Xiamen eastern area, transport infrastructure development was soon initiated after the plan had been approved. The initial estimated budget was three billion yuan (about 300 million pounds sterling). The detailed action plan of the eastern area has been also completed soon after Xiamen Urban Development Strategic Plan. The detailed action plan suggested that the eastern area should have the diverse functions of a comprehensive seaport and logistics park; high-tech industry; ecological seafront sports, tourism and recreation; a retail and trade centre in the South Fujian region; and more significantly, an operation base for economic, trade and cultural exchange and cooperation with Taiwan. Xiamen municipal government was ambitious to expand the city by the development of the eastern area.

As regards industrial development, the government has adopted major principles of the Xiamen Urban Development Strategic Plan, such as the industrial structure should be adjusted to concentrate on high-tech industries, while promoting business development in logistics, tourism, finance and banking, conventions and exhibitions. Based on the proposals of the Xiamen Urban Development Strategic Plan, the municipal government has suggested that Xiamen should also develop culture and education. These two elements have not been emphasized in the plan. It was mentioned in the Outline that Xiamen should be developed as 'a City of Sciences, a City of Cultures and a City of Education'. The government was targeting a comprehensive integrated increase of competitive capability.

Regional collaboration and communication have been mentioned and are regarded as one of the critical issues for the development of Xiamen. It was suggested in the Development Strategic Plan that 'win-win' opportunities throughout the region should be sought. The planning proposal was considered by Xiamen's municipal government as the supporting reference to establish the 'Cities Alliance' in South Fujian Province. In May 2003, the directors of the three municipal bureaux of urban planning administration met in Zhangzhou to promote regional development and planning coordination. It was the beginning of the 'Cities Alliance' in the region.

With the establishment of the 'Cities Alliance', there will be regular coordination meetings and official discussions among the mayors of the cities and the directors of the municipal bureaux whose responsibilities may need coordinating within the region. These include regional planning formulation and implementation policies, major infrastructure projects and public utilities development, and the regional coordination of industries and ecological environmental protection.

Conclusion

This chapter has analysed the purposes, objectives and process of a non-statutory plan, the Urban Development Strategic Plan in the case study of Xiamen. The application of the Urban Development Strategic Plan is an innovation in planning from the bottom-up from local government in contrast to the existing top-down urban planning system. Local governments are unable to wait for the reform of the existing planning system from top-down by the central government. In addition, the process of the present statutory plan, the City Comprehensive Plan, cannot help them deal with the rapid changes and the problems they are facing. It is argued by Paris (1982) that urban planning is not a form of rational action for urban development, but part of the social and economic change process instead of separate from the changes. Urban planning has to reflect and to cope with social, economic and political changes. The purpose of the urban development strategic plan is to meet the demands of the present changes and pressures from competition. Since the delivery of the Urban Development Strategic Plan in China, there have been debates on the functions of this new type of plan, typically if it can replace the traditional and statutory City Comprehensive Plan (Li 2003; Luo and Zhao 2003; Yin 2003; Zhang 2002; Zou 2003b). From the case of Xiamen, it is able to compose few commends that the new innovation in planning has achieved the initial planning targets. This new type of plan has helped local government to understand its competitive capability within its regional and local context. The plan proposed that spatial development schemes should provide available land for future demands, and the city image should be marketed to potential investors and developers. The plan had been completed in a short time. The main problem of the time delay of the statutory City Comprehensive Plan was avoided. It was illustrated in the Xiamen case that increasing competitive capability was a main objective of the Urban Development Strategic Plan.

It is also evident from the case study that the main implication of the Urban Development Strategic Plan for the function of land-use change is a more positive encouragement of the desirable rather than the prevention of the undesirable. The Urban Development Strategic Plan emphasizes the developmental objective to provide a stimulus for new activities whilst providing potential development possibilities to potential users; and proposes improving the institutional efficiency to provide a framework for the integration of different urban policies with spatial consequences.

The development of a city will be greatly influenced by the image of its neighbouring cities, even if they may be quite different. Such impact may be positive, as a city reaps the benefits of the attractive image and promotional efforts of its neighbours, or negative when a city suffers through spatial association with the poor image of its neighbour. Accordingly, the promotion of one city must take account of its neighbours. Xiamen Urban Development Strategic Plan has particularly emphasized regional coordination. It has been found that the proposal has been regarded as a support from the urban planning view for the establishment of the 'Cities Alliance'.

Chapter 8
Land Use System, its Reform and Planning Management

Introduction

The delivery of Chinese city and regional plans is operated through the development planning management process, which is the main task of the bureau of planning administration in local governments. A bureau of planning administration of a municipal or a county government is authorized by the City and Countryside Planning Act 2008 to intervene in urban development by the power of development planning management guiding by statutory city and countryside plans. This power refers to the administration of the formulation of plans and their approval process, and assessment and approval of the application for planning permits for a development project.

The term 'planning management' is used in the Chinese city and regional planning system for two reasons. First, there is no definition of 'development'. Second, in the centrally planned economy when the Chinese planning system was first initiated, the government decided the development tendency and dimension. The government also allocated all resources for development. In other words, the government had monopoly control of all the basic elements of resources for development. City and regional planning was regarded as the extension and physical realization of the five-year economic and social development plan (Huang 1999). City and regional planning was not authorized to control development, but to manage the process of allocating available land to meet the demands of the social and economic development projects decided and invested by the governments. There were no private property rights and the development rights likewise belonged to the government, therefore, it is not development control, but planning management.

The development planning management process is a processing through a complicated and time-consuming certificating system of so-called 'One Permission Notice and Two Permits' mechanism, which includes the 'Permission Notice' for the decision of land use location for a proposed development project; and the 'Planning Permit for Land Use' and the 'Planning Permit for Construction and Engineering' of the proposed project.

Within the certificate mechanism of 'One Permission Notice and Two Permits', the local planning bureau should assess and issue certificates according to the planning acts, statutory plans and relevant planning conditions, which are explained through controlling indexes, ordinances that are specified in the Regulatory Detailed Plans (Zhou and Huang 2003).

This chapter will explore and analyse the planning management in the Chinese city and regional planning system. However, it is useful to understand the land system and its reform since the late 1980s first of all. Chinese city and regional planning management mechanism has been closely associated with and impacted by the land use reform.

Chinese Urban Land Use Reform

The Features of Chinese Land System

China has operated a dual land ownership system since the 1954. According to Article Eight of the Land Administration Act 1986 of the People's Republic of China, and the amended Land Administration Act 1999, the land within urban areas is owned by the state; the use of urban land should be registered and permitted by issuing licence in local government which should be at least at a county level in the Chinese administrative hierarchical system. Land within rural areas is under collective ownership except those areas that are specified by the Land Administration Act. The use of rural collective land should be registered and permitted via issuing licence by the local county government. The different land ownership should be managed by a different mechanism. Any piece of land in rural areas is forbidden to have any urban development or industrial development. Before the land use reform in the late 1980s, the land uses both in urban and rural areas were free of charge and without any time limitation, or, in other words, it was possible to use a piece of land permanently if someone occupied it.

Policies of Chargeable Leasing for the Right of Land Use

It was after the open door policy, especially since the introduction of the 'Socialist Market Economy', that the urban land use system in China started its reform. The original urban land use system, which was introduced in the year of 1954 as an administrative distributing one, with the characteristics of free of charge and permanent use, has been changed.

In 1982, Shenzhen Special Economic Zone initially started its reform of land use in China. The Shenzhen municipal government began to charge for land use when cooperating with overseas investors. Nevertheless, it was not until May 1984, during the 6th National People's Congress that an experimental delivery of commercialization of housing development was promoted. Since then, real estate development has become an important contributor to GDP. Wuhan, Wushun and Guangzhou started their implementation of land use reform in 1984 immediately after the 6th National People's Congress.

It was in the year of 1986 with the promulgation of the Land Administration Act of the People's Republic of China that the chargeable land use by leasing its use rights started. It was also confirmed by the Land Administration Act that transferring the right to land use between urban land users was a legal action. The land market in urban areas was thus formally established. It has been specified by Article 4 of the Land Administration Act 1986 (NPC 1986) that any company, enterprise or institution or individual person from China and other countries, with a few exceptions specified by the Act, are able to transfer, rent out or mortgage their right of land-use within their usage period. However, the transfer of this right should be inspected and supervised by local municipal bureau of land administration. All duties and responsibilities associated with a piece of land should be transferred at the same time when usage rights of land are being transferred, rented or mortgaged.

Land use reform had changed the former unitary administrative distribution approach of land use to various methods of negotiation (private treaty), tender offers and auction. The significance of land use reform is that the right to land use has become a commodity with a certain price that is transferable in the market.

The State Council has regulated the maximum time period of leasing the right to land use as follows (State Council 1990a):

- 70 years for residential land use;
- 50 years for industrial land use;
- 50 years for education, science and technology, culture, health and sport land use;
- 40 years for commerce, tourism and recreation land use;
- 50 years for comprehensive development, or other land use.

It is defined in Circular 55 of *The Interim Regulations of the Assignment and Transfer of the Right to Use the State-owned Land in Urban Areas* issued by the State Council (1990a) that local municipal governments should take responsibility for leasing the right to land use to potential land users. The users should pay a lump sum fee for the lease of right to the local municipal government. Circular 55 of the State Council clarifies that the price charged for land leasing should be dependent upon the local level of socio-economic development, location of land, features of intended use, and leasing period. The land users must develop the plot of land according to the leasing contract and planning policies, and planning controlling indexes and ordinances. The user is able to enjoy the rights of transfer, rent, mortgage and undertake development on the land. Generally speaking, a piece of land distributing from tender offer or auction should be provided with the necessary infrastructures, e.g. roads, water and gas accessibility, by local government.

However, it is useful to realize that it took many years to implement the chargeable land use reform in most Chinese cities. It is not a smooth way to introduce the market mechanism. Many local governments were not very active to deliver the mechanism of leasing the right to land use in the early 1990s because GDP promotion and attraction of inward investment were the priority policy of Chinese local government. It was the concern of local governments that leasing the right to land use through auction may increase land prices, which may increase the potential costs for investors. It would then be a barrier for the attraction of inward investment.

Although the implementation of the Land Administration Act 1986 (NPC 1986) and the central government's circular *The Interim Regulations of the Assignment and Transfer of the Right to Use State-owned Land in Urban Areas* (State Council 1990a) in 1990 could be regarded as the milestone for Chinese urban land use reform, the dual land distribution systems, i.e. the commercialized right to land use leasing approach through tender offer or auction process and administrative allocation process of land with free or low fees, have always existed.

The private housing development and housing prices before the mid 1990s were not as high as those in the late 1990s, and especially in the 2000s. When the housing price was low, the income from leasing the right to land use was not significant to local governments. Low land use fees, sometimes even free, were used as one of the incentive packages for the attraction of inward investment by many Chinese local governments. Some local governments would have to reduce the price of land leasing to attract potential investors by ignoring the cost of land and destroying the irrecoverable resources (Dong 1999).

According to the research of Wang (2008), even in Shenzhen, one of the first cities to introduce the commercialized land use leasing system, 90 per cent of land has been still distributed by administrative allocation before 1999. The majority of Chinese urban land distribution had not been operated according to the market mechanism. It was not until 2001 when the Circular No. 15 of *Strengthening the Management of the State Owned Land Resources* (State Council 2001) and *Regulations of State Owned Land Auction and Tendering* (MOLR 2002) were promulgated to emphasize the policy that all the commercialized urban land uses including housing, retail, banking and industry must be distributed through the auction of the rights to land use, that the application of the market mechanism was considered by local governments. It was typically after the mid 1990s when housing prices in most Chinese major cities dramatically increased. Income from leasing the right to land use has become the most important resource for Chinese local governments. Increasing land use leasing price has been in the interests of the Chinese local governments since then.

Different Approaches of Land Distribution

Lease large size of land for development In order to attract foreign direct investment to China in the 1990s, it was specified by *The Interim Regulations for Leasing Tracts of Land Development and Operation by Foreign Investment*, the Circular No. 56 promulgated by the State Council issued in 1990, that investors from other countries were allowed to apply for land use leasing for large tracts of land. Investors and developers obtaining this type land use right must develop the land according to the local City Comprehensive Plan and the feasibility studies produced by consultants for the tracts of land. It was specified in Circular No. 56

(State Council 1990b) that the provision of necessary infrastructure, such as water supply, electricity, district heating, roads, telecommunication and other public utilities is the responsibility of the developers and investors in order to enable land to mature for further development. Depending upon the interests and intention of foreign investor(s) and developer(s) who develop the tracts of land, they could build industrial buildings and the supporting facilities, daily living and service facilities on the ground. It was also possible for the investors or developers to transfer their right to land uses to anyone who may be interested, or to engage their own business on the land, or to sell or rent out the buildings on the ground. Generally speaking, this approach of land use leasing for development normally covered a large area of raw land without any infrastructure. It was defined in Circular No. 56 that this approach of leasing right to land use would only be operated within the Special Economic Zones, the Open Coastal Cities and the Open Coastal Economic Zones (State Council 1990b). Circular No. 56 was terminated on 7 March 2008 since the majority of China was now generally opened to the outside world instead of just the experimental Special Economic Zones or Open Coast Cities.

The reform of the urban land use system provides opportunities to increase revenue for local governments. It was defined in Circular No. 172 of *The Interim Regulations for Income from Lease of Urban State Owned Right to Land Use* promulgated by the Ministry of Finance (MOF 1992) that 5 per cent of income from land use leasing should be submitted to the central government and 95 per cent of income could be kept by local governments. This policy has been amended by *Notice for the Income Management from Leasing State Owned Right to Land Use* issued by the General Office of the State Council (2006) that all the income from land leasing tending will be kept in local government.

Auction (tender offer) for obtaining right to land use   The auction or tender offer measure of the right to land use lease is the most attractive approach to lease land both for local governments and developers. Local government may have the opportunity to increase the leasing price for more income, while developers are able to consider the possibility of increasing heavy marginal profit for development on land. Delivery of this land leasing approach has amended the planning management mechanism. It is regulated by the City and Countryside Planning Act 2008 (NPC 2008) that before auction for leasing the right to land use, the nature of land use and development intensity should be confirmed by a regulatory detail plan. They should be formulated as important items of the contract for leasing the right to use start-owned land.

In the approach of an auction for leasing the use right to a piece of land, local governments, or an organization on behalf of local government, should produce a notice to announce that the use right of a piece of land will be leased. The notice should be available to developers (or investors) at least one month earlier before the auction date to leave time for them to operate an assessment before deciding to submit the proposal of development schemes and the proposed payment to the leasing land.

The notice of use right lease of a piece of land should include the necessary information about the land use functions, its planning controlling index and ordinances that will be associated with the land plot. It is possible for a potential developer or investor to undertake an initial analysis for marginal profit before a making a final decision on bidding for the land. If any developer or investor is interested in the piece of land, it is the task of developer and investor to provide a proposed development scheme and the price that they intend to offer for obtaining the right to land use.

Administrative allocation of land at low charging fee Investors and developers are able to apply for usage rights to a piece of urban land through the approach of administrative allocation from the local bureau of land administration, with the report of a project feasibility study, the approved documents of application for development planning permission, and the certificate of corporate enterprise. Normally, the approach of administrative allocation of land use is only for industrial development and public utilities construction. In addition to paying a development lump sum fee, the users of the land also pay a land tax every year.

However, in order to encourage investment in certain industries, e.g. high-tech and export-oriented industries, the Chinese government has made some favourable policies in the land tax for this category of land use. At the beginning of the 1990s, the highest tax was about 5–12 yuan per square metre per year including both development fee and land use tax. Nevertheless, the right to land use obtained through this approach is unable to be further transferred, rented and mortgaged. Moreover, it is impossible to change the land use function that was originally defined.

Leased Land without Development

The real estate speculation has been a critical issue in the Chinese cities. Some developers might obtain the use right to a piece of land for the purpose of re-selling the right to land use without actual physical development on the ground. It is regulated in the Land Administration Act 1986 (NPC 1986) that if a piece of land is without actual physical development after the right to land use has been held for two years, then the local government has the power to take the land back from the existing land usage rights keeper(s) or developer(s), and return it to the market for re-auction. In principle, local governments can take the land back without any payment. However, in practice, there is often some compensation to the existing land usage rights keeper(s) or developer(s). It can be done through the following two approaches:

1. On behalf of government, land auction administrative organization, e.g. Land Development Corporation, or Land Tender Centre can re-lease

the using right to the piece of land through auction in the market. After successful tender offer or auction, a certain percentage of the income would be paid to the existing right keeper(s) or developer(s) as payment for the 're-purchase' of the land by the government.
2. Negotiation for payment between the existing right using keeper and local government for taking back the piece of land. Land auction administrative organization will then provide the piece of land in the market after re-capturing land by agreed compensation to existing keeper of land.

For some special factors, such as illegal development on a piece of land, local government may confiscate the land. The land being confiscated by the government should be provided on the market for auction.

Land Acquisition and Allocation

Land acquisition and allocation for urban development have illustrated the features of the Chinese land system. In order to obtain land for urban development purposes, developers have to apply the right to land use leasing through different approaches of administrative allocation, tender offer and auction for a certain time period depending upon the land use functions.

There is no private land ownership, but either the state owned in the urban area or the collective owned in rural area. The land acquisition administration and the right to land use leasing administration are the task of and managed by local government. In the local government, except those amalgamating the land administration and planning, the Municipal Bureau of Land Administration (MBLA) should take responsibility for the process and decision on the urban land leasing and land tenure. It is not uncommon that there are usually contradictions between leasing the right to land use, and urban planning.

In terms of application of the right to land use, it can be divided into two processes of existing urban land, and the acquisition of agricultural land. Considering the severe shortage of cultivated land in China, the central government has made the decision to strictly control the total volume of cultivated land (State Council 2004). The approval power for the decision of transferring land use to non-agricultural function from existing agricultural use and acquisition of existing agricultural land is controlled by the State Council and provincial governments separately depending upon the size of land for transfer or acquisition.

The key issue during the application of the right to land use is to have the 'Permission Notice of Development Project Land Use Boundary'. This Notice should be utilized as one of the most important documents for the application of urban planning permit. The details of the application of the right to land use for both existing urban land and acquisition of agricultural land are illustrated in figures 8.1 and 8.2.

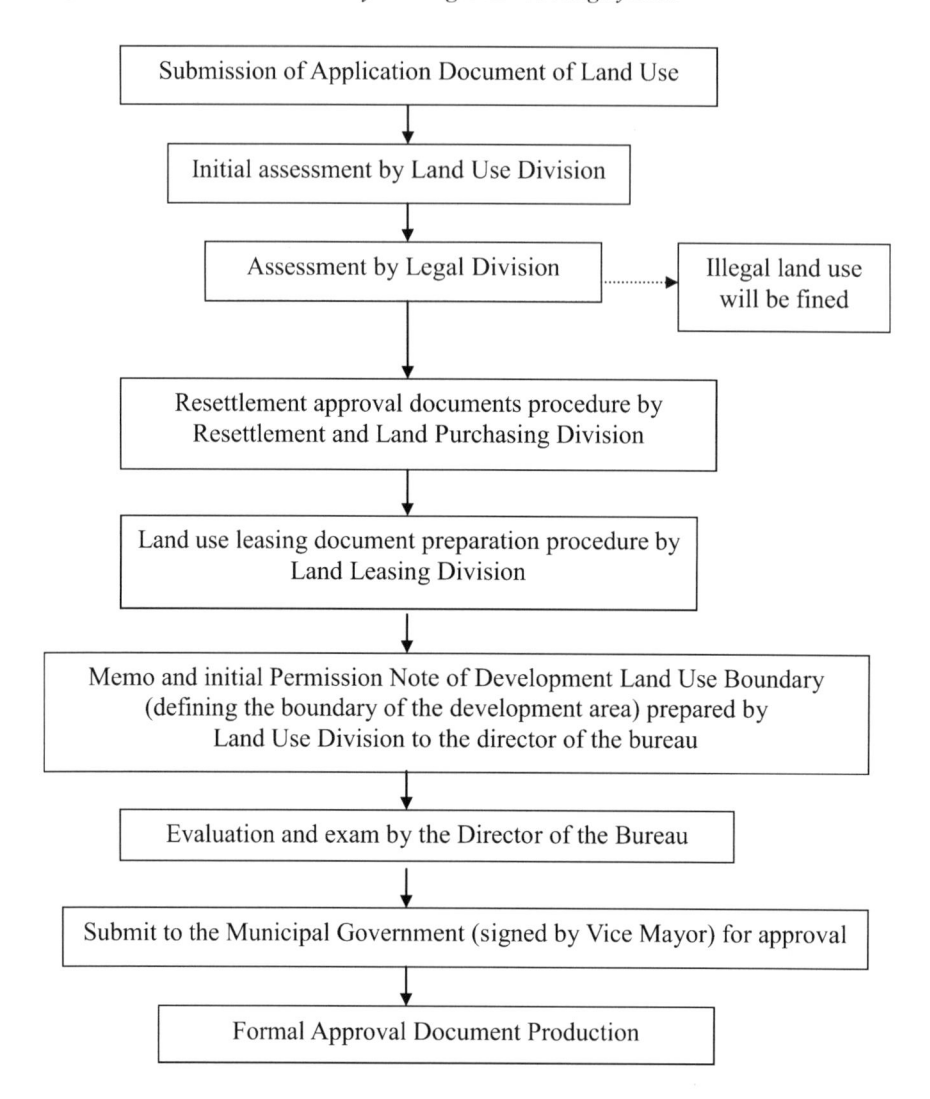

228　　　　　　　　*Chinese City and Regional Planning Systems*

```
┌─────────────────────────────────────────────┐
│ Submission of Application Document of Land Use │
└─────────────────────────────────────────────┘
                    ↓
    ┌─────────────────────────────────────┐
    │ Initial assessment by Land Use Division │
    └─────────────────────────────────────┘
                    ↓
        ┌───────────────────────────┐ ······▶ ┌──────────────┐
        │ Assessment by Legal Division │        │ Illegal land use │
        └───────────────────────────┘        │ will be fined    │
                                             └──────────────┘
                    ↓
    ┌───────────────────────────────────────────┐
    │ Resettlement approval documents procedure by │
    │ Resettlement and Land Purchasing Division    │
    └───────────────────────────────────────────┘
                    ↓
    ┌───────────────────────────────────────────┐
    │ Land use leasing document preparation procedure by │
    │ Land Leasing Division                              │
    └───────────────────────────────────────────┘
                    ↓
┌────────────────────────────────────────────────────────────┐
│ Memo and initial Permission Note of Development Land Use Boundary │
│ (defining the boundary of the development area) prepared by       │
│ Land Use Division to the director of the bureau                   │
└────────────────────────────────────────────────────────────┘
                    ↓
    ┌─────────────────────────────────────────┐
    │ Evaluation and exam by the Director of the Bureau │
    └─────────────────────────────────────────┘
                    ↓
┌──────────────────────────────────────────────────────────────┐
│ Submit to the Municipal Government (signed by Vice Mayor) for approval │
└──────────────────────────────────────────────────────────────┘
                    ↓
        ┌─────────────────────────────────────┐
        │ Formal Approval Document Production │
        └─────────────────────────────────────┘
```

Figure 8.1　Application process for using existing urban land

Land Use System, its Reform and Planning Management 229

```
                    ┌─────────────────────────────────────────┐
                    │ Submission of Application Document of Land Use │
                    └─────────────────────────────────────────┘
                                         │
                                         ▼
                    ┌─────────────────────────────────────────┐
                    │ Initial assessment by Land Use Division │
                    └─────────────────────────────────────────┘
                                         │
                                         ▼
                    ┌─────────────────────────────┐      ┌──────────────────┐
                    │ Assessment by Legal Division│┄┄┄┄▶│ Illegal land use │
                    └─────────────────────────────┘      │ will be fined    │
                                         │               └──────────────────┘
                                         ▼
        ┌──────────────────────────────────────────────────────────────────┐
        │ Land Purchase Documents Produced by Resettlement and Land Purchasing Division │
        └──────────────────────────────────────────────────────────────────┘
                                         │
                                         ▼
```

Documents prepared by Land Use Division within 30 days should including items:

- The proposed scheme of changing agricultural land use function;
- Proposed scheme of cultivated land replenishment (NB: regulated by Central Government that the total cultivated land must not be reduced. The side of transferring cultivated land must be compromised by similar size of cultivated land from other place);
- Map of cultivated land replenishment;
- Land use proposal of the development project;
- Opinion of the initial assessment of the project;
- Initial application report on behalf of the municipal government.

```
                    ┌─────────────────────────────────────────────┐
                    │ Evaluation and Exam by the Director of the Bureau │
                    └─────────────────────────────────────────────┘
                                         │
                                         ▼
        ┌────────────────────────────────────────────────────────────────────┐
        │ Submit to the Municipal Government (signed by Vice-Mayor) for approval │
        └────────────────────────────────────────────────────────────────────┘
                                         │
                                         ▼
        ┌────────────────────────────────────────────────────────────────────┐
        │ Submit to Provincial Government or Central Government for approval │
        │ depending upon the size of the proposed transfer of agricultural land │
        └────────────────────────────────────────────────────────────────────┘
                                         │
                                         ▼
        ┌────────────────────────────────────────────────────────────────────┐
        │ Approval by Provincial Government, or Central Government           │
        └────────────────────────────────────────────────────────────────────┘
                                         │
                                         ▼
        ┌────────────────────────────────────────────────────────────────────┐
        │ Issue of Initial Permission Note of Development Land Use Boundary  │
        │ (defining the boundary of the development area, prepared by Land Use Division │
        │ to the Director of the Bureau                                      │
        └────────────────────────────────────────────────────────────────────┘
                                         │
                                         ▼
                    ┌─────────────────────────────────────────────┐
                    │ Evaluation and Exam by the Director of the Bureau │
                    └─────────────────────────────────────────────┘
                                         │
                                         ▼
        ┌────────────────────────────────────────────────────────────────────┐
        │ Submit to the Municipal Government (signed by Vice-Mayor) for approval │
        └────────────────────────────────────────────────────────────────────┘
                                         │
                                         ▼
                    ┌─────────────────────────────────────┐
                    │ Formal Approval Document Production │
                    └─────────────────────────────────────┘
```

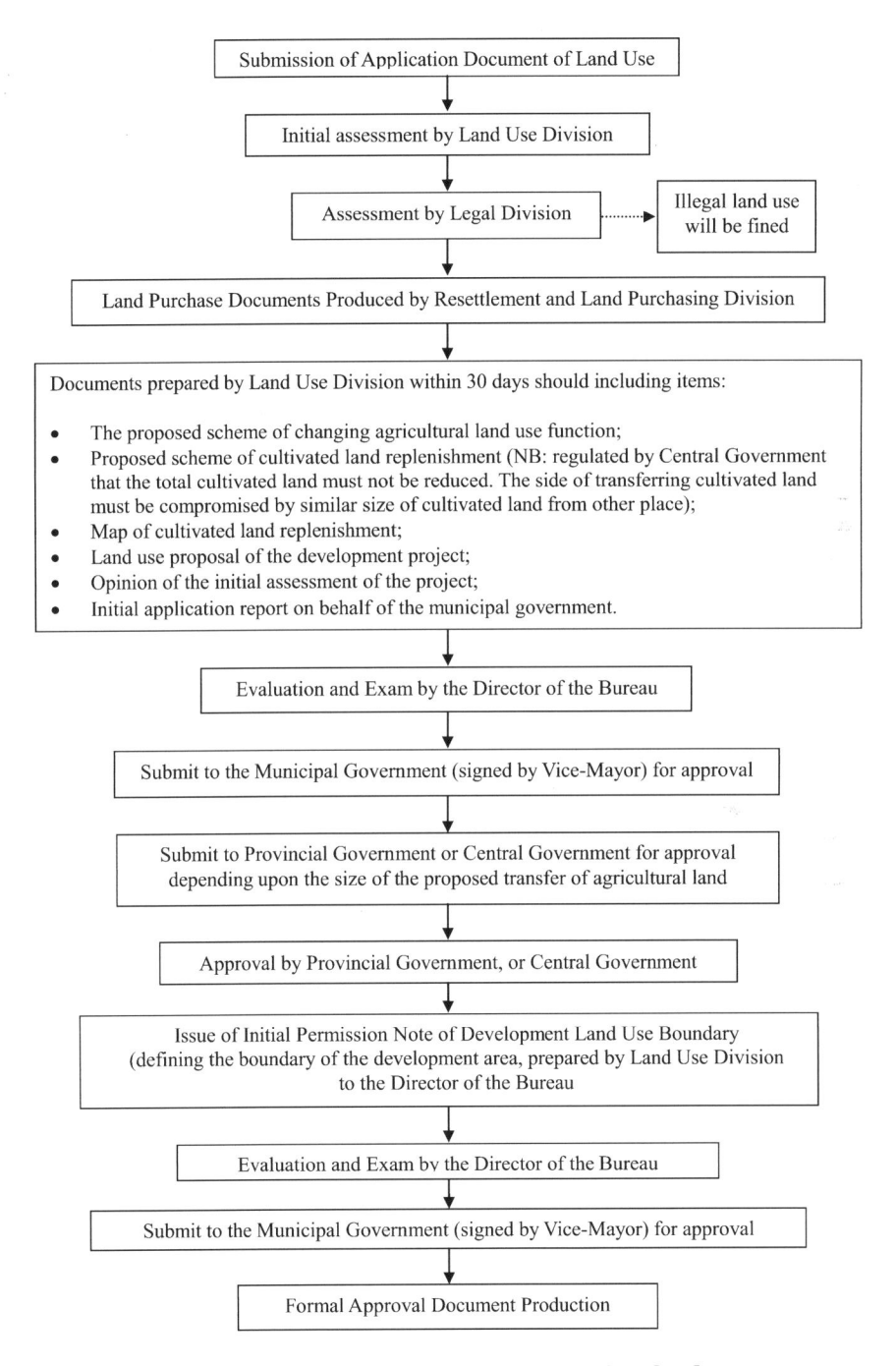

Figure 8.2   Application to transfer rural land to urban land use

The decision for leasing the right to land use of each piece of land should be made in local municipal liaison committee chaired by a vice-mayor who is responsible for urban and rural planning and development. The members of the committee consist of the directors of bureaus of land administration, urban planning administration, development and reform commission, and other municipal government bureaus that have responsibilities in economic development. After decision of land use leasing, the vice-mayor should sign the 'Permission Notice of Development Project Land Use Boundary' as a document for the approval of leasing the right to land use.

For some important projects, which may promote the development, or increase the local GDP, the land would not go through the auction process. They will be leased to the developer (investor) directly through the approach of administrative allocation of land use. This approach of land allocation has been mainly used to promote local industrial development. It is even possible for local governments to waive the land use fee for some important projects, typically the projects that can contribute to GDP and tax. The interests of local governments are then mainly for GDP and tax increasing instead of the income from leasing the right to land use.

Urban Land Development Corporation

The organization responsible for the tender offer, bidding and auction of the right to land use in local cities is the Urban Land Development Corporation, which is a quasi-government organization.

Since the mid 1990s, the income from land leasing has played a crucial role in the budget of local government. In order to generate more income from the leasing of the right to land use, to separate policy-making and management of land and operation of land as a business, many Chinese local governments have established urban land development corporations in order to operate the business of land purchasing, reserving and leasing of rights to land use to raise incomes that can be used for public services and social development by the local governments. Urban land development corporations have been authorized the power to monopolize the market of land use since all the urban land in China belongs to the state.

It is the responsibility of a land development corporation to purchase undeveloped land, typically existing cultivated land, and then to provide basic infrastructure on the land. When the land is mature for development, it will be leased to developers or any user in the market. The required financial capital for the purchase of land and initial infrastructure development on the land after purchasing from rural areas is provided by the local municipal government. The budget for land acquisition and development should be proposed by a land development corporation and approved by the local bureau of land and housing and local bureau of finance together.

Procedure of Granting Planning Permission to Development Project

The procedure of granting planning permission for a development project is generally divided into three stages. Although it is called a planning application, this process consists of at least nine local government bureaus including the bureau of planning administration (see Figure 8.3).

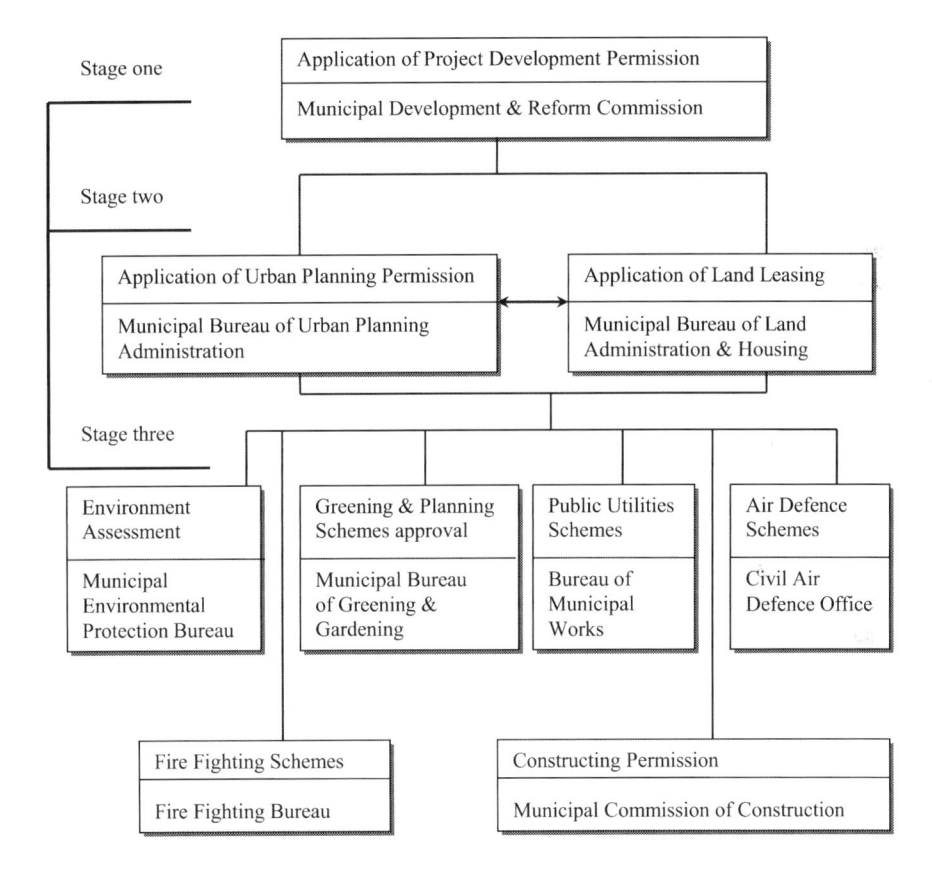

```
Stage one          Application of Project Development Permission
                   Municipal Development & Reform Commission

Stage two
                   Application of Urban Planning Permission   |   Application of Land Leasing
                   Municipal Bureau of Urban Planning         |   Municipal Bureau of Land
                   Administration                             |   Administration & Housing

Stage three
   Environment        Greening & Planning       Public Utilities    Air Defence
   Assessment         Schemes approval          Schemes             Schemes
   Municipal          Municipal Bureau          Bureau of           Civil Air
   Environmental      of Greening &             Municipal           Defence Office
   Protection Bureau  Gardening                 Works

   Fire Fighting Schemes                Constructing Permission
   Fire Fighting Bureau                 Municipal Commission of Construction
```

Figure 8.3   Application process of development projects

This process is complicated and time-consuming since it has to be decided by 10 different bureaus. It is not unusual that the whole application process can last for more than one year. It used to require the permission of at least 18 different local government bureaus or their divisions for the whole process of application of planning permission for a development project. In order to promote development, many local governments have tried to simplify the application process for planning management since the mid 1990s. A 'one-stop' administrative service system to reduce the costs of time and inputs by developers has become popular in many Chinese cities.

Certificating Mechanism of Planning Permission and its Amendment in Land Use Reform

Certificating mechanism of planning permission The implementation of plans and planning management functioning as a passive measure to deal with urban and rural development and construction is through the 'One Permission Notice and Two Permits' certificating mechanism. This mechanism was initiated in the era of the centrally planned economy when all urban land was used for free.

The 'One Permission Notice and Two Permits' certificating mechanism refers to three certificates that are issued by the local bureau of planning administration in the evaluation and approval process of application of a planning permit. 'One permission notice' is the certificate of decision for land use location of a specific development project. This was initiated from the concept that land in China was owned by the state, especially urban land. Any development on the state-owned land should apply a location permission of land use from local government. The 'two permits' refer to the Planning Permit for Land Use and the Planning Permit for Construction and Engineering.

Differences in mechanism of planning permission for different approaches of land distribution Since the land use reform in the late 1980s, a dual system for urban land use, i.e. assignment approach and auction for land leasing approach, has existed. The certificating system of planning management has to be amended to cope with the changes.

In the assignment approach of land use, the certificating mechanism of development planning management process has been kept the same as in the centrally planned economy. A development project is usually decided first. The process of application for land use through assignment by land authority and planning permission should be followed up by a developer, usually a government organization.

According to the availability of urban land at certain sites defined in a General Land Use Plan and a City Comprehensive Plan, a developer can propose a location for a development project through the process of applying for 'Planning Permission Notice for Location', which is issued by the local bureau of planning administration (Article 36, NPC 2008). The local bureau of planning administration should confirm the proposal of the project location, or select an alternative location for the proposed development or construction project according to the local City Comprehensive Plan, and other material for consideration, i.e. planning and environmental acts and the policies decided by the Circulars of the central government. After the approval of land use and location of the proposed development project, the developer should apply for the 'two permits', i.e. the Planning Permit for Land Use and the Planning Permit for Construction and

Engineering, according to the ordinances, controlling codes and indexes defined in the Regulatory Detailed Plan to a specific plot of land in order to carry out development and construction project.

However, in the auction process for leasing the right to land use, the location of a development project, which is associated with the leasing right to land use, has already been decided. It is unnecessary to apply for the planning Permission Notice for land use location. The process of auction of leasing the right to land use and then the contracting documents for the land use right leasing has already linked to a Regulatory Detailed Plan that defines land use boundaries, development functions, plot ratio, greening ratio, height of buildings, density of development and other controlling codes, indexes for the specific piece of land for leasing. The development on the land for leasing should follow all specific controlling ordinances and codes that should form an important component of the contract for land leasing with developer(s). After obtaining the leasing right of a piece of land through the auction process, developer(s) should employ planning consultants to prepare a construction detailed plan and architecture design according to these controlling indexes specified by the Regulatory Detailed Plan for the leased piece of land. The development or construction project is allowed to start only after the construction detailed plan is submitted and approved by the local bureau of planning administration.

Project development permission Before the reform of the urban land use system in China, there was a special stage in the application process of planning permission. This stage is still operated for the assignment approach of land use. It is for project development permission by local municipal development and reform commission (MDRC). The MDRC used to be the most important and powerful government department in the centrally planned economy because of its power in allocation and assignment of all resources and products for development. Its power has been reduced since China delivered its reform policies in the late 1980s, although it has still significant powers and responsibilities in economic and social development and the 'Five-Year Plan' formulation, and in the approval of major infrastructure and industrial development projects.

For development projects through assignment process of land allocation, the project applicants need to submit the following documents to the MDRC when applying for project development permission:

- Application report, which is required to explain reasons of the proposed development, scale and standards of the proposed project.
- Forecast and proposed budget and measures of fundraising, as well as the cost and benefits of the project.
- Feasibility study and design proposals of the development project.

In the application process of project development permit, after receiving the application report and other required documents of a development project, the MDRC will assess the project according to the policies and proposals of the local economic and social development plan (the Five-Year Plan). The MDRC should evaluate the availability of resources for the proposed project, typically those that require support of financial resources from the budget of local or central governments. For private development, or the project without financial support by the budget of local or central governments, developer(s) should have a deposit of at least 30 per cent of the total estimated cost of the proposed project when applying for project permission. The resources can come from any channel, e.g. a bank loan, private money, or mortgages.

For development projects with less than 8,000 square metres of building floor space, or of investment by government of less than 10 million RMB yuan, or self-funding projects, project proposal and its feasibility study report can usually be approved at the same time. However, if a project is more than 8,000 square metres of building floor space, or of investment by government of more than 10 million RMB yuan, the approval process of development permission is separated into two parts. Project proposal should be approved first. After the approval of a project proposal, the developer(s), either government organizations or private business persons, should organize a feasibility study and then submit the study report for approval by the MDRC.

After consulting with some bureaus of local government, e.g. urban planning administration, land administration and environmental protection, the MDRC will approve the proposed project development and issue the 'Project Permit for Application of Approval', which is a certificate to illustrate the approval of the proposed development.

Although devolution has been the main element of Chinese reform and transition to the market economy, the central government can still ask major development projects of infrastructure, or industrial and urban development, which are of national or regional social and economic significance, to be submitted for approval by the central government directly.

These projects with national or regional social and economic development significance are required to have a special assessment in terms of social, economic and environmental impacts by independent consultants. The social, economic and environmental impact report is needed as supporting and supplementary documents when applying for project development permission.

Application of 'Planning Permission for Land Use'

It is regulated that any development and construction project within the 'Urban Planning Area' has to apply for the 'Planning Permit for Land Use' being issued by the local bureau of planning administration (Article 31, NPC 1989a) as an important component within the process of application for the planning permission. Issuing 'Planning Permit for Land Use' should fully consider controlling indexes and policies decided by the Regulatory Detailed Plan that has followed the guidelines of the 'Urban Land Use Categories and Planning Land Use Standards' promulgated by the Ministry of Housing and Urban-Rural Development (see Chapter 6 for detailed information). The application of 'Planning Permit for Land Use' is complicated and time-consuming.

Despite the fact that the City Planning Act 1989 and City and Countryside Planning Act 2008 have defined the general requirements for the planning permission application, different cities may operate different approaches in the process of planning management. One difference is created by the different government administrative structure, typically in the administrative systems of planning and land management.

Chinese land is under a dual ownership system. In urban areas, it is defined as under state ownership, while it is collectively owned in rural areas, which means owned by all the local villagers. When rural land is proposed to be used for urban development, the ownership of the plot of land should be changed and transferred to the state ownership status from the rural collective status. Otherwise, it is not permitted to undertake any urban development. It is the task of the Ministry of Land Use and Resources and its lower hierarchical local government bureau of land administration to deal with land administration. From Chapter 3, we can see that the land use administration and city and regional planning are different responsibilities in two ministries of the central government. The Ministry of Land and Resources is responsible for land use, and the Ministry of Housing and Urban-Rural Development is responsible for urban–rural planning and development. However, at the local government level, the government's administrative frameworks are varied. Some cities follow the structure of the central government, e.g. Beijing, Shanghai, Xiamen and some other cities; other local municipal governments merge these two tasks into one government authority, e.g. Shenzhen, Wuhan and Qingdao. One authority structure of both land use and urban–rural planning and development can be more efficient and able to reduce conflicts during the development and construction process, typically for the assignment approach of land use. Even for the approach of auction of leasing the right to land use, it is easier to have coordinated policies between land use and urban–rural planning and development in one local government bureau. However, the disadvantage is that there is a risk of corruption since the authority controlling both land use and planning permission is very powerful if there is no appropriate supervision by the public.

Initial urban planning permission: Decisions of controlling policies and indexes While submitting the application of project development permission to the local municipal development and reform commission, applicants or developers are usually applying for initial planning permission by submitting the initial planning and design scheme of the proposed development project to the local bureau of planning administration,

The local bureau of planning administration will assess the proposed development scheme according to planning policies or local City Comprehensive Plan and relevant controlling indexes, norms and ordinances of Regulatory

Detailed Plans. This is an important step in the decision-making process of planning permission since most policies, development ordinances and controlling indexes to a specific piece of land for development should be decided at this time by the local bureau of planning administration. It shall significantly influence the final decision on the approval of planning permission. The local bureau of planning administration treats this step as the fundamental component of planning management.

However, it is because of the lack of appropriate communication with prospective developers and shortage of public notices which should explain the application process for planning permits in Chinese planning that many applicants may not fully understand the importance of the initial development scheme's assessment. Many applicants have not treated it as an important step that will impact formal application later. It is a common situation that the submitting documents required for approval for planning application may not be provided appropriately.

Discretion of planning officers in the process of decision-making   After receiving a planning application, a task planning officer should assess the proposed function of land use of a project and check if the initial design scheme fits the policies of a City Comprehensive Plan, and land use controlling indexes, e.g. height of buildings, buildings setback line, plot ratio, proposed provision of public facilities and infrastructures defined by the Regulatory Detailed Plans for the location of the proposed project and relevant national and professional planning norms. The task planning officer should make the initial decision and recommend any planning or design conditions if necessary to the application. There is the possibility for him or her to decide how to follow controlling indexes, codes and ordinances, or national and professional norms. After the decision by the officer, the application documents will then be submitted to the division chief.

The responsibility of a division chief is to review the recommended planning and design conditions proposed by the task planning officer. The division chief should examine the contents of the proposed development project and assesses its potential impacts on the entire urban built-up space from the perspective of scale and construction volume of the proposed development project. The division chief can add more or reduce required conditions and controlling ordinances to the proposed development project or change the proposed condition suggested by the task planning officer according to the City Comprehensive Plan and Regulatory Detailed Plans or national or professional planning norms.

The application documents will then be further assessed by a responsible director of the local bureau of planning administration. The responsible director should review the comments and suggestions by the planning officer and the division chief. It is possible for the director to add or reduce any planning controlling requirements to the application of the development project according to local urban plans, and national or professional planning norms.

It is not the end of the application process yet. Finally, the application documents should be submitted for discussion and final decision on a planning administrative panel meeting which are participated by planning directors and division chiefs of the local bureau of planning administration. The decision of planning conditions and controlling ordinances and indexes to the proposed development project that is decided at the panel meeting is the final planning controlling decision to the proposed development project.

If the proposed project is for industrial development, applicants should submit other documents, e.g. feasibility study report of the development project, which consists of information of applicable technologies, demands of transport, energy and public utilities (electricity, water supply and sewage, roads, gas and telecommunication), and an environmental impact analysis report.

During the approval process of planning permission, it is possible for planning officials at different levels to make the decision if necessary. There is the possibility of discretion in decision-making. It is because of the possibility of discretion in decision-making in planning, that the influence from local leaders (Cao 2009; Zhang 2004) and developers or investors cannot be easily avoided. It creates a risk of corruption if there is no appropriate supervision mechanism, typically supervision by the public.

The approval of 'General Layout and Construction Scheme Proposal of Development Project' and application of the 'Planning Permit for Land Use' After the initial application for planning permission has been approved by local bureau of planning administration, and the application of the development project is also approved by the local development and reform commission, the local bureau of planning administration will then issue the 'Permission Notice for Location' for the proposed development project. This notice is the principal document for applying land use allocation from the local bureau of land administration (MBLA). The 'Permission Notice for Location' for the development project should be co-issued by local bureau of planning administration and local bureau of land administration if they are the two separate departments in local government.

Since the reform of Chinese urban land use, the land uses of real estate, retail and business, and industries should be leased through an auction process. The City and Countryside Planning Act 2008 has changed the former requirement of application for 'Permission Notice for Location'. It regulates that except the assignment approach for land use, it is unnecessary to apply for a 'Permission Notice for Location' (Article 36, NPC 2008).

After land has been allocated for development either through auction approach for leasing the right to land use or assignment approach from the local bureau of land administration, developers or applicants should invite planning and design institutes to prepare construction detailed plan and urban design schemes for the proposed development project. The construction detailed plan and urban design schemes are produced in order to obtain approval by the local bureau of planning administration for 'General Layout and Construction Scheme Proposal'. At this stage of application for planning permission, the 'Permission Notice for Location' and the decided controlling indexes, codes and ordinances for the proposed project from the initial planning permission application should also be provided as materials for consideration.

A construction detailed plan usually specifies information about a development project in detail. It should define the boundary of project land use; location; stories, height and setback line of each building; location of greening and public spaces; roads, infrastructures and public facilities; and the corridors of high pressure electricity, microwave, or underground pipelines.

The required documents for a construction detailed plan for application of the approval of 'General Layout and Construction Scheme Proposal' are varied depending upon the sizes and types of the proposed development project. If it is a housing development, applicants or developers should provide following documents:

- For a project with less than 50,000 square metres floor space of development, the applicant should submit three copies of layout design at the ratio of 1:500, or 1:1000.
- For a project with development floor space between 50,000 to 200,000 square metres, the applicant should submit a construction detailed plan with at less two planning and design schemes.
- For a project with more than 200,000 square metres of floor space, the design should be taken by competition among the A Level professional planning and design institutes or firms. The developer should submit the models and plans.

It is also necessary for applicants to provide other detailed engineering and architecture design schemes and plans, e.g. design scheme and plan of public utilities, including pipeline layout map, roads and their vertical design schemes, water supply, sewage, drainage, telecommunication, electricity, gas and power sub-station within the proposed development site; plain, section and elevation schemes of individual buildings that should consist of the functions of each room, and list of floor space of each story.

For industrial, retail and business, or mixed land uses, or large housing development projects with more than 50,000 square metres of floor space, it is require to undertake a traffic impact analysis of the proposed development project, the report of which should also be provided for evaluation by the local bureau of planning administration.

If the construction detailed plan and the 'General Layout and Construction Scheme Proposal' of a proposed development project have been approved, the local bureau of planning administration should issue a 'Planning Permission for Land Use'.

City planning committee and power of local leaders in planning management process Since the 1990s, a city planning commission has been playing a significant role in decision-making in the process of planning management for development. It has become common in many cities that the decision of planning permission for important development projects should be decided by the local city planning commission. This commission can call in an approval process of planning permission for a development project if necessary.

A city planning committee is usually chaired by the mayor of a city directly. The members of the committee include the vice-mayor responsible for urban–rural development and planning, and directors of different municipal bureaus. The vice-mayor for urban–rural development and planning, as vice-chair of the committee, is involved in urban and rural planning and development projects on a regular basis.

For example, a 'Permission Notice of Development Project Land Use Boundary', which is issued by the local bureau of land administration to illustrate a piece of land being officially leased or assigned to a user, should be finally approved by the vice-mayor responsible for urban–rural development and planning on behalf of the committee. The vice-mayor has veto power in decisions for applications of land use and planning permission by the local bureau of planning administration and the bureau of land administration if there is any inappropriate actions, e.g. illegality, corruption or problems during the application process or any error in the approved documents. When an application is rejected by the vice-mayor, it is the end of the application process of planning permission and land use. However, this seldom happens. When something goes wrong, the vice-mayor usually asks for a review of the application case instead of rejecting it directly, except if there are very obvious severe or illegal problems, such as the proposed development project may generate severe impacts on the local environment, or negative impacts on local features, or corruption of planning and land administration officers during the approval of development planning permission or land leasing if such cases were identified and explored.

Application for the 'Planning Permit for Construction and Engineering'

After obtaining the right of land use either from assignment or leasing process, and then offering the 'Planning Permission for Land Use', applicants or developers should apply for the 'Planning Permit for Construction and Engineering' from the local bureau of planning administration in order to start engineering and construction work. It is the third stage of application of planning permission for development.

Concerning the possible impacts by the building engineering and construction to city layout pattern, environment, transport and the neighbourhood of a proposed development project, it is regulated by City and Countryside Planning Act 2008 (Article 32, NPC 2008) that:

> for new constructions, extensions of land alterations to buildings, structure, roads, pipelines and other engineering works within the city planning area, application

should be submitted to the city planning administrative authority together with related documents for approval. The city planning administrative authority will issue "Planning Permit for Construction and Engineering" for the said project.

When applicants or developers have been issued with the 'Planning Permit Note of Development Project Land Use Boundary', engineering design and special plans should be produced. The engineering design and special plans include schemes of fire fighting, civil engineering, architecture structure, landscaping and greening and others. These architectural and engineering design and special plans are required to be approved by various local government bureaus, which include the commission of construction for the permission of construction; bureau of civic engineering work for layout and engineering design scheme of water supply, sewage, gas, street lights, road access, and sanitation; fire fighting bureau for fire fighting facilities and dispersion arrangement in emergency; bureau of gardening and greening for the scheme of planting and greening.

Since the 1990s, environmental protection has become an important issue in development. It is usually required to submit the environmental impact assessment report for the proposed development project when applying for planning permission. The application for evaluation and approval of an environmental impact assessment report is ordinarily operated soon after the application of the 'Permission Notice for Location'. If there is any potential pollution created by the proposed development project, in principle, the veto power of the bureau of environmental protection can stop the application process for planning permission and land use. However, in practice, some development projects with potential pollution are able to be approved by local governments because development and GDP increase has been the priority task from the perspectives of local governments.

Implementation of Plans through Planning Management

Different Plans in Hierarchy and Planning Management (Development Control)

The statutory plans in the Chinese city and regional planning system have their own defined functions. A Provincial Integrated Urban System Plan expresses overall development strategy of the whole province, or a region. This type of statutory plan should function as an essential tool to guide regional development and urbanization. The purposes of the plans are to address contradictions and to coordinate different objectives among cities and towns within a provincial territory. A City Comprehensive Plan should provide general development policies of a city by defining the nature of a city, its growth targets, land use structure, and comprehensive arrangement for all types of development and construction. A City Comprehensive Plan consists of a set of different types of special plans, e.g. civil engineering plans, short-term construction plan, transport plan and others.

These two types of statutory plan define broad development policies and development schemes from a macro perspective. They are too general to guide any specific development project on the ground. A detailed plan in the lower hierarchy is required to follow the general policies defined in higher hierarchic plans and to guide control projects. A Regulatory Detailed Plan that follows the policies defined by a Provincial Integrated Urban System Plan and a City Comprehensive Plan, is the result of such demand.

A statutory Regulatory Detailed Plan is formulated as a tool to guide development and implementation of higher hierarchic statutory plans through the mechanism of planning management by application of planning permission. The establishment of this planning hierarchy is for a rational implementation of planning policies and development schemes. However, the actual development and plan implementation on the ground may be a different story.

In order to have a better understanding of Chinese development planning management and plan implementation in real practices, the author has selected Xiamen as a case study to exam the operation of planning management and its problems.

The survey was taken through the investigation in the application process of planning permission for development projects. The research methods of face-to-face interviews and focus group discussion were adopted for this research.

The deputy director general and division chiefs of Xiamen Municipal Bureau of Planning Administration, and the director, chief planner and senior planners from Xiamen Municipal Institute of Urban Planning and Design, were invited to participate in the focus group discussions. From the discussion, it was found that there were two main problems that had not been addressed properly in the relationship between the formulation of Regulatory Detailed Plans and planning management in the application process for development project planning permission. One was that the local features and social and economic context had not been carefully studied and analysed during the process of plan making. It was due to the lacking appropriate detailed analysis of local characteristics that some controlling indexes, codes and ordinances decided in Regulatory Detailed Plans were unreasonable and impractical. Some plans were even prepared without realizing that the controlling index and ordinances such as the height, or FAR, were different between two nearby buildings. This problem was created because of the limited time of plan preparation designated by the local bureau of planning administration. It was not unusual that a Regulatory Detailed Plan had to be completed and submitted to the local bureau of planning administration in 2–4 months. This not only happened in a few cities, but was common in many Chinese cities. In Tianjin, a metropolis with a population of 12.9 million and under direct jurisdiction of the State Council, a so-called 'Great Battle for Planning' was operated in 2008. Tianjin Municipal Bureau of Planning Administration intended to produce 119 different types of plans within 150 days (*Tianjin Daily*, 25 November 2008 at SINA News). In 2008, Chengdu organized a 'Great Battle for Planning' to provide planning services for the reconstruction after the disastrous earthquake in Wenchuang (Yin and Wang 2010). Yushu local government also introduced a 'Great Battle for Planning', in which 320 plans were completed in four months in order to start the re-development after the earthquake (Cun 2010). The limited time for plan making restrains the possibility of detailed analysis in planning. What planners can do is to complete plans and written documents without appropriate analysis of local characteristics.

The other problem is that the City Comprehensive Plan and the Regulatory Detailed Plans had not been treated as main materials for consideration by planning officers in the process of evaluating applications for planning permission of development projects as the consequence of the possibility of discretion.

City and countryside plans are produced to guide development. However, contradictions between planning policies defined by plans and major construction projects often appear owing to uncertainties in planning and development. It was found from surveys by the author that whenever there are large projects with large amounts of investment that will help to promote local economic development and GDP increases, the policies of city and regional planning would usually have to be adjusted or amended to meet the demands of developers or investors. This is because the main objectives of Chinese local governments are to pursue economic development, typically the increasing rate of GPD, which are also the criteria to evaluate the achievements of local governments. Similar to the mechanism of planning management of applying planning permission for development projects, the significant power of local leaders, e.g. party secretary, mayor(s), is able to force planning policies to meet the demands of their development targets.

It was also found by the author from the surveys that the main conflicts between the planning controlling indexes and application of planning permission for development projects were not in land use functions, but in some detailed controlling indexes, e.g. height of buildings, plot ratio, green spaces. It was a common problem that the proposed controlling index in the density, FAR (Floor Area Ratio) in Regulatory Detailed Plans had been the key items for bargaining and negotiation between developers and the local bureau of planning administration. These controlling indexes are also the items of a Regulatory Detailed Plan that developers attempt not to obey during development in order to seek maximum profit.

Developer and Investor in Process of Applying Planning Permission

Developers or investors are important stakeholders in the process of urban and rural development. The majority of applications for planning permission in Chinese cities are submitted by them. In order to understand the concerns of developers regarding planning policies, controlling indexes and planning management, the author carried out a survey in some Chinese real estate development companies.

The developers or investors in China can be divided into two main categories of government (or public) owned companies and the private ones including foreign investment or development companies.

The public development companies, typically those owned by the central government, enjoy maximum benefits and opportunities in urban and regional development. Of course, they may have to take some responsibility for the provision of public services as assigned by the government. It is because of their political status that the state-owned real estate development companies are able to establish links with local leaders. However, if there were not special reasons or for major projects that may significantly impact local economy, generally speaking local leaders do not make any decisions directly. They are usually forward applications to the local bureau of planning administration through the formal procedure of planning management.

In the process of application for planning permission, developers often attempt to negotiate with the local bureau of planning administration to change some detailed controlling indexes that may influence their return from investment, e.g. plot ratio, density, in order to increase their profits, or even to break controlling indexes and ordinances decided by Regulatory Detailed Plans. It is found by the author from the survey that this situation often happens in large development companies, especially the state-owned development companies or foreign investment companies.

The private developers, typically the developers of small and medium size companies, may have to obey planning controlling indexes and ordinances decided in the Regulatory Detailed Plan. The main reason for them to obey the controlling indexes and ordinance is the concerns of cost and benefit by against the planning policies. The small private real estate developers will usually avoid participating in the auction of leasing right to land use if they find the associated planning policies or controlling indexes are difficult for potential marginal profit of investment. However, if they decide to enter the market, they usually look forward to having a quick approval process of planning permission. It is a common view among the small private developers that the quicker the decision by the local bureau of planning administration, the larger the potential marginal profits achievable.

The main problems of the small private development companies in the process of application of planning permission appear in the stage of 'initial planning permission'. It was found that 80–90 per cent of the applications for the permission by the small private development companies were refused. The refusal rate declined to 20–30 per cent when re-submitting the application of permission with necessary documents or appropriate project planning and design schemes. This illustrates the confusion or misunderstanding of 'initial planning permission' by many real estate developers. The survey shows that the refusal rate of larger development companies, particularly the state-owned real estate development companies, was much lower.

The information in Table 8.1, which was provided by a municipal bureau of planning administration as required by the author in the survey, shows the cases for application of planning permission for development projects of a city between mid March and mid June 2002. This table is able to illustrate that most of the application cases were granted approval permission by the local municipal bureau

Table 8.1 Reasons of conditional approval or refusal for application of planning permission

	Numbers	Percentage
Total number of planning applications	312	100
Number of the approved applications with conditions	51	16
Number of the refused applications	69	22
Reasons of the approved applications with conditions	51	100
Shortage of required documents, or proposed planning and design schemes	16	31
Proposed project impacts the existing or planned greening land, or other public building	3	6
Violation of regulated indexes, e.g. density, plot ratio, height of buildings	12	24
Changing of the developers without formal agreed documents	3	6
Inappropriate proposed planning and design schemes	13	25
No payment of the land use right leasing fee	1	2
Changing the initial approved scheme during the process of the application	2	4
The City Plans has not completed their formulation	1	2
Reasons of the refused applications	69	100
Violation to the existing policies	10	14
Violation of the land use function defined in the City Plans	7	10
Reserved land by the government	2	3
Proposed project impact the existing or planned greening land, or general spatial structure	8	12
Adjustment of the existing City Plans or other government policies, re-application later	7	10
Expire of the decided submitting of supplementary documents for the approval	11	16
Violation of regulation and planning indexes	8	12
Shortage of the proposed planning schemes, drawings and other documents	1	1
Inappropriate proposed location of the projects	5	7
Poor planning and design schemes quality for the proposed project	7	10
Inappropriate developers, e.g. government departments for some projects	2	3
Others (not mentioned)	1	1

Source: Calculated by the author from the data provided by the Xiamen Municipal Bureau of Planning Administration (cases from mid March to mid June 2002).

of planning administration. About 16 per cent out of a total of 312 cases were conditionally approved. Three factors can be found for those cases of conditional approval:

- shortage of the necessary documents;
- inappropriate proposed planning and design schemes; which is the mistake of the planning design institute or planning consultants that produced the planning and design schemes;

- violation of planning controlling indexes, e.g. density, plot ratio, height of buildings.

Twenty-two per cent application cases were refused. By analysing the refused cases, it is possible to find that the refused cases are either of violation to policies decided in city and countryside plans and local development strategies, or of violation to controlling indexes and ordinances decided in Regulatory Detailed Plans. The poor quality of the planning and design schemes for the development project that should be submitted for application of planning permission was the other reason for refusal of planning permission.

Conclusion

China operates a dual land use ownership system that is divided into the state ownership in urban areas and the collective ownership in the rural areas. This feature significantly impacts Chinese city and regional planning, typically its planning management. The reform of land use in urban areas has changed the approaches of planning management characterized by the certificating mechanism of 'One Planning Permission and Two Planning Permits'.

The existing planning management mechanism is complicated and time-consuming. However, it is due to the discretion in decision-making in local government that helps the government to attract inward investment to promote local economic development and increase of GDP.

The planning and development decision power is controlled by local leaders, including the party secretary, mayors and a few planning officials. The leaders of local government have great discretion in planning decision-making. Similar discretion in decision-making is also for planning officers at different levels. In the process of application for the planning permission of a development project, it is not unusual to change planning control ordinances and index of land use and building control defined by plans, particularly in Regulatory Detailed Plans. Discretion in development planning management procedure is one of the factors that create the implementation of plans and development projects that depart from existing City Comprehensive Plan and Regulatory Detailed Plans. Failure to provide appropriate development control may create a risk of damaging the environment and worsening the daily living quality of local residents. This can be found from the evidence that during the last three decades, many Chinese cities have lost their local features that distinguish them from others.

It is the view of Cao (2009) that the existing mechanism of planning management creates weakness in plans delivery. It is impossible for higher hierarchical planning authorities to supervise planning policies and actual development projects on the ground because of the asymmetric information. This mechanism is able to control the development project under construction, but ignoring the stage of the project in operation. It lacks an important part of monitoring plan implementation (Zhang 2004).

The reform and improvement of the Chinese planning mechanism is necessary. Nevertheless, it is because the planning management mechanism is one component of the Chinese political and administrative system, that the reform of the planning management mechanism is unable to be separated from the whole system.

Chapter 9
Performance of Chinese City and Regional Planning System

Introduction

In this chapter, the performance of the Chinese planning system in the process of transition to the market economy and the associated impacts from the changing social and economic context is explored and analysed.

According to the performance theory initiated by the Dutch academics, e.g. Mastop and Faludi (1997), Lange et al. (1997), Damme et al. (1997), Needham et al. (1997), planning should be flexible with built-in opportunities and procedures that can be adjusted to allow for further subsequent elaboration, deviation and revision. The performance of planning should be concerned with the clarity of plans during the process of decision-making in the pluralistic stakeholders society by helping stakeholders to understand what they are doing and what they should do next. It is a concept in planning performance theory that to what extent a plan influenced the decision-making of stakeholders is the key indicator for the evaluation of planning performance. Moreover, values expressed by planners in planning process and the ability to mediate are critical to the capacity of city and regional planning to influence the decision-making of stakeholders in relation to development, and to persuade decision-makers about planning proposals. City and regional planning policies aim to influence the decision-making elements by implying various processes of negotiation, compromise and persuasion.

The evaluation of Chinese city and regional planning performance will be mainly based on, but not limited to, the following four principles:

- planning as a visionary and methodological framework;
- concepts of urban planning;
- growth promotion and public interest protection policies;
- a robust and resilient monitoring mechanism in implementation.

It is important to mention that these principles will be applicable in any case but the way they are applied will be varied depending on the different conditions and context.

Before evaluation of the performance of the Chinese planning system, it is useful to have a review of the theoretical framework of city and regional planning. The review will help to a better understanding of evaluating planning and its system.

Theoretical Framework for Evaluation of Chinese Planning System

Why is it necessary or helpful to explore the theoretical underpinnings of planning and implementation when it is clear to a person in the street that planning is a simple, common-sense activity? The reality is that the person in the street fails to understand that planning is complex and fiendishly difficult for the reasons that it:

- is concerned with the future and the future is uncertain;
- deals with complex, inter-related public policy problems;
- interferes in the free play of the market;
- creates conflicts of interest;
- involves value judgments; and
- presents policies that prove difficult to implement.

An appreciation of the theoretical underpinnings of planning and implementation will contribute to an understanding of the effectiveness of the planning system. Otherwise, why it is necessary or helpful to explore the theoretical underpinnings of planning when it is clear to the man in the street that planning is a simple, common sense activity forestall. The reality is that planning is affected by the social, economic and political conditions of a particular system of government, and it could be argued that it is not appropriate to review the nature of the Chinese city and regional planning system against theoretical concepts developed in the Western world, and if was still a wholly centrally planned economy that argument would have some force – concepts and ideas developed within the framework provided by a market economy operated in a democracy would not necessarily be relevant in a centrally planned economy. However, since the introduction of the open door policy in 1978 China has operated a market economy within a framework provided by a centrally controlled political system – 'a market economy with socialist Chinese characteristics' and the social, economic and political forces driving that city and regional planning system are not significantly different to those in the West.

The Nature of Planning

'Planning' has been regarded as a 'literature of controversy' (Dyckman 1979). It faces uncertainties in development and choices in the future. Different people, groups and even planners themselves have different views about planning because of different priorities and values. Thus, planning has many meanings depending upon the problem one wants to deal with, and the various social, political and economic milieus, as well as different professional aspects. In particular, planning stands for efforts to relate to the so-called operational decisions. Generally, city and regional planning is regarded as the course of actions in the future. It attempts to stimulate or control the local effects of social and economic growth. The city and regional plan is something to be used (Mastop and Needham 1997). People need to plan for the coordination of the various actions. For a long time, city and regional planning has been regarded as intervention by the government to protect public interests, equality and fairness, through guiding and controlling development or land use.

According to different planning aspects, there are various views. It is the point of view of some academics that planning is defined as the art of making social decisions rationally (Friedmann 1987). From a system theory point of view, planning is a human activity, which is concerned with optimizing the use of human abilities (Chadwick 1978).

Moor had a narrower definition. He suggested that all decisions about the allocation and distribution of public resources are planning (Moor 1978). This definition is related to the government's functions in planning, and enables the justification of planning in the course of legitimizing the government's intervention.

At the same time, it also obscures the fact that a wide variety of plans are routinely produced corporate plans that are prepared for private businesses; sectoral plans are prepared by public authorities for the economy, industrial development, education, health, social services, transport and land use (Bruton 1984). Indeed, many developing countries such as China also produce holistic national plans setting out proposals for economic and social development designed to bring about significant change over time in the way the country operates.

Experience has shown that plans, being for social and economic change at the national level or for corporate or public sector development, are rarely fully implemented in the way the policy-makers envisaged when the plan was produced. e.g. taking Cardiff as a typical case study, where 'planning in practice has been led by opportunity more often than carefully considered goals' (Harris 2006: 87). There is long-standing evidence to suggest that opportunity or changing circumstances, or politics and power, rather than planning, lead to developments on the ground which differ from plan proposals. Take the early example of the land use plan produced for Cardiff in the mid 1950s which made provision for extensive areas within the city's boundaries to retain their existing land use by zoning them as 'white land', i.e. areas where there would be no change to the then existing land use. Yet 10 years later much of this 'white land' had been developed for residential purposes, largely because the policy-makers had not foreseen the dramatic increase in demand for housing. Similarly a subsequent detailed land use plan for the redevelopment of Cardiff city centre proposed significant projects which failed to materialize, such as major traffic links, office developments and even a cathedral (Bruton 1983).

Given the nature of the planning and implementation process it is not surprising that the policies put forward in plans are rarely implemented in full for the reasons that:

- uncertainty about the future, and the complex inter-related nature of public sector problems, ensures that the policy-makers inevitably make wrong assumptions about what the future holds and how best to deal with those problems;

- attempts to bring about changes that would not otherwise come about without intervention in the 'free market' inevitably give rise to situations where conflicts of interest are generated and politics, power and bargaining become central to the process;
- policy-makers invariably fail to understand the complex inter-relationship between the formulation of policy and its implementation.

These complications in the process militate against the possibility of 'perfect implementation'.

A quotation from a recent and detailed case study of the application of the planning process in Cardiff summarizes the nature of the difficulties likely to be encountered in attempting to change the future through public sector planning:

> The principal economic discourse in plan-making in Cardiff has been one of attempting to secure an improved economic base for the city, diversifying its economy and aiming to attract prestigious inward investments. Yet early episodes in plan-making often co-incited with periods of economic uncertainty and decline. Such plans have therefore been characterized in large part by the making of relatively ambitious plans that have not been capable of being realized in the prevailing economic circumstances. One of the key functions of plans for Cardiff has therefore been to safeguard land to enable such projects to progress when conditions improve. This in turn has generated debates over the merit of such allocations and how to deal with pressures for alternative use of the safeguarded sites, especially in respect of employment land. Likewise, the planning of Cardiff has been predicated on the completion of its major road infrastructure, key parts of which have not been completed as assumed within specified plan periods. In addition, the transformation of the city's economy to a post-industrial form based on the dominance of services has been based more on dictated circumstances than deliberate design. (Harris 2006: 91)

Complexity and the inter-related nature of public policy issues; conflicts of interest and bargaining and a failure to implement policies and proposals as anticipated are central to the development of Cardiff over the last 25 years.

What is public policy? This book is concerned with public policy planning and implementation in China, but what is public policy? Simplistically it is whatever governments choose to do or not to do in the areas in which they are involved (Dye 1978), e.g. defence, welfare, economic planning, police, fire services, housing, transportation, land use planning, the environment. It is the idea of Wayne Parsons (1995) that public policy should tackle the issues that are defined as 'public', which refers to governmental or social regulation.

Hill and Hupe (2002) suggest that public policy involves actions (or inaction) and intention. It consists of a course of actions. Public policy initiates from a process over time. It is the views of Hogwood and Gunn that:

Any public policy is subjectively defined by an observer as being such and is perceived as comprising a series of patterns of related decision to which many circumstances and personal, group, and organisational influences have contributed. (Hogwood and Gunn 1984: 23–4)

Lowi (1972) adds to our appreciation of what public policy involves with his four-stage classification of public policies into:

- distributive policies, which generally benefit everybody;
- constituent policies, which are concerned with the 'rules of the game', such as election laws;
- regulative policies, which are designed to regulate behaviour, e.g. public health;
- redistributive policies which take from one group and give to another.

However, it should be recognized that the institutional framework within which public policy is developed and implemented determines much of what happens on the ground, which does not always accord with higher order policies. The diverse and complex nature of China, allied with the speed of social and economic change since the introduction of market reforms in 1978, has resulted in development on the ground that does not accord with officially approved plan proposals.

The following quotation from Goldsmith (1980) succinctly summarizes the views outlined above and provides a working definition of public policy as:

something to be affected in a particular way to secure a particular result, together with the statement as to how that result is to be achieved, and the necessary actions taken to implement the statement and achieve the desired results. (Goldsmith 1980: 23)

What is clear from the above review is that public policy:

- is complex;
- consists of a range of different types of policy which are invariably inter-related;
- can have unintended as well as intended outcomes;
- can involve social, physical and economic change;
- is political;
- involves the exercise of power;
- is influenced by the institutional framework within which it is developed and implemented; and
- is best understood as a process.

In this context, the next section explores the complexity of public policy issues.

Public Policy and Complexity

Public policy problems are complex and highly inter-related, e.g. in 2008 planning to accommodate population growth was related, amongst other things, to:

- a programme of strict birth control to limit population growth to manageable levels;
- economic development and industrial modernization to provide the wherewithal to support the growing population;
- global competition to provide food for the growing population and raw materials, especially oil and steel, to source economic development;
- the containment of social unrest which has developed as a consequence of agrarian and industrial restructuring;
- a severe degradation of the environment;
- an overriding concern to maintain political control in the hands of the CCP;
- international speculation against the yuan which could lead to a financial crisis and recession.

This example reinforces the view of Mason and Mitroff (1981: 4) that 'basically every real world policy problem is related to every other real world problem'. Acknowledgement of this situation is of vital importance to planners for it means that attempts to resolve a particular policy problem should consider the potential relationships between it and other policy problems – the unintended consequences. Thus in plan or policy formulation, it should be accepted that:

- any plan or policy is or should be concerned to resolve a number of problems and issues;
- these problems and issues will undoubtedly be inter-related to some degree;
- as a result policy to cope with one problem typically requires further policies to cope with all the other problems;
- few (if any) problems can be isolated for separate treatment. Each policy solution is likely to create additional dimensions to other related problems and the policy solutions developed to deal with those problems.

Thus public policy-makers are faced with a range of problems which are inter-related in a most complex way where a problem in one particular area can have knock-on effects in other areas (the unintended consequences) – effects which can be modified or magnified in such a way that the system becomes unstable and unpredictable. This emphasizes the need to appreciate that: (a) there can be a difference between policy objectives and the outcome of policy; and (b) policy objectives can be distorted in the process of implementation.

The outcome of policy In the real world the implementation of policies can have intended and unintended consequences, e.g. the introduction of land development reforms resulted in the unexpected transformation of Shangbu Industrial Zone in Shenzhen into a thriving commercial and retail district when the original plan proposed that it should continue as an industrial zone with a new commercial and retail centre being built elsewhere. This failure to implement in accordance with the then current plan arose from:

- the introduction of these land development reforms after the plan for Shenzhen had been prepared;
- a higher rental value being offered for commercial and retail property as opposed to industrial use;
- the rapid growth of the city, which had begun to absorb into the inner city area industrial estates that had initially been located at the edge of the city;
- the inability of the city government to enforce its own land use zoning controls.

These unintended consequences were eventually accepted and incorporated in later versions of the plan for Shenzhen.

Policy-making and implementation Policy intentions can be reshaped in the process of implementation as a consequence of changing circumstances or because of inadequacies in the process of implementation or because the implementing agency has a different agenda to the policy-making body.

Focusing on the intended consequences of public policies provides a starting point for analysing the actual consequences of policy implementation by asking:

- What were the intended consequences?
- What were the unintended consequences?
- How far have the intended consequence been achieved?
- To what extent have unintended consequences hampered the achievement of the intended consequences?

'Wicked Problems'

'Wicked problems', or problems of organized complexity, are not wicked in the sense of being evil. Rather they are 'wicked' in that the more one attempts to solve them the more complicated they become. Rittel and Webber (1973) identify nine characteristic properties of such problems, namely:

- 'wicked problems' have no one definitive formulation;
- formulating or understanding a 'wicked problem' is synonymous with solving it;
- there is no right or wrong solution to a 'wicked problem'. Solutions can only be good or bad relative to one another and the value system within which they are applied;

- there is no way of knowing when a 'wicked problem' has been solved. As a result they require constant monitoring, and there is always scope for improving the solution propounded;
- the possible range of methods which can be used to solve 'wicked problems' is unlimited;
- there are many explanations for 'wicked problems'. Depending on the explanation chosen, so the solution differs;
- it is never clear that 'wicked problems' are being dealt with at the proper level as they have no identifiable root cause, and can be considered as symptoms of other problems;
- once a solution to a wicked problem has been attempted it cannot be reversed. What is done is done!
- unlike the physical sciences there is no trial and error procedure which can be followed. Every wicked problem is unique.

Mason and Mitroff (1981: 12–13) amplify the ideas of Rittel and Webber in the context of the planner or policy-maker who seeks to serve a social system by changing it for the better. In doing so they identify a number of characteristics of public policy 'wicked problems', which in their words spell difficulty for the policy-maker. These characteristics are:

- Interconnectivity – each problem is strongly connected to other problems and sometimes these connections form feedback loops. The opportunity costs and/or side effects of policies put forward to deal with a specific problem will often depend on events out-with the scope of the specific problem, i.e. contingent factors.
- Complexity – given the numerous elements and inter-relationships which make up 'wicked problems' there are various points where analysis and intervention might focus and various approaches and policies for action might be applied.
- Uncertainty – 'wicked problems' exist in a dynamic and uncertain environment. Therefore risk taking, contingency planning and flexibility are essential ingredients in planning and public policy formulation.
- Ambiguity – there is no one correct interpretation of a wicked problem or a solution to it. The value system of the individual determines how the problem/solution is seen.
- Conflict and conflicts of interest among individuals or groups are inevitable in the competing claims for resources to resolve 'wicked' public policy issues. The interaction of these conflicts of interest determines how the solutions work out.
- Societal constraints – social, organizational, technological and political constraints and capabilities have a marked influence on the policies selected to resolve 'wicked' public policy problems, and the success or otherwise of those policies in resolving the problems.

Uncertainty in Planning

Uncertainty is endemic in planning for the future. Christensen (1985) in examining uncertainty in planning reinforces many of the points made by Mason and Mitroff (1981) and identifies four prototype conditions of planning situations. She argues that:

- traditionally planning practice has assumed that both the goals (or ends) of planning and the means of achieving those ends are known and accepted, i.e. the change being sought and the method(s) of bringing that change about are certain;
- in reality, the situation is much more complex with actual problems having different levels of uncertainty in both ends and means;
- if people agree on what they want and how to achieve it, then certainty prevails and planning is rational application of knowledge;
- if they agree on what they want to do but do not know how to achieve it, then planning becomes a learning process;
- if they do not agree on what they want but do know how to achieve alternatives, then planning becomes a bargaining process;
- if they agree on neither means nor ends, then planning becomes part of the search for order in chaos.

In amplifying this statement she suggests a framework for clarifying the conditions relating to variable (or uncertain) planning problems. Briefly she:

- establishes a two-dimensional matrix with the vertical dimension as the means or knowledge of how to do something, and the horizontal dimension as the goal or the desired outcome of planning (the former she labels 'technology'; the latter 'goal');
- dichotomizes the vertical and horizontal dimensions according to certainty/ uncertainty. Thus, the technology dimension is dichotomized into known/ unknown; the goal dimension into agreed/ not agreed.

Using the matrix Christensen identifies four prototype variations of conditions that can characterize planning, namely:

- known technology/agreed goal;
- unknown technology/agreed goal;
- known technology/no agreed goal;
- unknown technology/no agreed goal.

Whilst this framework could be criticized on the grounds that it is apparently simplistic, nevertheless it does point very clearly to the complex range of variable situations which impinge on planning problems and their resolution.

In the first situation – known technology and agreed goal – both 'ends' and 'means' are certain. In such conditions public policy can be implemented through standard procedures which can readily be replicated, e.g. the provision of roads, water, drainage and other public services can be matched to local conditions at the same time as meeting the agreed desirable level of provision. However, even in this apparently stable situation, periodic uncertainty is inevitable if circumstances change, e.g. the South East Asian financial crisis of 1998 produced potentially significant levels of uncertainty in China for a period, which would have destabilized any existing prototype condition of known technology and agreed goal.

The unknown technology and agreed goal situation generally arises when there is a public commitment to deal with pressing problems (an agreed goal) without the 'technology' (a proven solution) to deal with the problem, e.g. the agreed goal may be the regeneration of inner city areas but the means of achieving that goal may be far from clear. Thus, the first priority is to obtain the missing knowledge about the possible solutions. Practice frequently deals with such situations of uncertainty by a pragmatic, incremental, trial and error search for policies that work or a controlled experimental search for alternative solutions.

In circumstances of known technology and no agreed goal, a process of bargaining surrounds the issue at hand, e.g. the British Airports Authority and the airlines may know exactly what is needed to serve the British 'market' and where best to locate those airports. However, the residents of the areas likely to be affected by the development of new or existing airports may well resist the proposal. In such situations, the bargaining may take a variety of forms although the expected outcome of the process is to accommodate multiple but conflicting goals. Thus, each bargain is tailored to its particular participants, is unique and, as a result, replicable results are precluded.

Finally, unknown technology and no agreed goal represents situations where there are multiple or unarticulated goals and no known effective means of achieving those goals. In the words of Christensen (1985: 65):

> It is impossible to draw clear examples of uncertainty over both means and ends because these conditions are in chaos. Nevertheless, they are common. Goals are often nebulous and changing; facts are often ambiguous ... Moreover, the ways planning problems are defined focus our attention on some aspects of the problem and cause us to neglect others. These neglected aspects are left nameless and therefore untreatable.

However, it should be emphasized that real-world problems may not necessarily fit into any single category, given that political and institutional forces shape the way problems are defined and treated, and a problem may shift its location as agreement over ends and means shifts. Nevertheless, the framework shows very clearly that three of the four sets of conditions that characterize planning problems contain uncertainty – about ends, or means or both – whilst the set of conditions with agreed goals and proven technology is susceptible to periodic uncertainty as conditions change, 'in short, uncertainty characterizes all planning problems' (Christensen 1985: 66).

Public Policy and its Implementation

The purpose of planning is to bring about 'desired changes' in society that would not otherwise happen if left to the free play of the market. Thus, successful implementation of the policies and proposals for change is almost certainly the most important aspect of the public policy process. Yet since the 1970s there has been an acknowledgement that many of the measures for change introduced through the policy planning process have in practice brought about relatively little by way of fundamental or lasting change. This problem of failures of implementation was identified in Britain in the 1970s in urban regeneration, land development, employment, control of pollution and industrial restructuring. In the words of Blowers (1982: 141) 'there is now a general recognition that plans are rarely carried out according to plan'. It is the view of Barrett and Fudge (1981) that:

> Government appears to be adept at making statements of intention but what happens on the ground often falls a long way short of the original aspirations. Government either seems unable to put its policy into effect as intended, or finds that its intentions and actions have unexpected or counter-productive outcomes which create new problems. (Barrett and Fudge 1981: 3)

Why should this be? Where and in what way is the process of implementation failing? Hogwood and Gunn (1984: 197) suggest that in looking at 'policy failure' it is useful to distinguish between the failure to implement (or non-implementation) and unsuccessful implementation.

Non-implementation Non-implementation is where a policy is not put into effect as intended because the resources available for implementation were inadequate or those involved in implementing policies or proposals have been either unconvinced about the value of the policies, and/or uncooperative and/or inefficient, or because their best efforts could not overcome obstacles to effective implementation over which they had little or no control.

Unsuccessful implementation Unsuccessful implementation occurs when a policy is implemented in full and external circumstances are not unfavourable, but nonetheless the policy fails to produce the intended results. The reasons for such a failure could be attributed to:

- bad execution, where the policies have been ineffectively administered;
- bad policy, which is based on inadequate information, defective reasoning or unrealistic assumptions; or

- bad luck, where external circumstances were so averse that it was no one's fault.

This analysis makes the point that formulating policy and implementing that policy are not separate operations – they are part of the same process. What happens at the so-called implementation stage will influence the actual policy outcome. Conversely the probability of a successful outcome will be increased if thought is given at the policy design stage to potential problems of implementation.

Framework for Planning and Implementation

Bruton and Nicholson (1987), taking account of some of the theoretical aspects of strategic planning, have suggested an idealized framework for planning and implementation in the public sector. This framework provides a hierarchy of levels of decision-making, with the level above constraining the level below; gives the opportunity of defining national social and economic objectives; and allows the coordination of the decisions and actions of a variety of agencies concerned with the implementation of sectoral policies through more detailed physical development plans and the programming of resource allocation. Though the details of this framework will vary in application depending on particular circumstances and needs, the fundamental character of the various levels in the hierarchy is clear.

Implicit in this framework is the acceptance of the fact that it provides no more than a way of thinking. If such a framework were accepted, it would do no more than enhance the possibility of obtaining a more consistent and coherent interpretation of higher order policies through the implementation of more specific proposals for development and implementation. Fundamental to the successful implementation of policies and proposals at the different levels in the hierarchy is the need to allow the managers of change who are active at those levels to develop the most appropriate means of achieving the objectives set for them by higher order policies within the given situation facing them.

Thus the framework assumes that in different situations different types of plan and resource allocation mechanisms will be used. However, all will be concerned to contribute towards achieving social and economic change, and all will have to take account of the contingent factors impinging on particular situations.

Level 1, national objectives and policies At the top of the planning hierarchy, the nature of the social and economic change to be pursued throughout the policy formulation and implementation process should be clearly set out by government. In addition, more specific sectoral objectives and policies need to be stated in relation to issues such as the economy, income, housing and social welfare. The inter-relationships between these policies and city and regional development would also be established at this level, together with an allocation of resources to implement policies on a sectoral and regional basis.

An important function of these policy statements is to establish a commitment to an agreed set of national policies, in order to provide a clear 'input' to lower levels. These statements would ideally be produced for short periods of time (such as five years with a mid-term review) for which it is possible to foresee likely changes and the availability of resources necessary to implement the proposals.

Level 2, regional strategies   At the next level down, strategic policies are needed to set out the way in which national policies relating to both sectoral and regional development can be expected to affect major regional areas. Strategies at this level should be concerned to establish more specific socio-economic sectoral policies for the particular region, as constrained by the more generalized policies of Level 1 and contingent factors in the region the broad spatial strategy required to coordinate the investment into physical development associated with the implementation of these policies and a programme of development to implement these policies, which is related to the likely availability of resources.

Regional strategies therefore comprise two distinct elements. One component establishes the socio-economic strategy for the area, while the other translates that strategy into spatial terms. The socio-economic element seeks to establish broad strategies for such sectors as employment, industrial development, housing, transport and education, within the constraints set by national level policies. Simultaneously, they are also influenced by regional and local politics, which could give rise to conflicts of interest, bargaining or the exercise of power, together with a technical assessment of the feasibility of policy implementation. The second element, the spatial strategy, then translates the socio-economic strategy into policies for physical development, identifying for example growth centres for additional population and employment, or the location of infrastructure improvements. Ideally, national coverage would be aimed for with a plan timescale of some 10–15 years.

Level 3, local spatial implications   Such regional strategies, by themselves, cannot ensure that either their socio-economic or spatial policies are achieved. It is only through a series of short-range programmes of development that higher level objectives can be achieved. The third level in the hierarchy, which would form a sub-unit of the region within which it is located, therefore seeks to translate the general land use proposals and broad policies established in the level above into detailed local plans where appropriate. Full coverage of the region would not be expected. These plans would serve as a guide to developers and establish a basis for the co-ordination of public and private investment into development and re-development. They therefore potentially constitute an important lead-in to implementation and would be needed especially in areas with significant development pressures and/or where change is to be concentrated in the near future. In line with this emphasis on the progression of general policy towards implementation, the timing of proposed changes and the resources required would be worked out in detail; also, the timescale of such plans would be shorter than for those in Level 2, looking ahead a maximum of 10 years. Again the form of these plans would be constrained by policies in Level 2 and influenced by more local contingent factors.

Level 4, resource allocation   Given that the failure to deal adequately with the resource aspects of strategic policies is a key factor in explaining the relative lack of success in implementing such proposals in the UK and elsewhere, it is essential that account is taken of this factor. One solution would be the use of a series of sectoral rolling programmes for public investment into, for example, economic/ industrial development, housing, transport and other elements of physical infrastructure and development requiring recurrent investment. Such programmes could be produced and rolled forward annually, and would outline spending plans for a relatively short period ahead, such as five years. These programmes would be constrained by the general policies and resource allocation proposals contained in the higher levels of the hierarchy, and would most likely be formulated for the regional areas defined in Level 2. However, given the range of potential policies and proposals, and the range of contingent factors in any specific situation, it must be accepted that a variety of ways of handling Level 4 would be developed.

Level 5, implementation   Following resource allocation, the process moves into implementation. Here, the proposals contained in Levels 1 to 4 are converted into development on the ground through the coordination and control of public and private investment in the built environment. It is important to ensure through careful programming that the required combination of resources is available when required, and that development and expenditure are constantly monitored to ensure the early identification of any departures from schedule, together with the associated 'under-spend' or 'overspend'.

The idealized hierarchical framework outlined above explicitly takes a policy-centred or 'top-down' view of the process of implementing policy, and assumes that lower-level implementers are constrained by (or act as agents for) higher-level policy-makers. It is in the nature of idealized models to abstract from or simplify reality so as to allow analysis, and in this respect the scheme developed here is no exception.

However it is important to acknowledge that the hierarchical concept of organization, whereby policy emanates from the top and is transmitted down the hierarchy and translated into more specific formulations, rules and procedures as it goes to guide or control action on the ground, is rarely found in reality.

Performance of Chinese City and Regional Planning System

A Visionary and Methodological Framework for Users

From the previous chapters, it is possible to find that the city and regional planning in China has been operated as public policies and spatial proposals for development. The top-down hierarchical planning system is designed to carry out policies of the central government and to influence the development of the local municipal governments. Nevertheless, it is because of devolution and marketization in the transition to a market economy that the role of the central government has been declining greatly. Local governments, including agencies responsible for infrastructure provision, as well as planning, housing and other aspects, are playing a crucial role in development. This has greatly influenced the other stakeholders, e.g. developers, local residents and other interest groups. Even so, this situation will have to change gradually, as in the pluralistic stakeholder society, with more investment from non-government resources, so that the role of government in development will be incrementally reduced. From the analysis of the existing city and regional planning system and practices, it is evident that the basic concepts and notions of the Chinese city and regional planning have not considered the interests of pluralistic stakeholders in planning and development. Chinese planners mainly react to the pressures from politicians of local governments, who regard increase of GDP and economic development as a priority. City and regional planning is not concerned to seek the consensus of all the stakeholders in society, other than some government departments with the resources for development.

Chinese city and regional planning has been used to provide advice to the governments to define the nature of the dimension of a region, a city or a town, to forecast the population growth, to identify the development policies, and then as the basic and traditional function of planning, to propose the spatial structure for development. To achieve planning targets, planners have used the notion of a systematic and rational planning approach, and their knowledge and methodologies, e.g. cost–benefit analysis, during the formulation of plans.

Legacy of Rational Planning and a Process of Elite Actions

Chinese modern planning was initially introduced from the former Soviet Union within the system of the centrally planned economy. The existing Chinese city and regional planning concepts and the practices have been impacted by the legacy of the centrally planned economy and the rational approach to planning.

In the centrally planned economy, planners normally could safely ignore market forces, which did not exist when producing the plans. The intention and decision of the governments should be regarded as the main consideration in the planning process. This was practicable since the investment in development and construction were the tasks of the governments; especially with investment as a type of allocation of revenue from the central government.

In the market economy, the situation has been changed to be more complex because of the pluralistic stakeholder society and the uncertainties. Investment is derived from multiple resources in the market instead of from the government only. Within such a context, the poor understanding of the market by planners creates many conflicts between planning policies and development.

However, Chinese planners are still regarded as experts by themselves or even ordinary citizens. The planners, as experts, should be capable of defining the direction of city development because of their educational backgrounds and working experiences. They should be able to achieve the required task of foreseeing city development tendencies and objectives. Planners should have responsibility to predict and to control the future development of a city. From the perspectives of planners, they often complain that the planning policies proposed by them have been often ignored in implementation process. No one has thought whether the proposed planning policies or development schemes have considered the uncertainties and the interests of the different stakeholders in a market economy.

City and regional planning has been attempting to function more appropriately through the use of 'scientific' and 'objective' methods to guide development. Scientific planning has been a popular subject in journals (Chen 2003; Duan 2003; Ma 2003; Shi 2003; Zou 2003a). Some have suggested that city and regional planning should seek a perfect technical method (Zhang 2001).

It is found from the survey that many planners do not understand what kind of role planners and planning should have in a market economy. Planners produce plans from a standpoint of planning rationality, yet the rationality of planners, as one of a number of diverse disciplines, is not enough to deal with the complex inter-relationship of many diverse interests, especially when the implementation of planning initiatives mainly depends upon private investment.

Chinese city and regional planning has been a process of elite action. The formulation of city and regional plans predominantly depends upon planners' investigations into local social and economic development, and the knowledge of planning methodologies. Despite the fact that public participation in plan making through public hearing is specified in the City and Countryside Planning Act 2008, the implementation of public participation in plan making has not been treated as priority. The most prominent consulting to the planning policies is mainly to the leaders of local government and local government bureaus with development resources.

The process of elite action explains the reason why Chinese planning has mainly tried to improve its function of encouraging economic development but ignoring the basic values of city and regional planning in the provision of 'public good', equal opportunities for all and fairness to all.

Local Responsibility of City and Regional Planning in Chinese Transition to the Market Economy

Chinese city and regional development is influenced both by the government and the market, but dominantly by the government. Government is able and has tried to impact or lead development by planning policies or investments in the infrastructures.

Although China is moving towards a market economy, the government is still playing a significant function, especially in city and regional development and its policy-making. The main reason is that there is shortage of mechanisms for the accessibility of developers and local communities to the city and regional planning process. The majority of local communities and small developers do not have the opportunities to express their opinions and suggestions in both plan making and implementation. Instead, they follow the actions of local governments. If the local governments are able to implement their development projects by following planning policies, city and regional planning policies can then be delivered. However, in practice, many planning policies are in fact violated by local governments themselves.

It is not unusual for local government to adjust, revise or violate the local approved planning policies, typically the planning policies that may impact development tendency or general spatial structure of a city by ignoring any formal revision process as indicated from leaders of local governments. When development has been the predominant concern and the increase of GDP the priority, and planning and development of a city has been devolved to local governments as one of their responsibilities, planning policies proposed by planners may have to consider the interests of the developers and investors, typically the large investments with significance for local GDP increases under pressure from local governments. The local bureau of planning administration is one bureau under local government's jurisdiction. Planning officers have to obey the decisions of local government leaders.

It is because of the great influence of the government on other stakeholders in city and regional development in present-day China, if planners can persuade local governments to support and to follow the proposed city and regional planning policies, then city and regional development is able to have a plan-led approach. However, it should be borne in mind that with further decline in investment in infrastructure by the government and the political reform, the existing functions of the government should be changed sooner or later.

A Mechanism for Debate, Communication, Negotiation and Conflict Arbitration in Planning Process

City and regional planning has to deal with the pressures imposed by different sources of power. This is a common problem worldwide. The definition of power suggested by Cooke (1983) is that power represents the capacity of getting others to do something they may not otherwise do. City and regional planning, under the pressure of power, has to attempt to mediate among the various sources of power and interest groups, as well as the local communities.

Mediation, negotiation and conflict arbitration are critical functions of planning in the market economy. There are usually conflicts due to many interest groups. It is critical for planning to establish a certain mechanism of mediation, negotiation and conflict arbitration to manage and mobilize changes in the pluralistic stakeholder society (Healey 1998). In the period of the centrally planned economy in China, the function of mediation, negotiation and conflict arbitration in city and regional planning was not very important. The development and planning commissions in both the central and local governments would make decisions on development and resource allocation. City and regional planning in the centrally planned economy was to allocate land use to the proposed development projects defined by the social and economic development plan. However, in the market economy with Chinese characteristics, the role of the development and planning commissions (renamed as the Development and Reform Commission 10 years ago) in the allocation of resources has declined gradually. The city and regional development will have to depend upon different resources in the market instead of government only. The function of mediation, negotiation and conflict arbitration of planning becomes critical.

However, the functions of mediation, negotiation and conflict arbitration of city and regional planning in China have been mainly operated within governmental departments during plan formulation, and then among the governments' departments, developers and investors in the stage of plan implementation. Nevertheless, it is because of a lack of understanding of the market, and inappropriate planning concepts of most Chinese planners that the mediation, negotiation and conflict arbitration functions of planning have not worked well in China.

It is very common that when negotiating with other government departments to deliver their development projects within the city and regional planning policy framework, these government departments are very difficult to be mediated to follow city and regional planning policies. They usually insist on delivering their development projects from either their professional perspectives or their interests in the development.

Regarding the involvement of local communities in the planning policy-making process, although public participation in plans formulation has already operated in plan making process, and some urban planning exhibitions have been held for the local citizens after the completion of the plan's production in some cities, generally speaking, the communication function of planning with the communities is very weak. The interests and voices from some stakeholders have been ignored, especially the local communities.

It is regulated by City and Countryside Planning Act 2008 (NPC 2008) that: 'Before submitting for approval, planning organisation should publish city and countryside plan drafts for consulting with experts and the public through the public hearing or other approaches. Public participation should be less than 30 days' (Article 26, NPC 2008). However, local communicates are not told the whole story, but only part of the policies. It is regulated in the City and Countryside Planning Act 2008 (ibid.) that: 'City and Countryside plan making administrative authority should notice in public the approved city and countryside plans on time, except the contents that are forbidden to be noticed in public by the laws and government's security regulations' (Article 8, NPC 2008).

Key elements of a City Comprehensive Plan are still regulated by the security law and security regulations of the government to be confidential. It is not possible to obtain the whole contents of the City Comprehensive Plan of a city without authorization, but a brief summary of the policies is available. It is not possible for local citizens and developers to obtain detailed information of local plans, but only the general policies.

The citizens are only told what has been proposed in the plans for the cities in which they are living. Their opinions have had slight influence on plan making and delivery. It is still difficult for local communities to participate in decisions and policy-making.

Promoting Economic Development

During the last three decades, the promotion of development has been the starting point of the planning. It is the priority of government and it has to be seen by planners as the most important aspect of the planning function (Needham et al. 1997). This is typically true in China since the transition to the market economy. Jingxia Wang (2003), former president of China Academy of Urban Planning and Design, argues that planning should emphasize the services to economic development in the transition to a market economy.

Economic development promotion has been the principal policy of the Chinese government since the reform and open door policy in the late 1970s. It was particularly after Deng Xiaoping's trip to China's southern provinces in early 1992 that he redefined the meaning of socialism, and put forward his famous 'three types of advantages', namely, the advantage of promoting the growth of the productive forces in a socialist society; the advantage derived from increasing the overall strength of the socialist state; and the advantage gained by raising the people's living standards.

Deng's new socialism definition and the 'three types of advantages' stopped the debate on which way to go ahead, whether socialism or capitalism, and ensured a market-oriented economic reform which was finally clinched in the 14th National Congress of Chinese Communist Party in October 1992. The concept of 'development as an absolute need and priority' as raised by Deng Xiaoping has become the principal policy of the Chinese government. Economic development took over all other issues.

The function of city and regional planning is not merely a form of rational action applied to city and regional development, but it has to be processed in a social, political and economic context. When there are social and economic changes, planning should be part of the changing process instead of being separate from it. Paris (1982) has argued that city planning should reflect and contribute to the changes. The former chief planner of the Ministry of Construction, Madam Chen (2001), argued that urban planning had to adapt itself to the social and economic changes, instead of the other way round that the social and economic changes meet the demands of the traditional planning approaches. When development is defined as the government's priority, city and regional planning has to reflect the government's policies and development objectives in its targets and to apply the most efficient and effective methods to achieve the defined development objectives of the government.

Chinese city and regional planning has tried to help local governments to coordinate the state's policies and interests with the local governments' interests. As a requirement of national policies, and using their professional knowledge, Chinese planners should consider and apply the development proposals and policies decided in the Five-Year Economic and Social Development Plans produced by both national and local governments in city and regional plans. City plans should seek the most efficient social, economic and environmental benefits of a city while protecting the local nature and characteristics, as well as bringing the local economic merits and advantages into full play.

In the case of Xiamen Comprehensive Plan, the planners applied a project oriented pragmatic planning approach under the pressure of inward investment. The emergence of non-statutory Urban Development Strategic Plan in Chinese cities is the result of great pressure in competition for economic development and city expansion. The function of the promotion of economic development in city and regional planning has been fully achieved.

Protecting Public Interests

The change and adjustment of an urban planning policy or scheme to meet the demands of investment projects as required by developers and investors, especially international investment projects, have been criticized as indicating 'capture' by developers (Wang 1999b). Some academics criticize Chinese planning as satisfying developers' and investors' requirements (Xu and Song 1998), or planning as being absolutely obedient to developer or investor (Liu 2000; Wang 1999b). However, these criticisms may not fully explain the entire picture. City and regional planning has to work within the existing social and political context.

These criticisms do explain a tendency in Chinese city and regional planning. It is difficult for planners to turn away investment when producing planning policies and proposals. City and countryside planning policies should be produced within the existing social and political context. In the market economy, developers and investors are crucial stakeholders in the development. It is typical in China where 'development is an absolute need and priority' that planning policies may have to help to attract inward investment. However, the critical issue within such context is that planning policies should also consider the interests of local communities, the public and sustainability. Nevertheless, it is not difficult to find that Chinese planning policies often ignore the interests of local communities, disadvantaged groups and sustainability. The interests of the disadvantaged and policies for poverty reduction had not been regarded as a priority, and are not even considered as planning policies when making city and regional plans.

As public policies, city and regional planning is required to include policies to service the needs of local communities, and to target protection of the long-term benefits of local residents, while serving business interests at the same time. In particular, it should guide public investment in those areas that are of little interest to private capital owing to the low rates of return, diffused benefits, and the large investment required, such as investment in infrastructure and public facilities. Although this function of planning is particularly important in a competitive environment since the appropriate infrastructure provided for inward investment is regarded as one of the critical criteria for assessing the 'investment environment', the demands of public goods for local communities should not be forgotten. The former Chinese Construction Minister, Yu Zhengsheng, used to explain that the main task of planning as a functioning tool of government should not only propose to 'erect buildings on the ground', but to protect the public interests, to control development and to provide a good living quality through development planning management (Yu 2000).

While indicating the provision of 'public goods', there is one issue that should be considered by planners, namely what kind of 'public goods' should be provided? There are some criticisms in the literature of the definition of 'public goods'. Some academics argue that when defining the 'public good', it should avoid the problems of decision and definition by a small group, e.g. politicians, civil servants and experts, as a kind of elitist and inconsistent process (Simmie 2001).

Moreover, there is a risk associated with the provision of 'public goods', particularly those 'public goods' of infrastructure, that of building them in advance of development, providing them before any real demand, for the purpose of demonstrating politicians' and government's 'political achievements'. This has been a common problem in China. The leaders of a local government have a term of four years. The objectives of local leaders are to secure their achievements in the economic and urban development during their four-year term as a priority. They need to achieve their political aspirations through the achievement of city and regional development and economic growth. They attempt to influence the city and regional planning by emphasizing short-term and ambitious targets. It has been a common problem that local leaders seek to build large infrastructure projects without any cost–benefit analysis, and with no concern about the financial affordability. From a planning perspective, after the decisions by the leaders of local governments, planners have to obey and follow these decisions. The development of Zhuhai Airport in the 1990s was such an example. The project brought Zhuhai municipal government to the threshold of bankruptcy since few airlines use Zhuhai Airport. As suggested by Flyvbjerg (1998), the rationality and reality are all defined by those in power.

Forwarding to Sustainable Development

Sustainable development seeks diverse objectives but focuses on delivering those objectives now and in the future. Economic growth is only one, though an important objective as it will secure a better living quality. Chinese city and regional planning system has provided an important role in economic development during the last three decades. Nevertheless, it should be realized that economic growth should not be at the social and environmental costs, including enlarging inequality. The objectives of economic development should also target poverty reduction, and improvement of the existing environment, while protecting non-renewable resources. From this perspective, Chinese city and regional planning has not made the contribution that it should.

Since the Rio World Summit 1992, the Chinese government has published the Sustainable Development Agenda, which consists of all aspects of social and economic development, and environmental protection. The Agenda has been regarded as national development policy. The problem is that there have not been any real concerns on sustainable development, not an action plan or detailed practical proposals put forward through city and regional planning policies and regulations for local governments. There were either only a few sentences about sustainable development in principle, or even without any word about sustainability in city plans. In fact, city and regional planning has been criticized as giving too great an emphasis to economic objectives (Wang 1999b; Wu 2000), but not other issues, typically in social and environmental sustainability.

It was not until the twenty-first century that some city and regional plans begin to consider ecological issues as a main launch to sustainability. Although ecology and environment are only part of sustainable urban and regional social and economic resources, it is progress for planning to include the principles of sustainability.

On 25 November 2009 the State Council of China made the announcement that, by 2020, the emission of carbon dioxide per GDP unit will be reduced 40–45 per cent compared to that in the year of 2005 (Xinhua News Agency 2009). This ecological and sustainable development has become one of the main development objectives in China. Low carbon eco-city developments, which consist of social, economic and environmental aspects, have become mainstream.

According to the survey by Chinese Society of Urban Studies, 97.6 per cent of prefectural cities and above have already specified low carbon eco-city as the local development strategy. Among which 53 per cent have started different types of low carbon eco-city development; 28 per cent are formulating plans; and 19 per cent are at the strategic stage. However, these policies and understanding have been achieved at the severe cost of damage to the ecological environment and scarce natural resources, including land during the development of the last 30 years.

Sustainability is not only important in policy-making, but more significantly in actions taken. City and regional development without considering sustainability will create serious problems. This has already happened in China during the last three decades. One example is city sprawl and expansion, which have been exceedingly difficult to control by city and regional planning. A considerable amount of cultivated land has been occupied for urban and industrial development. It has been a severe problem in many cities. Deng (1997) indicated that from 1992 to 1993, during the period of 'Developing Zone Fever',[1] there were more than 10,000 developing zones in China, which occupied 15,000 km^2 of land. The majority of these zones were only defined for development, but nothing

---

1   The intention to establish developing zones everywhere.

was actually developed on the ground at all. As a consequence, a large amount of cultivated land was occupied and demolished without any development. The situation arose for two reasons. One was the ambitious development targets of local governments. The second was the failure of planning to guide and to persuade decision-makers to consider sustainable development to protect natural resources for future generations, perhaps without benefiting the existing generation. According to Deng's research (1999), during the same period of time, real estate attracted investment of around 73.2 billion yuan in 1992. The price of housing was distorted owing to speculation. Until July 1993 when the central government tried to intervene and control the real estate market, more than 1,000 development zones and several billion yuan investment had been made but without returns.

The reasons for such a disordered development situation without any concerns of sustainability are complicated. However, the main reasons can be summarized as follows:

- The ambitious development targets raised by the leaders and planners in local governments because of seeking GDP increase.
- City and regional planning has tried but failed to persuade decision-makers to adopt realistic development policies and objectives.
- The planning is easily captured by developers and investors.
- Shortage of public participation in the planning process. Local communities are unable to function as a type of power block in the decision-making procedure.
- The inappropriate delineation of property rights, especially in land resources. The land is defined as state-owned, but local governments operate and obtain the revenues from leasing the right to land use, while many state-owned enterprises and existing land users generate profits from the land they occupy.

These factors show that in addition to planning performance problems, the government's function and its institution in the development process are also problematic. Typically, if there were any association between government departments and private developers, the chaos and the cost will be a further increase. The 'malposition' of local government that not only functions as a policy and rules regulator, but also an operator in the market, is the main reason for this situation. Local governments are not only responsible for promoting economic development and providing policies to regulate development, but also involved directly in the investment and business operations. The need of increasing GDP and attracting inward investment as priority tasks of local government means that they have limited choices. This is a typical problem in the transition caused by the contradictory roles of the government as a development promoter and a regulator. This phenomenon is the underpinning factor that planning is unable to function appropriately in guiding sustainable development and in providing equal opportunity for all. As a component of the administrative system in China, the reform of the Chinese city and regional planning system is unable to separate from and has to associate with political and administrative reform in China.

A Rigid and Complicated Systematic Top-down Plans Production

Chinese city and regional planning system, as a hierarchical arrangement of policy and its formulation that allows the relationships between policy options to be pursued at each level, but within a framework, is a rigid and complicated, sometimes overlapping, system.

The Chinese planning system, which consists of the Five-Year National Economic and Social Development Plan, National Development Priority Zone Plan and National Integrated Urban System Plan at level 1, Provincial Integrated Urban System Plan and Provincial Development Priority Zone Plan at level 2, a City Comprehensive Plan at level 3, a Regulatory Detailed Plan at level 4 and planning management (development control) through certificating mechanism of application for planning permission at level 5, is a top-down framework. This hierarchic framework with different levels of decision-making, but with the level above constraining the level below, gives the opportunity of defining development policies for the coordination of the decisions and actions of a variety of governmental organizations.

Higher level plans give expression to development policies and schemes, but only in principle and general terms. The lower level plans should follow policies decided in higher level plans. At the lowest levels 4 and 5, a Regulatory Detailed Plan is used as the main tool for urban development planning management.

However, within this framework, there are potential contractions and overlapping between different types of plans at one level. At levels 1 and 2, several different types of plan, e.g., Five-Year Economic and Social Development Plan, Development Priority Zone Plan, Integrated Urban System Plan, General Land Use Plan, are produced by diverse ministries with different national and regional planning responsibilities. Planning policies decided by separate plans may create contradictions and overlapping tasks among them. At level 3, a City Comprehensive Plan should be in accordance with the Economic and Social Development Plan, but contradictions have always existed between a City Comprehensive Plan and a General Land Use Plan in most Chinese cities because of the differences in planning targets and objects.

City and regional planning has two powers of 'positive' and 'negative'. The 'positive' power is to produce policies and decisions in relevant city and regional plans, but this 'positive' power is weak. The main reason is that planners have neither resources, nor the real power to implement the plan. Whatever a plan proposes, its effectiveness is dependent upon it being used (Mastop and Faludi 1997). Pickvance expressed a similar opinion. He argued (1982) that what planning may do is to use the 'negative' power to refuse permission for development which does not fit the planning policies. This is the power of planning management, the term that is used in the Chinese city and regional planning system, and what is called development control in the UK.

Planning management plays a significant role in the planning process. It is a practical process of city and regional land and spatial allocation and uses. The objective of the process is to deal with the conflicts in actual development among the different stakeholders. Planning management is a process of bargaining and negotiation with developers and investors. The role of 'gatekeeper' in city and regional planning is to protect the interests of local communities, to seek the possibility of a 'win-win' compromise solution. More critically, planning management should function as an important means of monitoring and supervising the changes that are taking place on the ground once plan implementation begins, and to measure the changes that are taking place against the policies set out in the plans. If development on the ground departs in a significant way from the original policies contained in the plans, then the plans should be reviewed and changed to meet the new/unexpected conditions. Without a perfect and efficient monitoring and supervision system, it is very difficult to achieve a balance between the long-term objectives and short-term benefits, between the general interests of a city or a region and the interests of the districts or local communities, and between the interests of the developers and investors and that of the public.

Resilience of Plans Delivery

Although Chinese planning system is of a feature of the rigid top-down plan production, it provides certain flexibilities in its operation, typically the national policy of 'development is an absolute priority' in the Chinese open door and reform process. The interest of the Chinese local governments has been to promote economic development and attract more inward investment, especially from large and well-known international companies. This investment can both increase employment to accelerate local economic development and create the reputation of a 'perfect investment environment' in a highly competitive market. It is also able to illustrate the political achievement of local leaders. Because of the assessment criteria of GDP increases, it is unrealistic to expect that local governments would ignore inward investment. Chinese local governments are confronted with the pressure of competition for investment projects, which will contribute to local economic development and increasing GDP.

Although planners prefer potential investment and available inward projects that can fit the nature of the city and without damage to the environment, the attraction of investment to promote local economic growth and increase GDP has always been the main objective of city and regional planning in China. Planners have to compromise and obey local development priority. It is the reason that planners will not be permitted to refuse inward investment and allow it to be allocated to other competitors. Chinese planners are very realistic during the implementation of city and regional plans. It has been a common practice that planners are involved in discussion and negotiation with investors and developers to explain the merits of planning proposal and scheme to influence the location decision of investment projects. Some investment projects may not be appropriate to a specific area, planners can report to the leaders of local government, and to change policies to conform to the decision, if it is decided by local governments. Sometimes, planners are placed in a very difficult position due to the conflicts between their values and the pressures to promote economic development in a competitive world, typically when potential project might change local characteristics or create risk to local environments. Planners often argue and insist development should be plan-led. It is the wish of planners that the investment and relevant projects should follow the proposals and policies defined in city and regional plans. They worry about how development may impact on or create damage to the environment. They are concerned with the problems of wasting scarce resources, i.e. cultivated land. They attempt to avoid failing to control city sprawl and expansion, decline in the quality of the environment, and damage to the historical heritage and areas. At the same time, they have to change their policies and proposals to meet political and economic demands when necessary, especially under the pressure of the power from local governments. It is because of the political pressures that if there are any significant projects, available land has to be allocated immediately to meet the demands of the development, planners may have to change planning policies and land use schemes after discussion and negotiation with the leaders of local government and investors (or developers) of important projects. The change in planning policies and design schemes may even overlook the defined city and countryside plans amendment process decided in the City and Countryside Planning Act 2008. This is the main factor why a pragmatic planning approach has been used in practice without any knowledge of this planning theory. Planners have no other choice. Although city and regional plans, typically a City Comprehensive Plan, have worked as a basic reference to negotiate and bargain with the inward developers and investors for important development projects, some policies and proposed schemes may have been changed in the negotiation and bargain between local government and developers or investors because of the significance of the large projects to local economy and GDP. The changes include the layout of road networks, public utilities, or even changes in land use. It happens in many cities that proposed greening areas had been used for office blocks and residential development to increase the return of investment in some rehabilitation projects due to pressure from developers and politicians.

It shows that Chinese city and regional planning is enormously complicated, non-transparent, and time-consuming when assessing the Chinese city and regional planning system, including the government structure, plans' formulation and approval procedure, and governance mechanism in plan implementation. It is because of this complicated system that there has to be an alternative to deal with the rapid development in practice. It is the reason that Chinese local governments have applied the non-statutory Urban Development Strategic Plan as a flexible planning approach, which avoided the time-consuming and comprehensive process of statutory city comprehensive planning process, to cope with economic development and rapid changes, even though it is not specified in the City and Countryside Planning Act 2008. Flexibility is a term in planning to explain the adjustment to meet demands of the market. Public interests could sometimes be sacrificed to satisfy the market. It is a dilemma in city and regional planning. Planning has to be flexible in a market economy owing to uncertainties in development. Nevertheless, some basic values and principles of city and regional planning, e.g. sustainability, public interests and equal opportunities, should also be considered.

It is very difficult to avoid conflicts between flexibility and planning values of public interests, equal opportunities for all and others. However, an eclectic or compromise solution between planning flexibility and planners' values and principles can be a potential solution.

Conclusion

In this chapter, the performance of the Chinese planning system is evaluated. Numerous problems have been highlighted as a result of the evaluation. They include inappropriate concepts and understanding of city and regional planning in the market economy.

An investigation of planning practice illustrates that Chinese city and regional planning has endeavoured to improve its function to encourage development and to try its best for keeping pace with economic development. Nevertheless, the primary values of planning, e.g. sustainability, public interests, equal opportunities, have often been ignored. Chinese planning fails to achieve its basic function of equal opportunity to all and democracy in terms of an open and transparent planning process.

The main problem of the planning institution is its designated and complicated contents and the procedures of plan formulation and approval. The requirement for the approval process of a City Comprehensive Plan by the higher hierarchical governments may not be necessary. This is especially so when city and countryside development have been devolved as the responsibility of local governments. Since the reform and the transition to a market economy, devolution has been one of the principal features. At the National Conference of Urban Planning in October 1980, a decision was made that the major task of local governments should be to plan, build and manage the cities. This is part of the devolution process, which has been the principal core of the reform in China. Nevertheless, a City Comprehensive Plan approval process is still regulated by the central government in that all City Comprehensive Plans of metropolitan and capital cities of the provinces, and other designated cities should be submitted to the central government for examination and approval. This plan approval regulation seems unrealistic and unreasonable, while the procedure is time-consuming. It creates a problem for city planning to cope with the rapid social and economic changes. It forces local governments to find alternative solutions. They may either submit their City Comprehensive Plan for approval without reflecting on the actual development situation, or apply other effective and efficient approaches. The application of the Urban Development Strategic Plan is such an approach. The practice seems a departure from the system. This situation creates a problem of trust in and integrity of the Chinese city and regional planning system and practices. The alternative solution to address the contradictions between local development policies and national policies, and some projects with nationally and regionally significant issues in city and regional development and planning can be addressed by the 'call in' mechanism authorizing to the central government. This should be defined by planning acts.

It is found from the investigation of the author that through the government's policies, city and regional planning has continued to be regarded as an essential instrument to guide development in the transition to a market economy. The critical challenge is how planning policies can consider the interests of other stakeholders. Without this consideration, it will be difficult to implement planning policies, especially since investment in development is dependent upon diverse resources. The role of planners is required to adjust to meet the changes in the transition to a market economy.

According to the investigation, the role of government in development is significant. Private developers, typically the small size developers, normally follow the actions of the government. Under such a situation, the role of local governments is crucial in city and regional planning.

With devolution and China's transition to the market economy, competition for scarce inward investment and development projects among local governments has been serious. It is the impulse of local governments to apply the concepts of urban entrepreneurialism. This requires local governments to reduce various controlling and restraining policies and regulations to promote local economic development (Short and Kim 1998). At the same time, urban entrepreneurialism normally requires project-led development to explore and attain full use of opportunities to encourage local economic development. The application of the pragmatic planning approach and the Urban Development Strategic Plan can meet this demand.

The market economy is a pluralistic stakeholder society. There are many different interests and conflicts. Coordination and collaboration in planning as mediation and negotiation functions have to be introduced in the formulation and delivery of plans.

It is found by the author that the general policies defined in a City Comprehensive Plan can mainly be followed in development, typically in the defined land uses, through the mechanism of planning management. Within this process of planning management, controlling indexes, codes and ordinances of regulatory detailed plan have been used as the main tool to control development. However, the changing controlling indexes, codes and ordinances have been a critical problem in the process of planning management. The discretion in planning is the main cause of this problem.

Reform of the Chinese city and regional planning system is certainly needed. Nevertheless, the city and regional planning system is one component of the overall political and administrative system. The reform in the planning system is unable to be independent from the general one.

Bibliography

ADB (Asia Development Bank). 2007. *Reducing Inequalities in China*, Asian Development Bank Seminar, published online 9 August 2007, available at: www.adb.org/conf-seminarpapers/2007/10/30/2393.csr.yu/ [accessed: 10 April 2008].

Ashworth, G. and Voogd, H. 1990. *Selling the City: Marketing Approaches in Public Sector Urban Planning*, London and New York: Belhaven Press.

Bachmann, J. and Leung, A. 2008. Towns against a Background of Cities: Recent Evolution of Town Urbanization Policy in the People's Republic of China, *International Planning Studies*, 13(1), 13–30.

Barrett, S. and Fudge, C. 1981. *Policy and Action: Essays on the Implementation of Public Policy*, London: Methuen.

BBC News. 2008. How China is Ruled, available at: http://news.bbc.co.uk/1/shared/spl/hi/in_depth/china_politics/government/html/1.Stm [accessed: 28 March 2008].

Blowers, A. 1982. *Urban Change and Conflict: An Interdisciplinary Reader*, London: Harper & Row.

Blowers, A. 1994. *Planning For A Sustainable Environment*, London: Town and Country Planning Association/Earthscan.

Bo, W., An, H. and Li, J. 2011. Construction of Development Priority Zones and Chinese Regional Coordinated, Development: Promotive or Aggressive, *China Population, Resources and Environment*, 21(10), 121–8.

Bradsher, K. 2007. Trucks Power China's Economy, at a Suffocating Cost, *New York Times*, 8 December, available at: www.nytimes.com/interactive/2007/12/08/world/asia/choking_on_growth_7.html [accessed: 19 April 2008].

Bruton, M., Bruton S. and Yu, L. 2005. Shenzhen: Copy with Uncertainties in Planning, *Habitat International*, 29(2), 227–43.

Bruton, M.J. 1983. *Bargaining and the Development Process*, Cardiff: UWIST, Department of Town Planning.

Bruton, M.J. 1984. *The Spirit and Purpose of Planning* (2nd edn), London: Hutchinson.

Bruton, M.J. and Nicholson, D.J. 1987. *Local Planning in Practice*, London: Hutchinson Education.

Cao, H. 2009. Logical Reform of the Urban-Rural Planning Management System, *Planner*, 25(7), 82–5.

Cao, J. 2002. *From a Closed Society to an Open One: Review and Prospects of 20 Years' Reform*, speech at Huaxia Times, 12 April 2002.

CAUPD (China Academy of Urban Planning and Design). 1995. *Xiamen City Comprehensive Plan 1995*, Beijing: China Academy of Urban Planning and Design, P.R. China.

CAUPD (China Academy of Urban Planning and Design). 2006. *Urban System Plan along Liaoning Coastline, Liaoning Province, 2006–2020*, Beijing: China Academy of Urban Planning and Design, P.R. China.

Chadwick, G. 1978. *A Systems View of Planning* (2nd edn), Oxford: Pergamon Press Ltd.

Chen, A. 2005. Assessing China's Economic Performance since 1978: Material Attainments and Beyond, *Journal of Socio-Economics*, 34(4), 499–527.

Chen, B. 2003. Re-consideration of the Scientificity of Urban Planning, *City Planning Review*, 27(2), 81.

Chen, G. 1984. *Interpretation and Assessment of Laozi*, Beijing: China Book Press.

Chen, J. and Wang, G. 2006. The HE Policy Approach to Coordinated Regional Development, *City Planning Review*, 30(12), 15–19.

Chen, S., Yao, S. and Zhang, Y. 1999. Comprehensive Thoughts of Urbanisation in China, *Economic Geography*, 19(4), 111–16.

Chen, X. 2001. Urbanisation and the Urban Development Problems, *Urban Planning Forum*, 4, 1–3.

Chi, W. 2004. *Research of Guanzi*, Beijing: High Education Press.

China State Bureau of Statistics. 2008. *China Statistic Year Book 2008*, Beijing: China Statistic Year Book Press.

Christensen, K. 1985. Coping with Uncertainty in Planning, *American Planning Association Journal*, 51(1), 63–73.

Cooke, P. 1983. *Theories of Planning and Spatial Development*, London, Melbourne, Sydney and Auckland: Hutchinson & Co. Ltd.

Cun, S. 2010. Great Battles for Re-construction Planning and Design this Winter and Spring, available at: http://news.cntv.cn/20101112/105844.shtml [accessed: 4 March 2011].

Dai, F. and Duan, X. 2003. Strategic Planning: Why Do We Need It? A Practice in Guangzhou and Some Implications for Reforming the Chinese Planning System, *City Planning View*, 27(2), 28–34.

Dai, S. 2009. Research on Land Finance and the Usage of the Land of Local Government, *Journal of Fujian Normal University (Philosophy and Social Sciences Edition)*, No. 4, 21–6

Damme, L., Pen-Soetermeer, M. and Verdaas, K. 1997. Improving the Performance of Local Land-Use Plans, *Environment and Planning B: Planning and Design*, 24, 833–44.

Deng, L. 2002. New Changes in Taiwan's Economic Growth, *Journal of Xiamen University (Arts & Social Sciences)*, 153(5), 31–9.

Deng, L. and Du, Li. 2006. Research of Addressing Unbalancing Development between Regions by the Approach of Development Priority Zones, *Economist*, 4, 60–64.

Deng, X. 1992. *Selected Works of Deng Xiaoping*, Vol. 3, 374–6. Beijing: People Press.

Deng, W. 1999. On Fundamental Transition of Urban Planning Following the Way of Economic Growth, *Urban Planning Forum*, 5, 17–20.

DGBAS (Directorate-General of Budget, Accounting and Statistics, Executive Yuan). 2009. *Republic China Statistics Year Book 2009*, Taibei: Directorate-General of Budget, Accounting and Statistics, Executive Yuan.

Dong, J. 2012. Creating New Capital: Republican Nanjing's City Design and Planning Politics: An Analysis of the Capital's Planning (1928–1929), *Nanjing Social Sciences*, 5, 141–8.

Dong, K. 1999. Market Mechanism and Urban Land Use under the Intervention by Government, *Urban Planning Forum*, 3, 57–60.

Dong, L. 1992. Review and Prospect on Paid Urban Land Use System in China, *Yunnan Geographic Environment Research*, 4(2), 16–29.

Dong, Z. 2008. Initiate, Existing Situation and Approaches to Address the Problems from Land Fiscal Income for Local Government, *Journal of Socialist Theory Guide*, No. 12, 13–15.

Du, H. 2008. Study on the Legislation of Regulatory Detailed Plan: Local Practice of Guangdong Province, *Journal of Huazhong University of Science and Technology (Urban Science Edition)*, 25, 76–80.

Duan, J. 1999. Some Thoughts on Urban Planning System Structure, *Planners*, 15(4), 13–18.

Duan, J. 2003. Strengthening the Scientificity of Urban Planning: Regression of Noumenon, *City Planning Review*, 27(2), 83.

Duan, J. 2008. Problems and Solutions of Regulatory Planning in China, *City Planning Review*, 32(12), 14–15.

Dyckman, J. 1979. *Introduction to Readings in the Theory of Planning: The State of Planning Theory in America*, mimeo.

Dye, T.R. 1978. *Understanding Public Policy* (3rd edn), Englewood Cliffs and London: Prentice-Hall.

ELRD (Economic Law Research Department, the Commission of Legislative Affairs of National People's Congress). 2008. *Interpretation of City and Countryside Planning Act 2008*, Beijing: Intelligent Right Publication Press.

Esherick, J. 1987. *The Origins of the Boxer Uprising*, Berkeley: University of California Press.

Eversheds. 2007. A New Employment Contract Law is Passed, issued 12 July 2007, available at: www.eversheds.com/uk/Home/Articles/index.page?ArticleID=templatedata%5CEversheds%5Carticles%5Cdata%5Cen%5CChina_Focus%5CNew_employment_contract_law_is_passed [accessed: 7 April 2008].

Faculty of Urban Planning of Tongji University. 1985. *Urban Planning Principle*, Beijing: China Building Press.

Faludi, A. 1973. *Planning Theory*, Oxford, New York and Toronto: Pergamon Press.

Faludi, A. 1986. *Critical Rationalism and Planning Methodology*, London: Pion Limited.

Fei, X. 1984. *Small Town but Large Issue*, Nanjing: Jiangshu People's Press.
Flyvbjerg, B. 1998. *Rationality and Power: Democracy in Practice*, Chicago and London: The University of Chicago Press.
Friedmann, J. 1987. *Planning in the Public Domain*, Princeton: Princeton University Press.
Fu, Z. 2002. The State, Capital, and Urban Restructuring in Post-reform Shanghai, in *The New Chinese City: Globalisation and Market Reform*, edited by J. Logan. Oxford and Malden: Blackwell Publishers Ltd, 106–20.
General Office of the State Council. 2000. *Circular No. 25: Notices of Strengthening and Improving City and Countryside Planning*, Beijing: General Office of the State Council, P.R. China.
General Office of the State Council. 2006. *Circular No. 100: Notices for the Income Management from Leasing State Owned Land Use Right*, Beijing: General Office of the State Council, P.R. China.
Goldsmith, M. 1980. *Politics, Planning and the City*, London: Hutchinson.
Hall, P. 2005. A Global Urban Agenda, Lecture presented to the Urban Land Institute, London, 15 June, available at: www.uli.org [accessed: 22 April 2008].
Hall, P. and Pfeiffer, U. 2000. *Urban Future 21: A Global Agenda for 21st Century Cities*, London: Spon.
Han, Q., Gu, C. and Yuan, X. 2011. Spatial Coupling of the Urban Master Planning and Main Functional area Planning, *City Planning Review*, 35(10), 44–50.
Harding, M. 1996. *Weather to Travel*, London: Tomorrows Guides.
Harris, N. 2006. Planning and Spatial Development in Cardiff: A Review of Statutory Development Planning in the Capital, in *Capital Cardiff 1975–2020: Regeneration, Competitiveness and the Urban Environment*, edited by A. Hooper and J. Punter. Cardiff: University of Wales Press, 71–96.
Harvey, D. 2005. *Brief History of Neoliberalism*, Oxford: Oxford University Press.
Healey, P. 1998. Collaborative Planning in a Stakeholder Society, *Town Planning Review*, 69(1), 1–21.
Heilig, G. 1999. *Can China Feed Itself?* IIASA: Laxenburg Austria, available at: www.iiasa.ac.at/Research/LUC/ChinaFood/index.htm [accessed: 24 June 2008].
Hill, M. and Hupe, P. 2002. *Implementing Public Policy: Governance in Theory and in Practice*, London, California and New Delhi: SAGE Publication.
Hogwood, B.W. and Gunn, L.A. 1984. *Policy Analysis for the Real World*, Oxford: Oxford University Press.
Hu, A. and Ma, W. 2012. Social and Economic Transformation in Modern China: From Two-sector to Four-sector (1949–2009), *Journal of Tsinghua University (Philosophy and Social Sciences)*, 27(1), 16–29.
Hu, J. 2007. *High Holding the Great Banner of Socialism with Chinese Characteristics for Attaining the Objective of Building a Moderately Prosperous Society in All Respects* (Report to the 17th Chinese Communist Party Central Committee on Oct. 15, 2007), available at: http://cpc.people.com.cn/GB/6416 2/64168/106155/106156/6430009.html [accessed: 10 October 2011].

Huang, G. 1999. Three Principles of Urban Planning in China, *Planners*, 15(4), 7–11.
Hubbard, P. and Hall, T. 1998. The Entrepreneurial City and the New Urban Policies, in *The Entrepreneurial City: Geographies of Politics, Regime, and Representation*, edited by T. Hall and P. Hubbard. Chichester: John Wiley & Sons Ltd, 1–23.
Hutchings, G. 2001. *Modern China*, Cambridge, MA: Harvard University Press.
Hutton, W. 2007. *The Writing on the Wall: China and the West in the 21st Century*, London: Little, Brown.
Jiang, Z. 2002. *Objectives of Building a Well-off Society in an All-Round Way (Jiang Zemin's report at 16th Party Congress)*, available at: http://news.xinhuanet.com/english/2002-11/18/content_632550.htm [accessed: 10 October 2011].
Kahn, J. 2007. In China a Young Activist Imperils Himself, *New York Times*, 26 August, available at: www.nytimes.com/interactive/2007/10/14/world/asia/choking_on_growth_3.html [accessed: 19 April 2008].
Kahn, J. and Landler, M. 2007. China Grabs West's Smoke-Spewing Factories, *New York Times*, 21 December, available at: www.nytimes.com/2007/12/21/world/asia/21transfer.html [accessed: 19 April 2008].
Kahn, J. and Yardley, J. 2007. As China Roars, Pollution Reaches Deadly Extremes, *New York Times*, 26 August, available at: www.nytimes.com/2007/08/26/world/asia/26china.html [accessed: 19 April 2008].
Khakee, A., 1996. Urban planning in China and Sweden in a Comparative Perspective, *Progress in Planning*, 46(2), 29–96.
Kirkby, R. 1985. *Urbanisation in China: Town and Country in a Developing Economy 1949–2000 A.D.*, New York: Columbia University Press.
LACNPC (Legislative Affairs Commission of National People's Congress). 2008. *Definitions of City and Countryside Planning Act*, Beijing: Intelligent Property Press.
Lange, M., Mastop, M. and Spit, T. 1997. Performance of National Policies, *Environment and Planning B: Planning and Design*, 24(6), 845–58.
Li, B. and Xu, H. 2004. New Trends of Spatial Plan in China: With Urban System Plan at the Province Level as a Case, *City Planning Review*, 28(12), 9–14.
Li, B., Guo, J. and Huang, Y. 2006. A Historical Study on the Paradigm of Early-modern City Planning of Shanghai, 1943–1949, *Urban Planning Forum*, 184(6), 83–91.
Li, G., Wang, X. and Wang, Y. 2006. The Problems and Solutions of Regional Planning in China, *Areal Research and Development*, 25(5), 10–13.
Li, J. and Ren, W. 2011. A Few Key Points on the Implements of National Development Priority Zones Plan, *Economic Research Guide*, 27, 240–43.
Li, K. 2012. Promoting Urbanisation Collaboratively as an Important Strategic Alternative for Modernisation, *Administration Reform*, 11, 4–10.
Li, X. 2003. Some Consideration on Urban Spatial Strategic Studies, *City Planning Review*, 27(2), 28–34.

Li, X. and Men, X. 2004. Study on the Reform of Management System of Urban and Rural Planning, *Planner*, 20(3), 5–8.

Li, X., He, L. and Zhang, J. 2009. Using the Urban-Rural Planning Law to Reform Regulatory Plans, *Planner*, 24(8), 71–80.

Li, Z. 1999. A Few Issues of Regulatory Detailed Plan, *Planner*, 15(4), 69–72.

Liang, W. 2006. Environment Livability Index in Regulatory Planning: A Case Study of Beijing, *City Planning Review*, 30(5), 27–31.

Liaoning Provincial Department of Statistics. 2008. *Liaoning Provincial Statistic Year Book*, Shenyang: Liaoning Provincial Statistic Year Book Press.

Lin, J. 2005. *The Development and Management of Megacities in China*, Beijing: Development Research Center of the State Council PRC, available at: www.scj.go.jp/ja/int/kaisai/mega2004/detail/pdf_pre/dr_lin.pdf [accessed: 22 June 2008].

Liu, C. 2008. Adjustment of Fiscal Relations and Changing Functions of Local Government, *Journal of Finance and Economics*, 134(11), 16–27.

Liu, C. and Lu, Z. 2012. Urban Planning Transformation Driven By Development Priority Zone Planning, *Planner*, 28(8), 13–17.

Liu, J. 1999. Some Shortages and Amendments to Urban Planning in China, *Planners*, 15(4), 93–5.

Liu, J. 2000. Some Shortages and Amendments to Urban Planning in China, *Planners*, 15(4), 93–5.

Liu, J., Xing, H. and Zhang, J. 2008. Under Market Economy Conditions the Urban-rural Planning and Management of Provincial Domain Facing Questions and the Institutional Innovation, *Urban Studies*, 18(6), 56–61.

Logan, J. 2002. Three Challenges for the Chinese City: Globalisation, Migration, and Market Reform, in *The New Chinese City: Globalisation and Market Reform*, edited by J. Logan. Oxford and Malden: Blackwell Publishers Ltd, 3–21.

Lowi, T. 1972. Four Systems of Policy, Politics and Choice, *Public Administration Review*, 32(4), 298–310.

Lu, K. 2009. Rigidity or Flexibility: The Dilemma of Regulatory Plan Administration, *Planner*, 25(10), 78–80, 89.

Luo, Z. and Zhao, M. 2003. On the Strategic Study of Urban Development and Urban Strategic Planning, *City Planning Review*, 27(1), 19–23.

Ma, W. 2003. The Assessment of Urban Planning Needs Scientific Criteria, *City Planning Review*, 27(2), 80–81.

Mackerras, C. 2001. *The New Cambridge Handbook of Contemporary China*, Cambridge: Cambridge University Press.

Martin, W. 1983. The nationalist revolution: from Canton to Nanking 1923–28, in *The Cambridge History of China Vol 12: Republican China 1912-49 Part 1*, edited by Fairbank, J. Cambridge: Cambridge University Press.

Mason, P.O. 2005. *China's Booming Economy*, published online 16 September 2005, available at: http://news.bbc.co.uk/1/hi/programmes/newsnight/default.stm [accessed: 27 March 2008].

Mason, P.O. 2008. *China's Migrant Workforce* (published online 26 March 2008), available at: http://news.bbc.co.uk/1/hi/programmes/newsnight/default.stm [accessed on: 27 March 2008].

Mason, R.O. and Mitroff, I. 1981. *Challenging Strategic Planning Assumptions: Theory, Cases and Techniques*, New York and Chichester: Wiley.

Mastop, H. and Faludi, A. 1997. Evaluation of Strategic Plans: The Performance Principle, *Environment and Planning B: Planning and Design*, 24(6), 815–32.

Mastop, H. and Needham, B. 1997. Performance Studies in Spatial Planning: The State of the Art, *Environment and Planning B: Planning and Design*, 24(6), 881–8.

Metro. 2008 (23 April). Food Shortages are like a Silent Tsunami, Cardiff, *Metro*.

MOC (Ministry of Construction P.R. China). 1978. *The Opinions of Strengthening the City Construction Management*, Beijing: Ministry of Construction, P.R. China.

MOC (The Ministry of Construction P.R. China). 1984. *Regulation of Urban Planning*, Beijing: Ministry of Construction, P.R. China.

MOC (Ministry of Construction P.R. China). 1991. *Circular No. 12: Regulations of Plans Formulation*, Beijing: Ministry of Construction, P.R. China.

MOC (Ministry of Construction P.R. China). 1992. *Circular No. 22: Planning Management for Leasing of State Owned Urban Land*, Beijing: Ministry of Construction, P.R. China.

MOC (Ministry of Construction P.R. China). 1994. *Circular No. 36: Regulations for Urban System Plan Formulation and Approval*, Beijing: Ministry of Construction, P.R. China.

MOC (Ministry of Construction P.R. China). 2000. *Circular 25: Notice of Strengthening and Improving City and Countryside Planning*, Beijing: Ministry of Construction, P.R. China.

MOC (Ministry of Construction P.R. China). 2005. *Circular No. 149: Regulations of Plans Formulation*, Beijing: Ministry of Construction, P.R. China.

MOF (The Ministry of Finance). 1992. *Circular 172: The Interim Regulations for Income from Lease of Urban State Owned Land Use Right*, Beijing: The Ministry of Finance, P.R. China.

MoHURD (Ministry of Housing and Urban-Rural Development). 2010. *Circular No. 3: Regulations of Provincial Urban System Plan Formulation and Approval*, Beijing: Ministry of Housing and Urban-Rural Development.

MOLR (Ministry of Land and Resources, P.R. China). 2002. *Regulations of State Owned Land Auction and Tendering*, Beijing: Ministry of Land and Resources, P.R. China.

MOLR (Ministry of Land and Resources). 2009. *Circular No. 43 Regulation of Formulation and Approval for Land Use General Plan*, Beijing: Ministry of Land and Resources, P.R. China.

Moor, T. 1978. Why Allow Planners to do What They Do? A Justification from Economy Theory, *American Planning Association Journal*, 44(4), 387–8.

National People's Congress. 2011. *121.7 Million Hectares of Chinese Arable Land: A Further Reduction of 8,200,000 Hectares*, Chinese Government Webpage, available at: www.gov.cn/jrzg/2011-02/24/content_1810106.htm [accessed: 4 February 2013].

NBS (National Bureau of Statistics). 2001. *China Statistics Year Book 2000*, Beijing: China Statistics Year Book Press.

Needham, B., Zwanikken, T. and Faludi, A. 1997. Strategies for Improving the Performance of Planning: Some Empirical Research, *Environment and Planning B: Planning and Design*, 24(6), 871–80.

NPC (National People's Congress P.R. China). 1986. *Land Administrative Act 1989*, Beijing: National People's Congress, P.R. China.

NPC (National People's Congress P.R. China). 1989a. *City Planning Act 1989*, Beijing: National People's Congress, P.R. China.

NPC (National People's Congress P.R. China). 1989b. *Environmental Protection Act*, Beijing: National People's Congress, P.R. China.

NPC (National People's Congress). 1999. *Land Administration Act 1999*, Beijing: National People's Congress, P.R. China.

NPC (National People's Congress). 2008. *City and Countryside Planning Act 2008*, Beijing: National People's Congress, P.R. China.

OECD. 2005. *Policy Brief: Economic Survey of China, 2005*, Paris: OECD, available at: www.oecd.org/dataoecd/10/25/35294862.pdf [accessed: 29 March 2008].

Paris, C. 1982. Introduction by the Editor, in *Critical Readings in Planning Theory*, edited by C. Paris. Oxford, New York and Toronto: Pergamon Press Ltd, 3–11.

Parsons, W. 1995. *Public Policy*, Cheltenham and Northampton: Edward Elgar.

Pickvance, C. 1982. Physical Planning and Market Forces in Urban Development, in *Critical Readings in Planning Theory*, edited by C. Paris. Oxford, New York and Toronto: Pergamon Press Ltd, 69–82.

Pigou, A. 1932. *The Economics of Welfare* (4th ed.), London: Macmillan.

Population Division of the Department of Economic and Social Affairs of the United Nations Secretariat. 2007. *World Population Prospects: The 2006 Revision and World Urbanization Prospects*, New York: United Nations.

Qiu, B. 2003a. City Strategic Plan Should Emphasis the Research on Urban Economy, *City Planning View*, 27(1), 6–11.

Qiu, B. 2003b. The Limitations of the City Planning Act from the View of the Rule of Law, *City Planning Review*, 26(4), 11–14.

Qiu, B. 2006. Predicament in the Development of Small Cities and Towns of China and its Solutions, *Urban and Rural Development*, 1, 6–10.

Qiu, B. 2010. Challenges and Strategies for China's Urbanization, *Urban Studies*, 1, 7–13.

Rao, H. 1989. About the Scale Efficiency of Cities, *China Social Sciences*, 4, 3–18.

Ratcliff, E. 2003. The Great Green Wall of China, *Wired News*, April, available at: www.wired.com/wired/archive/11.04/greenwall.html [accessed: 21 April 2008].

Ren, Z. 2000. Briefing the Historical Mission of Chinese Comprehensive Plans, *Planners*, 4, 84–8.
Rittel, H. and Webber, M. 1973. Dilemmas in a General Theory of Planning, *Policy Sciences*, 4(2), 155–69.
Shao, L. 2001. Impulse Mechanism and Development Models of Chinese Urbanisation, *Shanghai Urban Planning*, 1, 15–17.
Shi, N. 2003. The Scientificity of Urban Planning comes from Scientific Planning Practice, *City Planning Review*, 27(2), 82–3.
Shi, Y. 2008. The Study of Interactive Relations among Development Priority Zone Planning, Urban-rural Planning, and General Land Use Planning, *Macro Economic Research*, 8, 35–40.
Short, J. and Kim, Y. 1998. Urban Crises/Urban Representations: Selling the City in Different Times, in *The Entrepreneurial City: Geographies of Politics, Regime, and Representation*, edited by T. Hall and P. Hubbard. Chichester: John Wiley & Sons Ltd, 55–75.
Simmie, J. 1993. *Planning at the Crossroad*, London: UCL Press.
Simmie, J. 2001. Planning, Power and Conflict, in *Handbook of Urban Studies*, edited by R. Paddison. London, Thousand Oaks, New Delhi: SAGE Publications Ltd, 385–401.
SINA News. 2010. *Drawing Magnificent Blueprint and Building Beautiful Hometown Together*, available at: http://news.sina.com.cn/o/2008-12-14/122 014878324s.shtml [accessed: 4 March 2011].
State Council. 1990a. *Circular 55: The Interim Regulations of the Assignment and Transfer of the Right to Use State-owned Land in Urban Areas*, Beijing: State Council, P.R. China.
State Council. 1990b. *Circular 56: The Interim Regulations for Leasing Tracts of Land Development and Operation by Foreign Investment*, Beijing: State Council, P.R. China.
State Council. 2001. *Circular No. 15 of Strengthening the Management of the State Owned Land Resources*, Beijing: State Council, P.R. China.
State Council. 2002. *Circular No. 13: Notice of Strengthening City and Countryside Planning Control*, Beijing: State Council, P.R. China.
State Council. 2004. *Circular No. 28: Decisions for Further Reform and Strict Control of Land Administration*, Beijing: State Council, P.R. China.
State Council. 2007. *Circular 21: Decisions on Preparation of National Development Priority Zones*, Beijing: State Council, P.R. China.
Sun, Q. 2006. Urban Planning, its Institutional Background and the Urban Spatial Morphology in Modern Shanghai, *Urban Planning Forum*, 184(6), 92–101.
Sun, S. 2003. Strengthening Immediate Planning, and Stimulating the Revolution of Urban Planning Idea and Methods, *City Planning Review*, 27(3), 13–15.
Tang, L. 1993. Unitary Plan in England, *Overseas Urban Planning*, 7(3), 15–18.
Tang, L. 2006. Towards the Effective Control And Guidance of Planning: Review and Prospect of Regulatory Planning, *City Planning Review*, 30(1), 28–33.

Tang, W. 2000. Chinese Urban Planning at Fifty: An Assessment of the Planning Theory Literature, *Journal of Planning Literature*, 14(3), 348–65.

Tian, L. 2007. Regulatory Planning in China: Dilemma and Solution, *City Planning Review*, 31(1), 16–20.

US Embassy Beijing. 1996. *China's Food Security – for a Prosperous China: A View from 'Science and Education for a Prosperous Future'*, available at: www.usembassy-china.org.cn/sandt/fddeb.12.htm [accessed: 17 June 2008].

Wang, F. 2000. From the Urban Planning System to the Urban Planning Mechanism: Evolution of Urban Planning in Shenzhen, *City Planning Review*, 24(1), 28–33.

Wang, F. 2001. Urban Planning in Shenzhen: Evolution and Its Evaluation, *City Planning Review*, 25(1), 28–33.

Wang, G. 2002a. *The Consideration of the Chinese Urbanisation*, Speech to the Ministry of Sciences and Technology, P.R. China, 15 October 2002.

Wang, J. 2003. Exploring an Urban Planning Theory and Practice with Chinese Characteristics, *City Planning Review*, 27(2), 17–21.

Wang, J., Ge, C., Yang, J., et al. 1999. Taxation and the Environment in China: Practice and Perspectives, in *Environmental Taxes: Recent Developments in China and OECD Counties*, Paris: OECD Publication Service, 61–108.

Wang, K. 1999a. The Impacts and Changes of Principal Ideologies to Chinese Urban Planning during the Last 50 Years, *Planners*, 15(1), 28–33.

Wang, K. 2002b. From Guangzhou to Hangzhou: Appearance of the Development Strategic Plan, *City Planning Review*, 26(6), 57–62.

Wang, K. 2007. History and Reality of National Urban System Planning, *City Planning Review*, 31(10), 9–15.

Wang, L. 1993. Urban Planning System in Britain, *Overseas Urban Planning*, 7(3), 9–14.

Wang, L. 1999b. Urban Planning Decision and the Authority of the Planning, *Shanghai Urban Planning*, 4, 2–5.

Wang, X. and Zhou, M. 2000. Analysis and Comparative Study of Urban Land Payable Use, *China Real Estate*, 3, 7–10.

Wang, Y. 2002c. Urbanisation Approach in Zhejiang Province, *Urban Problems*, 2, 17–19.

Wang, Y. 2008. *For a Higher Standard: 30 Years' Experiences of Chargeable Land Use System Reform in China*, available at: www.clr.cn/front/read/read.asp?ID=148363 [accessed: 19 May 2009].

Wei, Y. 2000. *Regional Development in China-States, Globalisation, and Inequality*, London and New York: Routledge.

Wells, S. 2005. *China's Fuel Demands*, published online 8 March 2005, available at: http://news.bbc.co.uk/1/hi/programmes/newsnight/default.stm [accessed: 27 March 2008].

Wen, J. 2010. China will Expand Import from USA, China Daily Webpage, available at: www.chinadaily.com.cn/hqgj/2010-03/22/content_9625013.htm [accessed: 3 February 2013].

Wu, F. 2003. The (Post-) Socialist Entrepreneurial City as a State Project: Shanghai's Reglobalisation in Question, *Urban Studies*, 40, 1673–98.

Wu, W. 2000. The Milieu and the Direction of the Urban Planning Reform in the New Times, *Planners*, 16(5), 80–83.

Xiamen Government. 2002. 'Implementation Outlines of the Xiamen Development Strategy', available at: http://www.xm.gov.cn/database/nt/view/newsdetail.asp?nid=18921 [accesed: 15 May 2007].

Xiamen Municipal Development and Reform Commission. 2001. Xiamen Economic and Social Development Plan of 2001, Xiamen: Xiamen Government.

Xiamen Statistics Bureau. 2010. *Xiamen Economic and Social Development Statistics Annual Bulletin 2009*, Xiamen: Xiamen Statistics Press.

Xie, Q., Parsa, G. and Redding, B. 2002. The Emergence of the Urban Land Market in China: Evolution, Structure, Constraints and Perspectives, *Urban Studies*, 39(8), 1375–98.

Xinhua News. 2003. *Communiqué of the Third Plenum of the 16th Central Committee of the Communist Party of China*, available at: http://news.xinhua net.com/newscenter/2003-10/14/content_1123116.htm [accessed: 24 August 2011].

Xinhua News Agency. 2003. *Creating More Jobs, Fundamental Way to Relieve Urban Poor*, available at: http://news.xinhuanet.com/english/2003-03/07/con tent_763937.htm [accessed: 10 October 2011].

Xinhua News Agency. 2006. Report of Ministry of Land and Resources: Declining Eight Million Hectares of China's Arable Land in 10 Years, Xinhua News Agency, 16 March 2006, available at: http://news.xinhuanet.com/house/2006-03/16/content_4308627.htm [accessed: 4 February 2013].

Xinhua News Agency. 2009. China Announces Greenhouse gas emission reduction Target, 26 November, available at: http://news.xinhuanet.com/politics/2009-11/26/content_12545939.htm [accessed: 4 February 2011].

Xinhua News Agency. 2013a. Possible Food Struggle Threatens China's Urbanisation, XinhuaNet, available at: http://news.xinhuanet.com/english/china/2013-01/26/c_132130119.htm [accessed: 4 February 2013].

Xinhua News Agency. 2013b. China to Accelerate Modernization, Boost Developmental Vitality in Rural Areas, XinhuaNet, available at: http://news.xinhuanet.com/english/china/2013-02/04/c_132149103.htm [accessed: 4 February 2013].

XMG (Xiamen Municipal Government). 2002. Implementation Guideline for Fastening Xiamen Development as a Bay Pattern City (Approved by the Fifth Session of the Ninth Chinese Communist Part Xiamen Committee on 27 November 2002), available at: www.xm.gov.cn/database/nt/view/newsdetail.asp?nid=18921 [accessed: 15 November 2003].

Xu, H., Wang, D. and Lv, X. 2011. Compilation Model of Regulatory Plan with the New Urban Rural Planning Law, *Planner*, 27(1), 94–9.

Xu, J. and Song, X. 1998. New Mechanism for Plan Implementation in Metropolises in China, *Planners*, 14(3), 86–9.

Xue, P. 1995, Urban Planning, Construction and Development Control in Singapore, *Overseas Urban Planning*, 9(4), 31–6.
Yan, L. 2008. Regulatory Planning after the Issue of Urban and Rural Planning Law: An Analysis of Regulatory Planning from Technology and Public Policy Aspects, *City Planning Review*, 32(11), 46–50.
Yang, J. 1993. Exploratory Study on the Regulatory System in Urban Planning, *City Planning Reviews*, 23(4), 5–11.
Yang, T. 2007. *Interpretation of Zhouli*, Shanghai: Shanghai Guoji Press.
Yang, X. 1989. Regulatory Detailed Plan for Wenzhou Old Town Renewal, *City Planning Review*, 19(6), 55–9.
Yao, Y. 2007. Thinking on the Compilation of Regulatory Detailed Planning in View of Development Control, *Modern City Research*, 9, 10–14.
Yardley, J. 2007. Beneath Booming Cities, China's Future Is Drying Up, *New York Times*, 28 September, available at: www.nytimes.com/2007/09/28/world/asia/chokimg_on_growth_2.html [accessed: 19 April 2008].
Yeh, A. and Wu, F. 1999. The Transformation of the Urban Planning System in China From a Centrally-Planned to Transitional Economy, *Progress in Planning*, 51(3), 167–252.
Yin, J., Zhang, J. and Luo, X. 2006. Chinese City Development and Entrepreneurialism of Local Governments in the Transition, *Urban Problems*, 4, 36–41.
Yin, Z. 2003. Understanding and Opinions on City Development Strategic Research, *City Planning Review*, 27(1), 28–9.
Yin, Z. 2008. Thinking about Compiling Urban and Rural Planning Scientifically and Democratically, *City Planning Review*, 32(1), 44–5.
Yin, Z. and Wang, M. 2010. A Great Battle for Planning after Disaster of the Earthquake, available at: http://kc.cdcc.gov.cn/Info/InfoView.aspx?ID=373, Chengdu Association of Surveying and Design [accessed: 4 March 2011].
Yu, L. 1994. Urban Planning and the Market Economy, *Urban and Rural Construction*, 9, 20–23.
Yu, L. 1995. The British Development Planning System and Its Characteristics, *Overseas Urban Planning*, 9(1), 27–33.
Yu, L. 2005. Chinese Urban Planning in Transition: Market Economy and its Way Forward, *Global Built Environment Review*, 5(2), 21–38.
Yu, L. 2009. Study on Development Objectives and Implementing Policies of Chinese Eco-city, *Urban Planning International*, 24(6), 102–7.
Yu, L. 2010. Some Concerns for Chinese Eco-city Development from International Experience, *Building Sciences*, 13, 30–35.
Yu, L. and Zhu, L. 2009. Chinese Local State Entrepreneurialism: A Case Study of Changchun, *International Development Planning Review*, 31(2), 199–220.
Yu, Y., Wang, F. and Fan, Y. 2010. Reforming Regulatory Plans through Dynamic Implementation: Chongqing Example, *Planner*, 26(10), 16–21.
Yu, Z. 2000. *Tasks of Urban Planning*, speech at the Beijing Urban Planning Conference on 2 November 2000.

Yuan, Q. and Hu, Y. 2010. The Role and Development of Regulatory Planning, *Planner*, 26(10), 5–10.

Zeng, F. 1992. About the Strategic Plan, *Overseas Urban Planning*, 6(1), 19–24.

Zhai, Z., Duan, C. and Bi, Q. 2007. The Fundamental Characters of Floating Migrants in Beijing, *The Flag Manuscript*, 2, 28–31.

Zhang, B. 2002. Where is the Tendency: Generation, Development and Future of Strategic Plan, *Urban Planning Review*, 26(6), 63–8.

Zhang, H. and Meng, T. 2007. Empirical Research of Chinese Fiscal System Changes between 1949 and 2004: From the Perceptive of Financial Pressures and Competition, *Economic System and Reform*, 4, 100–104.

Zhang, J. 1993. *Full Interpretation of Shang Book*, Guiyang: Guizhou People's Press.

Zhang, W. 2001. Urban Planning in Market Economy, *Urban Planning Forum*, 4, 18–30.

Zhang, X. 2004. A Preliminary Study on Urban Planning Management and Making, *Planner*, 20(10), 31–2.

Zhang, X. and Wu, Y. 2009. Theories and Methods of Provincial Subjective Function Zone Planning, *Resource Development & Market*, 25(12), 1092–6.

Zhao, M. and Le, Y. 2009. Regulatory Detailed Planning under the City and Countryside Planning Act: From Technical Document to Statutory Evidence, *City Planning Reviews*, 33(9), 24–30.

Zhou, G. 2002. Urbanisation and Cities with Historical and Cultural Interests, *City Planning Review*, 26(4), 7–10.

Zhou, J. and Huang, J. 2003. Of Planning Administrative Power and Urban Planning Management System, *Planner*, 19(1), 61–4.

Zhou, Y. 2012. Regulatory Planning Implementation Review Since the Enactment of City and Countryside Planning Act, *Planner*, 28(7), 45–50.

Zhu, J. 2004. From Land Use Right to Land Development Right: Institutional Change in China's Urban Development, *Urban Studies*, 41(7), 1249–67.

Zou, D. 2003a. The Scientificity of Urban Planning, *City Planning Review*, 27(2), 77–9.

Zou, D. 2003b. Considering the Situation, Planning as a Whole, Designing the Future, Exploiting and Going Ahead, *City Planning Review*, 27(1), 17–18.

Index

'901' project (Taiwan) 153, 154, 155, 156, 157, 164

agricultural production 2, 8, 13, 19, 27, 27–31, 32, 33, 44, 59
air pollution 2, 18, 19, 33, 34, 62
ancient China 2, 4, 37, 41–5
Ashworth, G. and Voogd, H. 200–201, 206
auction (tender offer) (leasing rights) 188, 223–4, 225, 226–7, 230, 233, 235, 237, 243

Bachmann, J. and Leung, A. 20, 59, 60
banking system 18, 73, 101
Beijing 3, 7, 20, 21, 24, 34n9, 36, 62, 86
  water shortages 32, 35, 61
Beijing-Tianjin-Hebei Province ('Greater Beijing Area') 9, 64, 65, 113
Book of Changes, The see *Yijing*
Boxer Rebellion (1900) 5
Bruton, M.J. and Nicholson, D.J. 258
'Building Construction and Engineering Permit', *see* 'Planning Permit for Construction and Engineering'

carbon emissions 68, 90, 136, 268
Cardiff 249, 250
CCP (Chinese Communist Party) 7, 8, 9, 14, 21, 37, 82, 83, 84, 85
Central Military Affairs Commission 84–5
centrally planned economy 16, 23, 27, 38, 41, 48, 49, 56, 63–4, 81, 94
  planning system 47, 55, 79, 95, 97, 194, 221, 261, 263–4
  urban planning 50, 51, 55, 57, 59, 79, 157, 168, 213
Chen, A. 17, 18–19
Chen, G. 42
Chen, Madam 265
Chen Xiaoli 59

Chen Xiwen 31
Chengdu 241–2
Chiang Kai-shek 7, 8
China (1840–1911) 4–6
China, ancient era 2, 4, 37, 41–5
China, People's Republic of (PRC) 1–2, *2*, 3–4, 8–10, **13**, 37–9, 50
China, Republic of (ROC) 2–3, 6, 37
Chinese Communist Party Central Committee 21, 24, 54, 57, 73, 76–7, 82, 104
Chongqing 86, 87
Christensen, K. 155, 165–6, 216, 255–7
cities 20–21, 41–4, 45–7, 54, 57, 60–63, **61**, 81, 86–8
  eco-city 78, *78*, 86, 268
  large 20, 46–7, 48, 51–2, 60, **61**, 66, 67–8, 107, 129, 139, 191
  medium 20, 48, 51, 60, **61**, 68, 107, 129, 139, 191
  mega-cities 21, 32, 61, 62–3, 67, 68, 75, 115
  metropolises 24, 52, 60, **61**, 66, 86, 128, 191
  small 20, 60, **61**, 66, 67, 68, 107, 129, 139, 191
Cities Alliance 105, 119–21, 123–4, 136, 219, 220
  Quanzhou 119, 120, 121, 122, 123
  Xiamen 119, 120, 121, 122, 123, 219, 220
  Zhangzhou 119, 120, 121–3, 219
City and Countryside Planning Act 2008 (NPC 2008) 95, 106, 140, 143, 221, 225, 235, 237, 239–40, 262, 264
  City Comprehensive Plan 110, 137, 145, 171
  Integrated Urban System Plan 106, 108, 136, 171

Regulatory Detailed Plan 171, 175, 182, 185, 186, 193
City Comprehensive Plans 56, 57, 76, 110–11, 137–9, 144, 145, 160–63, *161*, 195–6, 240–241, 264–5, 273–4
　General Land Use Plan 142–4, 270
　Integrated Urban System Plan 139–40
　Regulatory Detailed Plan 171, 175, 177, 180, 182–5, *184*, 186, 242, 245
　Xiamen 151, 152–4, 155–6, 157–60, 162–3, 164, 167, 168, 199, 203, 219, 266
city marketing 75–6, 200–201, 219
city master plan 48, 49, 50, 51, 52, 79
City Planning Act 1989 (NPC 1989a) 52, 53–4, 55–6, 79, 80, 106, 110, 137, 143, 154, 163, 235
co-urban areas *64*, 64–5, **65**, 67, 68, 204, 212
　Beijing-Tianjin-Hebei Province 9, 64, 65, 113
　Liaoning Coastal Area 64, 127, 128, 130, 131–3, 134, 135
　Pearl River Delta 63, 64, **65**, 113, 118, 204, 205, 206, 208
　Yangtze River Delta 3, 63, 64, 65, 113, 118, 123, 204, 206, 208
coal 2, 19, 33–4, 36, 67, 135
Confucianism 43
Control Detailed Plan 56–7, 58, 80
Controlling Development Priority Zones 111, 114, 116, 117, 118
CSR (Corporate Social Responsibility) 25
cultivated land 2, 28–9, 32, 49, 57, 92–3, 142, 143–4, 196, 268–9
　shortages 31–2, 33, 65, 67, 75, 142, 227
Cultural Revolution (1966–76) 8, 11, 14, 37, 50, 51, 79, 83, 89, 100

Deng Xiaoping 8, 9, 14, 16, 37, 41, 54, 60, 73, 87, 265
desertification 6, 27, 35–6, 38
development 115–16, 265
development corridors, Xiamen 166–8

development planning management process, *see* planning management
Development Priority Zones 105, 106, 111–13, 115, 116, 117–18, 123, 136, 270
development projects 21, 98, *99*, 185, 245–6, 264, 268–9, 272, 274
　industrial development 47–8, 90, 102
　planning application *231*, 231–40, 241–2, 243–5, **244**
　planning management 186, 187–8, 192–3, 221
　Xiamen 162–3, 164–5, 168
devolution 51, 54, 69–71, 79, 99, 100–102, 103, 118–19, 155, 165, 261, 273

eco-city development *78*, 78, 86, 268
ecological environment 18, 77–8, 94, 107, 108, 114–15, 117, 129, 214–16, 268
Ecological Environment Threshold Analysis 214–16
economic development 9–10, 16–19, 21–5, **22**, 31–3, 36, 64–5, 76–8, 79, 267–8
　'open door and reform policy' 16, 17, 71, 100, 141, 172, 265
education policies 8, 15, 17
employment 15, 20–21, 24, 25–7, 71, 125, 150, 203
employment rights 25, 26
entrepreneurialism 54, 75, 76, 80, 102, 165, 193, 274
environment 2, 9, 18, 31, 32–4, 35, 36, 76–8, 93–4, 129, 240, 267–8
Ethical Code of Zhou's Rituals 42–3

FDIs (foreign direct investment) 8, 64, **65**, 65, 71, *72*, **74**, 74–5, 135, 203–5, **204**
　Xiamen 149–51, 203–4, 208
Fengshui 41, 45
fiscal system **69**, 69–70, 90, 94, 100–101, 103
'Five Dots and One Belt', Liaoning Coastal Area 127, *127*, 130, 134, 135
Five-Year Plan, *see* National Social and Economic Development Plan

floating population 20–21, 23–4
flooding 27, 32, 36, 42, 191–2
food security 19, 27–31, 44, 93
foreign investment, *see* FDIs
fundraising 69, 73, **74**, 102

General Land Use Plan 95, 142–4, 232, 270
'General Layout and Construction Scheme Proposal' 237, 238
global warming 36, 68, 77, 90
globalization 37, 62, 71, **72**, 75, 76, 80, 99
Great Leap Forward (1958–60) 8, 11, 14, 37, 49, 50, 79
'Greater Beijing Area', *see* Beijing-Tianjin-Hebei Province
Guangzhou 3, 57, 173, 196, 222
Guanzhong 42

Hall, P. 62–3
harmonious society 25, 26, 27, 76–7, 107, 111n1, 115, 130, 136
health care policies 8, 11, 15, 17, 38
Heilig, G. 12, 28, 29–30
Hogwood, B.W. and Gunn, L.A. 250–51, 257
housing 23–4, 38, 50, 62, 65, 91–2, 101, 189–90, 224, 238, 269

industrial development 21, 47–8, 49, 90, 102, 114, 140–41, 237
  Liaoning Coastal Area 134, 135–6
  Xiamen 154, 166–7, 203, 215
industrialization 8, 9–10, 23, 48, 50, 59, 79, 93, 140–141
Integrated Urban System (megalopolis) Development Plan, Liaoning Coastal Area *127*, 127–33, *133*, 134–5, 136
Integrated Urban System Plan 107, 139–40, 270
  Liaoning Coastal Area *127*, 127–33, *133*, 134–5, 136
  National 105–6, 108, 136, 139–40, 171, 270
  Provincial 105–11, 171, 182, 240, 241, 270

Japan 5, 7, 37, 46
Jiang Zeming 155

Kirkby, R. 59
Kuomintang (Nationalist Party) 6, 7, 8, 37

Labour Contract Law (2007) 26
labour force 10, 15–16, 20–21, 23–7, 125, 131–2, 135, 150, 203, 208–9; *see also* unemployment
Land Administration Act 1986 (NPC 1986) 95, 142, 143, 222–3, 224, 226
land allocation 49, 92–3, 142–4, 186–8, 224–6, 230
land leasing 93, 101, 172, 186, 187–8, 194, 222–5, 226, 227
  auction 188, 223–4, 225, 226–7, 230, 233, 235, 237, 243
land shortage 31–2, 33, 65, 67, 75, 142, 227
land use control 171, 177–81, **178–9**, 186, 187–8, 225, 233, 236
'Land Use Planning Permit', *see* 'Planning Permission for Land Use'
Laozi 44
large cities 20, 46–7, 48, 51–2, 60, **61**, 66, 67–8, 107, 129, 139, 191
leasing rights (land use) 93, 101, 172, 186, 187–8, 194, 222–5, 226, 227
  auction 188, 223–4, 225, 226–7, 230, 233, 235, 237, 243
legal system 89
Li Xiaolong 196
Li Xiaolong and Men Xiaoying 110, 111
Liaoning Coastal Area 125–6, *126*
  co-urban areas 64, 127, 128, 130, 131–3, 134, 135
  'Five Dots and One Belt' 127, *127*, 130, 134, 135
  industrial development 134, 135–6
  Integrated Urban System Plan *127*, 127–33, *133*, 134–5, 136
Lin, Jiabin 61–2
Liuwudian, Xiamen 211, 212, 217
loans 18, 101, 102
local governments 51–2, 54, 57, 74–5, 85–9, 103–4, 195, 261, 267, 274
  devolution 69, 70–71, 79, 99, 101–2

land leasing 101, 223–4, 225, 225–6
municipal 57, *99*, 99–100, 145, 195, 261
planning system 97, 172, 245, 263
provincial 97–8, 100, 105–11, 145, 171, 182, 240, 241, 270
low carbon eco-cities 78, *78*, 86, 268

Mao Zedong 3, 7, 8, 14, 37, 84
market economy 17, 27, 38, 54, 56–7, 63, 79, 130, 172, 261, 274
marketization 47, 53, 54, 72–4, 80, 99, 103, 150, 207, 261
Marxism 6–7, 37
Mason, R.O. and Mitroff, I. 252, 254, 255
MCR (Mega City Region) 62–3
 Pearl River Delta 63, 64, **65**, 113, 118, 204, 205, 206, 208
 Yangtze River Delta 3, 63, 64, 65, 113, 118, 123, 204, 206, 208
MDRC (municipal development and reform commission) 233–4, 235
medium cities 20, 48, 51, 60, **61**, 68, 107, 129, 139, 191
mega-cities 21, 32, 61, 62–3, 67, 68, 75, 115
metropolises 24, 52, 60, **61**, 66, 86, 128, 191
MoEP (Ministry of Environmental Protection) 35, 93–4
MoHURD (Ministry of Housing and Urban-Rural Development) 91–2, 93, 95, 98, 105, 108, 109, 235
MoLR (Ministry of Land and Resources) 30, 92–3, 95, 142–4, 235, 270
MoT (Ministry of Transport) 90–91
municipal governments 57, *99*, 99–100, 145, 195, 261

Nanjing 6, 7, 46–7
National Development Priority Zones 112, 270
National Integrated Urban System Plan 105–6, 108, 136, 139–40, 171, 270
National Social and Economic Development Plan (Five-Year Plan) 8, 25, 39, 48, 49, 66, 90, 140–141, 209, 266, 270

nature 44–5
NDRC (National Development and Reform Commission) 68, 89–90, 105, 112, 117, 136
NPC (National People's Congress) 31, 66, 83–4, 89, 153, 222
nuclear power 34, 94

OECD (report) 16–18
'One Permission Notice and Two Permits' 186, 221, 232, 245
'open door and reform policy' (1978) 5, 8–9, 14–15, 19, 21, 37, 41, 50–51, 59–60, 66
 economic development 16, 17, 71, 100, 141, 172, 265
Opium War, First 5, 45
Optimizing Development Priority Zones 111, 113, 117

Pearl River Delta 63, 64, **65**, 113, 118, 204, 205, 206, 208
'Permission Notice for Location' 186, 187, 232, 237–8, 240
pipelines 191
planning 247, 248–50, 255–7, 263, 265, 266–7, 272–3
planning application *231*, 231–40, 241–2, 243–5, **244**
planning management 56, 171, 186, 192, 193–4, 221, 232, 240–42, 245–6, 267, 270–71
 development projects 186, 187–8, 192–3, 221
planning norms 54, 188–92, 194
planning permission *231*, 231–40, 241–2, 243–5, **244**
'Planning Permission for Land Use' 234–5
'Planning Permit for Construction and Engineering' 186, 221, 232–3, 239–40
planning system 55, 79, 95–7, *96*, 117–18, 248–50, 260–65, 266–7, 269–72, 273–4
 centrally planned economy 47, 55, 79, 95, 97, 194, 221, 261, 263–4
pollution
 air 2, 18, 19, 33, 34, 62

water resources 2, 18, 29, 32, 33, 35, 61–2
population 10–14, 31, 32, 37, 55, 60, 66, 67, 93, 104, 142
poverty 11, 14–15, 17, 20, 22, 24, 33–4, 77–8, 266
prefectures 78, 85, 87, 88, 97, 268
Prioritizing Development Priority Zones 111, 113–14, 116, 117
Prohibiting Development Priority Zones 111, 114–15, 116, 117, 118
Provincial Department of Housing and Urban-Rural Development 106, 107–8, 109, 110–11
Provincial Development Priority Zones 106, 112–13, 270
provincial governments 97–8, 100, 105–11, 145, 171, 182, 240, 241, 270
Provincial Integrated Urban System Plan 105–11, 171, 182, 240, 241, 270
public participation 77, 264–5
public policies 250–51, 252–4, 257–60, 266

Quanzhou 205, *206*, 212
 Cities Alliance 119, 120, 121, 122, 123

railways 27, 91, 120, 125, 191
rational planning 53, 54, 79, 156–7, 159, 160, 168, 169, 198, 211, 218, 261
regional planning 75, 76, 90, 105, 136, 260–261; *see also* planning system
'Regulations for Urban Plans' Formulation' (MOC 2005) 55–6, 137–8, 144, 180, 196
Regulatory Detailed Plan 171, 172, 173, 174, 175–7, *177*, 180–82, 192–4, 221, 241, 270
 City Comprehensive Plans 171, 175, 177, 180, 182–5, *184*, 186, 242, 245
 land use control 171, 177–81, **178–9**, 186, 187–8, 225, 233, 236
renewable energy 34, 136
Rittel, H. and Webber, M. 253–4
rural land 72, 88, 222, 227, *229*, 235
rural–urban migration 16, 17, 20–21, 23–4, 55, 65–6, 67, 71, 93

SARs (Special Administrative Regions) 3, 85, 86, 87
'Scientific Outlook on Development' 82, 111, 115, 127
Shang Yang 44
Shanghai 3, 24, 27, 36, 45–6, 61, 62, 86, 172–3, 202
Shenzhen 25–6, 193, 202–3, **204**, 222, 224, 253
shortages
 cultivated land 31–2, 33, 65, 67, 75, 142, 227
 water resources 2, 27, 32, 35, 61, 66, 215
small cities 20, 60, **61**, 66, 67, 68, 107, 129, 139, 191
small towns 66–7, 108, 120, 130, 133, 134
social harmony 19, 21, 25, 26, 27, 76–7, 107, 115
South Fujian Province, *see* Cities Alliance
Soviet Union 6–7, 8, 37, 47, 48, 63, 81, 86, 261
Special Economic Zones 8, 50, 51, 87, 100, 118, 193, 195, 203, 225
 Xiamen 121, 122, 146, 147, 152, 153, 199, 203, 207–8
state-owned enterprises 15, 16, 24, 63, 72, 73, 100, 102–3, 269
state-owned land 52, 146, 227, 232, 235, 245
 land rights leasing 175, 176, 186, 192, 223, 224, 225, 269
state-run union 26–7
Sun Yat Sen 6, 7, 14, 37
'Sunan Type' (urbanizaton) 60, 66
sustainable development 19, 25, 31–2, 55, 77, 107, 129–30, 135, 136, 267–9
systematic planning 53, 54, 79, 156–7, 160, 199, 261, 270–71

Taiwan 5, 7, 8, 86, 123, 147, 155, 204, 205, 209
 '901' project 153, 154, 155, 156, 157, 164
tax revenues **69**, 69–70, 100–101, 103
techno-rationalism 53
tender offer (leasing rights), *see* auction
Tianjin 61, 62, 86, 131, 241

tourism 71, **72**, 121, 135
  Xiamen 151, 157, 162, 164, 166, 167, 203, 205, 209, 210
transport corridors 114, 125, 132, 133
transport plans 120, 190–91
TVEs (township and village enterprises) 66

uncertainty 14, 37, 57, 155, 255–7
unemployment 15–16, 17, 20, 24, 131, 134–5
Urban Construction 48, 49
Urban Development Strategic Plan 57–8, 196, 197–9, 216, 219–20, 266, 272–3, 274
  Xiamen 196, 199–200, 201, 202, 205, 207–10, 211–14, 216–19, 220
urban entrepreneurialism 75, 76, 80, 102, 193, 274
urban land 88, 188–90, 195, 222, 227, *228, 229*, 232–6, 245; *see also* land leasing
Urban Land Development Corporation 230
urban planning 19, 32, 41, 47–52, 54–5, 98, 157
  centrally planned economy 50, 51, 55, 57, 59, 79, 157, 168, 213
urban planning areas 48, 52, 137, 145–6
urbanization 15, 17, 20–21, 23, 28, 33, 58–63, 64–8, 93, 195

'village in city' 146

water resources 29, 31, 32, 38, 42, 61, 67, 91, 95, 122, 215
  pollution 2, 18, 29, 32, 33, 35, 61–2
  shortages 2, 27, 32, 35, 61, 66, 215
Wenzhou 56–7, 173–5

'Wenzhou Type' (urbanization) 60, 66
Wuhan 25, 26, 222

Xiamen (Fujian Province) 121, 146–51, *147, 148, 149*, **158**, 201–2, *206*, 206, 241
  '901' project 153, 154, 155, 156, 157, 164
  Cities Alliance 119, 120, 121, 122, 123, 219, 220
  City Comprehensive Plan 151, 152–4, 155–6, 157–60, 162–3, 164, 167, 168, 199, 203, 219, 266
  development corridor 166–8
  Ecological Environment Threshold Analysis 214–16
  FDIs 149–51, 203–4, 208
  industrial development 154, 166–7, 203, 215
  Liuwudian 211, 212, 217
  Special Economic Zone 121, 122, 146, 147, 152, 153, 199, 203, 207–8
  tourism 151, 157, 162, 164, 166, 167, 203, 205, 209, 210
  Urban Development Strategic Plan 196, 199–200, 201, 202, 205, 207–10, 211–14, 216–19, 220
'Xiamen PX (Paraxylene) Event' 77
Xiamen Urban Development Strategic Plan *198*

Yangtze River Delta 3, 63, 64, 65, 113, 118, 123, 204, 206, 208
Yijing (*The Book of Changes*) 44–5
Yuan Shikai 6

Zhangzhou 119, 120, 121–3, 219